AF941203

Treatability of Perchlorate-Containing Water by RO, NF, and UF Membranes

The mission of the Awwa Research Foundation (AwwaRF) is to advance the science of water to improve the quality of life. Funded primarily through annual subscription payments from over 1,000 utilities, consulting firms, and manufacturers in North America and abroad, AwwaRF sponsors research on all aspects of drinking water, including supply and resources, treatment, monitoring and analysis, distribution, management, and health effects.

From its headquarters in Denver, Colorado, the AwwaRF staff directs and supports the efforts of over 500 volunteers, who are the heart of the research program. These volunteers, serving on various boards and committees, use their expertise to select and monitor research studies to benefit the entire drinking water community.

Research findings are disseminated through a number of technology transfer activities, including research reports, conferences, videotape summaries, and periodicals.

Treatability of Perchlorate-Containing Water by RO, NF, and UF Membranes

Prepared by:
Gary Amy, Yeomin Yoon, Jaekyung Yoon,
and **Myoungsuk Song**
Civil, Environmental, and Architectural Engineering
University of Colorado, Boulder, CO 80309
John Pellegrino
National Institute of Standard and Technology
Boulder, CO 80303
Sun Liang, Karen Scott, and **Joon H. Min**
Metropolitan Water District of Southern California
La Verne, CA 91750

Jointly sponsored by:
Awwa Research Foundation
6666 West Quincy Avenue
Denver, CO 80235

and

U.S. Environmental Protection Agency
1200 Pennsylvania Avenue, N.W.
Washington, DC 20460

and

East Valley Water District
1155 Del Rosa Avenue
San Bernardino, CA 92413

Published by the

Awwa Research Foundation

Advancing the Science of Water®

IWA Publishing

DISCLAIMER

This study was jointly funded by the Awwa Research Foundation (AwwaRF), the U.S. Environmental Protection Agency (USEPA), and East Valley Water District (EVWD) under Cooperative Agreement No. R 826422-01. AwwaRF, USEPA, and EVWD assume no responsibility for the content of the research study reported in this publication or for the opinions or statements of fact expressed in the report. The mention of trade names for commercial products does not represent or imply the approval or endorsement of either AwwaRF, USEPA, or EVWD. This report is presented solely for informational purposes.

Published by IWA Publishing, Alliance House, 12 Caxton Street, London SW1H 0QS, UK

Telephone: +44 (0) 20 7654 5500; Fax: +44 (0) 20 7654 5555; Email: publications@iwap.co.uk;
Web: **www.iwapublishing.com**

AwwaRF report number 90932F. Originally published by AwwaRF for its subscribers in 2003.
IWA Publishing version published 2004.

ISBN 1 84339 862 1

CONTENTS

TABLES

FIGURES

FOREWORD

The Awwa Research Foundation is a nonprofit corporation that is dedicated to the implementation of a research effort to help utilities respond to regulatory requirements and traditional high-priority concerns of the industry. The research agenda is developed through a process of consultation with subscribers and drinking water working professionals. Under the umbrella of a Strategic Research Plan, the Research Advisory Council prioritizes the suggested projects based upon current and future needs, applicability, and past work; the recommendations are forwarded to the Board of Trustees for final selection. The foundation also sponsors research projects through the unsolicited proposal process; the Collaborative Research, Research Applications, and Tailored Collaboration programs; and various joint research efforts with organizations such as the U.S. Environmental Protection Agency, the U.S. Bureau of Reclamation, and the Association of California Water Agencies.

This publication is a result of one of these sponsored studies, and it is hoped that its finding will be applied in communities throughout the world. The following report serves not only as a means of communicating the results of the water industry's centralized research program, but also as a tool to enlist the further support of the nonmember utilities and individuals.

Projects are managed closely from their inception to the final report by the foundation's staff and large cadre of volunteers who willingly contribute their time and expertise. The foundation serves a planning and management function and awards contracts to other institutions such as water utilities, universities, and engineering firms. The funding for this research effort comes primarily from the Subscription Program, through which water utilities subscribe to the research program and make an annual payment proportionate to the volume of water they deliver and consultants and manufacturers subscribe based on their annual billings. The program offers a cost effective and fair method for funding research in the public interest.

A broad spectrum of water supply issues is addressed by the foundation's research agenda: resources, treatment and operations, distribution and storage, water quality and analysis, toxicology, economics, and management. The ultimate purpose of the coordinated effort is to assist water suppliers to provide the highest possible quality of water economically and reliably. The true benefits are realized when the results are implemented at the utility level. The foundation's trustees are pleased to offer this publication as a contribution toward that end.

Perchlorate originates as a contaminant in ground and surface waters. Based on recent health effects data, perchlorate was placed on the Contaminant Candidate List in 1998. The objective of this study was to determine the removal/ rejection of perchlorate ion by pressure-driven membranes at bench and pilot scale, to evaluate the effects of water quality parameters, and to provide/develop transport parameters/models that can be used in engineering correlations. Results of this research will be useful as drinking water utilities plan to meet a future regulation.

Edmund G. Archuleta, P.E.
Chair, Board of Trustees
Awwa Research Foundation

James F. Manwaring, P.E.
Executive Director
Awwa Research Foundation

ACKNOWLEDGMENTS

PROJECT ADVISORY COMMITTEE

The advice and help provided by the Project Advisory Committee (PAC) and Awwa Research Foundation (AwwaRF) project officer, Traci Case, are sincerely appreciated. The PAC consisted of Ken Thompson (CH2M Hill), Jim Lozier (CH2M Hill), Shankar Chellam (University of Houston), and Zahra Panahi (City of Riverside Public Utilities).

PARTICIPATING UTILITIES

This work has been supported by AwwaRF, with participation by a number of drinking water utilities, including several AwwaRF subscribers. The authors of this report are indebted for the cooperation and participation of the following water treatment utilities that were involved in this project: Metropolitan Water District of Southern California (MWDSC), San Gabriel Basin Watermaster (SGBW), City of Redlands, City of Riverside, Southern Nevada Water Authority (SNWA).

EXECUTIVE SUMMARY

INTRODUCTION

Perchlorate ion (ClO_4^-) is an oxidizing agent that originates as a contaminant in ground and surface waters from the dissolution of ammonium, magnesium, or sodium salts associated with manufacturing processes (e.g., rocket fuel). Under typical pH conditions found in natural waters, ClO_4^- exists as an anion in association with perchloric acid ($HClO_4$), a strong acid. Perchlorate has a molecular weight of 100 and exhibits an ioninc radius of ~1.4×10^{-10} m, larger than that of Cl^- but smaller than that of SO_4^{2-}.

With the recent emergence of perchlorate (ClO_4^-) contamination as an important parameter in drinking water quality, this study involved a careful assessment of high-pressure membrane technology for ClO_4^- removal. In particular, reverse osmosis (RO), nanofiltration (NF), and *tight* ultrafiltration (UF) membranes were tested to evaluate the ability of several classes of pressure-driven (ΔP) membranes to reject ClO_4^- by various hypothesized mechanisms, including solution-diffusion for RO/*tight* NF, steric (size) exclusion by NF/*tight* UF, and electrostatic (charge) exclusion by charged membranes. Through rigorous bench-testing of a broad materials-spectrum of membranes, both water quality (e.g., pH) and membrane operational factors (e.g., percent recovery) and their effects on ClO_4^- rejection were investigated in this study.

MEMBRANE CHARACTERIZATION AND TESTING PROTOCOLS

Each membrane was characterized according to molecular mass cut-off/molecular weight cut-off (MMCO/MWCO), surface/pore charge (zeta potential), and membrane material; this information augmented that supplied by the manufacturer (e.g., polymeric material). Both natural waters (with ambient and/or spiked levels of ClO_4^-) and synthetic (model) waters were studied at the bench-scale with a cross-flow test unit using flat-sheet specimens of membranes. This approach highlighted short-term rejection as a function of water quality and selected operational parameters; here, the goal was to define membrane-treatment conditions that maximize percent-rejection of ClO_4^- while minimizing the percent reject (concentrate) stream under economically-reasonable productivity conditions. The data obtained from the bench-scale cross-flow flat-sheet filtration tests were modeled by application of a non-equilibrium thermodynamic model, based on *hindered* diffusion embodying membrane properties (e.g., charge). Pilot-scale single elements tests were also performed with natural waters to demonstrate long-term perchlorate rejection trends, and the effects of fouling on rejection.

REPORT ORGANIZATION AND STUDY BREADTH

This report is organized according to twelve (12) chapters, including: introductory material (Chapter 1); general background information (Chapter 2); study objectives and hypotheses (Chapter 3); general methods and protocols (Chapter 4); bench-scale evaluations to address the influence of membrane properties on perchlorate rejection (Chapter 5), assess perchlorate rejection mechanisms (Chapter 6), develop a quantitative perchlorate rejection model based on *hindered* diffusion (Chapter 7), and discern the effects of operational variables on perchlorate rejection (Chapter 8); tradeoffs between full-stream NF versus split-stream RO treatment in terms of water quality (perchlorate and corrosivity) and costs (Chapter 9); special studies to highlight use of surfactant-modified UF for enhanced perchlorate rejection and the effects of perchlorate as an oxidizing agent on membrane flux and rejection (Chapter 10); pilot-scale tests in the form of single elements tests to examine long-term perchlorate rejection trends and the impacts of fouling (Chapter 11); and conclusions and recommendations for future research (Chapter 12).

OBJECTIVES AND HYPOTHESES

The primary objectives of the study were to:

1. determine the removal/rejection of perchlorate ion (ClO_4^-) by pressure-driven (ΔP) membranes, including RO, NF, and *tight* UF membranes;
2. evaluate the effects of water quality parameters, pH, ionic strength (conductivity), and co- and counter ions, on process performance;
3. study membrane operational conditions such as percent recovery and the J_o/k ratio (the ratio of initial flux (J_o) to back-diffusional mass transfer coefficient (k)) on perchlorate rejection, with the goal of maximizing system (feed water) recovery and minimizing the volume of retentate;
4. develop a transport model that can be used to simulate/predict perchlorate rejection; and
5. evaluate ClO_4^- removal performance in single element tests (with RO, NF, and UF) for assessing long term rejection of perchlorate from natural waters.

The following hypotheses have been tested for perchorate rejection:

1. membrane properties enhancing ClO_4^- rejection include lower MMCO and a greater (negative) surface charge; and
2. water quality conditions promoting ClO_4^- rejection include high pH and lower conductivity (ionic strength).

RESULTS

One of the significant findings of this research was that known and hypothesized rejection mechanisms (e.g., solution diffusion, steric (size) exclusion, and electrostatic interaction (repulsion) for solute transport (rejection) of a target ion (ClO_4^-) were demonstrated by several measurement techniques. Another significant finding of this research was that perchlorate transport was predicted well by a non-equilibrium thermodynamic model, based on hindered diffusion embodying membrane properties (e.g., charge).

Changes in membrane surface charge, becoming more negative as pH increases and conductivity decreases, were found to enhance perchlorate rejection because electrostatic repulsion significantly occurred between the perchlorate anion and negatively charged membranes. Changes in pore size for a given membrane, becoming smaller in the presence of co- and counter ions, were found to not significantly enhance perchlorate rejection because the ionic radius of the perchlorate ion is still much smaller than the pore size of NF and *tight* UF membranes. However, for the range of different membranes studied, RO and relatively small pore size NF membranes showed higher perchlorate rejection than relatively large pore size NF and *tight* UF membranes, indicating that steric exclusion affected perchlorate transport. Because RO membranes operate under a different rejection mechanism (solution-diffusion versus steric and/or electrostatic) rejection, their membrane properties are less influential.

It was shown that electrostatic repulsion did not significantly occur between an uncharged inorganic species, As(III) in the form of H_3AsO_3, and negatively charged membranes. As (III) rejection only slightly varied over a pH range of 4 to 8.0 for two (negatively) charged NF membranes since the As (III) exists primarily as an uncharged species below pH 9.13 based on its pK_a of 9.13. However, As (III) rejection significantly increased at pH 10 where As (III) is anionic.

Changes in membrane surface charge, becoming more negative as pH increased and/or conductivity decreased, were found to reduce hindered diffusion and partitioning for negatively charged NF and UF membranes. Measured hindered diffusion was used to predict perchlorate transport dominated by a diffusion (versus a convection) driving force. The predicted perchlorate flux by both diffusion and convection decreased with increasing pH and increased with increasing conductivity. The predicted perchlorate flux by convection was much greater than that by diffusion regardless of pH and conductivity conditions. These results are consistent with the theory that solute flux is dominated by convection for a relatively large pore size membrane.

An increase in recovery and J_o/k ratio resulted in a decrease of perchlorate rejection for RO and NF membranes. For the RO membranes, an assumption was made that an increase of recovery and J_o/k ratio significantly affected solution diffusion due to concentration polarization. For the NF membranes, an increase of concentration polarization of solutes with increasing recovery and J_o/k ratio also significantly affected membrane surface charge and slightly affected membrane pore size, which significantly increased perchlorate transport. Perchlorate rejection was significantly higher in perchlorate-spiked pure water versus high-TDS natural waters (e.g., Colorado River water), particularly for NF and tight UF versus RO.

Based on results of four different membranes at varying recoveries with natural waters, calcite ($CaCO_3$) scale was dominant at the membrane surface and detected by various measurements. Antiscalants, characterized by phosphonate groups, were effective in the inhibition of calcite formation.

The concentration of various solutes at the membrane surface can be predicted by a semi-empirical constant (K) related to concentration polarization as influenced by recovery. Considering results for each ion (perchlorate and other anions, and cations) and four membranes, most K values of ions with RO membranes approached 1, while K values for NF membranes were smaller than RO membranes.

In an evaluation of costs and water quality of full- versus split-stream treatment with respect to perchlorate rejection and corrosion potential and effects on water cost associated with design and operational parameters, split-stream RO treatment was shown to be more cost effective than full-stream NF treatment for an illustrative analysis based on a specific RO and NF membranes. Also, split RO stream treatment reduced the corrosion potential of RO full stream treatment, providing a cost-effective alternative to, and better water quality including rejection of perchlorate than, NF full stream treatment. Considering design and operational parameters such as maintenance cost, interest rate, power cost, chemical cost, plant life, and membrane replacement, each parameter affected the product water cost.

Perchlorate rejection by a UF membrane modified with a cationic surfactant was greater than expected, based mostly on steric/size exclusion as a result of a decrease in the membrane pore size imparted by the surfactant conditioning. However, a limitation in use of a cationic surfactant to modify a negatively charged UF membrane for perchlorate rejection was that a significant degree of flux-decrease also occurred. As expected, perchlorate transport was not affected by anionic surfactant conditioning, which did not significantly influence membrane pore size and/or membrane surface charge.

In tests performed to study perchlorate as an oxidizing agent, no membrane degradation was observed, either in terms of flux or rejection properties.

In single element tests representing pilot testing, RO and NF membranes showed effective perchlorate rejection from a natural water regardless of the influent concentration, however, a *tight* UF membrane proved ineffective for removing perchlorate. Tests were performed with and without reject-stream recycle, with recycle employed to simulate an element situated later in an array. While RO results can be explained by simple solution-diffusion as a rejection mechanism for ions including perchlorate, electrostatic rejection appeared to be the dominant mechanism of rejection for the NF membranes. For long-term tests, the specific flux declined by 25% for NF without recycle, which indicates fouling, and perchlorate rejection decreased slightly after fouling. With 50% recycle, the NF membrane did not foul within 30 days of operation under similar condition compared to without recycle. Also, for long-term tests with and without recycle, the RO membrane did not show any signs of fouling for all conditions tested, and consistent perchlorate rejection was observed. EC reduction correlated well with perchlorate rejection, and can be used as a good indicator for perchlorate rejection.

RECOMMENDATIONS TO UTILITIES

Recommendations to utilities are provided through insight into membrane selection and operation as a function of treatment objective (perchlorate rejection) and water quality (other anions and cations). A major water quality constraint is TDS (conductivity) imparted by various co- (e.g., SO_4^{2-}) and counter- (e.g., Ca^{2+}) ions and, in the case of (negatively) charged membranes, lower pH conditions. A major operational tradeoff is embodied by recovery and the J_o/k ratio; as recovery and J_o/k increase, productivity increases but rejection decreases. The choice between RO and NF is a key decision, with RO virtually required for high TDS waters; an important related decision is whether to employ spilt-stream RO versus full-stream NF. An important issue not addressed in this study is reject (concentrate) stream disposal; one interesting possibility would be to investigate the synergistic possibility of employing a biological processes for treating a high-concentration perchlorate brine stream.

CHAPTER 1
INTRODUCTION

With the recent emergence of perchlorate (ClO_4^-) contamination as an important parameter in drinking water quality, this study involved a careful assessment of membrane technology for ClO_4^- removal. In particular, reverse osmosis (RO), nanofiltration (NF), and ultrafiltration (UF) membranes were tested to evaluate the ability of several classes of pressure-driven (ΔP) membranes to reject ClO_4^- by well-understood mechanisms. These mechanisms included solution-diffusion for RO/*tight* NF and steric (size) exclusion by NF/*tight* UF, as well as by less-understood mechanisms such as electrostatic exclusion by charged membranes. Through rigorous bench-testing of a broad materials-spectrum of membranes, both water quality (e.g., pH, total dissolved solids (TDS), dissolved organic carbon (DOC), temperature, etc.) and membrane operational factors (e.g., J_o/k ratio and percent recovery) and their effects on ClO_4^- rejection were investigated in this study. Each membrane was characterized according to molecular mass cut-off/molecular weight cut-off (MMCO/MWCO), surface/pore charge (zeta potential), and membrane material (attenuated total reflection-fourier transform infrared (ATR-FTIR) spectroscopy); this information augmented that supplied by the manufacturer (e.g., polymeric material). Both natural waters (with ambient and/or spiked levels of ClO_4^-) and synthetic model waters were studied with a cross-flow test unit using flat-sheet specimens of membranes. This approach highlighted short-term rejection as a function of water quality and selected operational parameters (e.g., J_o/k and recovery). The goal was to define membrane-treatment conditions that maximize percent-rejection while minimizing the volume of the reject stream under economically-reasonable productivity conditions, as represented by specific flux (L/m^2-day-kPa or gal/ft^2-day-psi). The bench-testing program was preceded by a comprehensive literature survey to help finalize testing conditions, including membrane type, water qualities, and operational conditions.

The data obtained from the bench-scale cross-flow flat-sheet filtration tests were modeled by application of a non-equilibrium thermodynamic model. The model has five transport parameters: the molecular transport coefficient (ω), osmotic pressure gradient ($\Delta\Pi$), molecular reflection coefficient (σ), the average bulk fluid interfacial concentration between the feed and permeate side (C_{avg}), and the solvent flux (J_v). The transport parameters were determined by independent measurements (and calculation with minimum assumptions). For example, the molecular transport coefficient (ω) was obtained by diffusion cell measurements under varying pH and conductivity conditions.

IDENTIFICATION OF RESEARCH NEEDS

Public Health

An Environmental Protection Agency (EPA) report (U.S. EPA, 1998) summarizes the effects of perchlorate on human health based on tests with laboratory animals (rats). In that report, the target tissue of perchlorate toxicity is the thyroid gland in both adult and postnatal rats. Hormonal effects occurred at the lower range of exposures tested, from 0.01 to 1.0 mg/kg-day. Also neurobehavioral effects and effects in the offspring occurred at higher concentrations. Thyroid tumors were observed in previous studies with rats exposed in long-term bioassays at high doses (Caldwell *et al.,* 1995, Mannisto *et al.*, 1979, and Shigan, 1963). However, there is currently no National Primary Drinking Water Regulation for perchlorate. Perchlorate has been placed on the Contaminant Candidate List since March 1998; as such, it has been listed as a contaminant that requires additional research and occurrence information before regulatory determinations can be considered. The EPA and the California Department of Health Service (CDHS) have recently developed an analytical (ion chromatographic) procedure of ClO_4^- with a minimum detection limit (MDL) of 2 μg/L. Based on a U.S. EPA assessment of ClO_4^- toxicology, CDHS has defined a *provisional action level* of 4 μg/L (U.S. EPA, 1998 and Hautman and Munch, 1999).

Occurrence of Perchlorate in Drinking Water and Impact on Water Utilities

Perchlorate began to be discovered at various manufacturing sites and in well waters and drinking water supplies during the months following the 1997 development of a low level (2 μg/L) detection method. There are 14 states with confirmed releases in ground or surface waters. There are 44 states that have confirmed perchlorate manufacturers or users based on EPA Information Request responses. In California, most of the locations where perchlorate has been detected are associated with facilities that have manufactured or tested solid rocket fuels for the Department of Defense or the National Aeronautics and Space Administration. A ClO_4^- occurrence survey in California showed detectable perchlorate found in 69 of 232 wells samples, with 24 exhibiting levels above 18 μg/L; samples derived from highly contaminated sites contained up to 8,000 to 9,000 μg/L.

RELATED TECHNOLOGIES FOR PERCHLORATE REMOVAL

Ongoing research is investigating various treatment technologies for perchlorate rejection, including anaerobic biodegradation, ion exchange, and granular activated carbon, among other processes. A number of these processes have recently demonstrated the capability to meet the current perchlorate *provisional action level* of 4 μg/L for drinking water. These include i) an adsorption process using biologically activated carbon adsorption systems (Brown *et al.*, 2000), ii) perchlorate removal by electro-dialysis (Booth *et al.*, 2000), and iii) biological treatment of perchlorate (Logan *et al.*, 2000). Meanwhile, it has been recognized that pressure driven (ΔP) membrane technologies can provide a viable treatment opportunity to meet *the provisional action level* for perchlorate among the technologies (Yoon *et al.*, 2000).

Clearly, there has been ample demonstration of removal of other anions such as nitrate (NO_3^-) by RO. Moreover, recent work has shown rejection of nitrate (NO_3^-) by UF membranes (Morel *et al.*, 1997 and Morel *et al*, 1991) and As (III) and As (V) rejection by RO, NF, and UF membranes (Urase *et al.*, 1998 and Brandhuber and Amy, 1998), with NO_3^- and As (V) reflecting some general chemical similarity to ClO_4^-. In another study (Waypa *et al.*, 1997), bench tests of As (III) and As (V) filtration by two different RO and one NF membrane showed removals of > 97.5%, > 96% and 94%, respectively. Several studies were performed to investigate the separation potential of RO, NF and/or UF membranes in transport of common ions (e.g., Cl^-, SO_4^{2-}, Na^+/K^+, and Ca^{2+}/Mg^{2+}) (Rautenbach and Groschl, 1990, Eriksson, 1988, and Cadotte *et. al.*, 1988).

It is expected that membrane treatment, particularly RO, will be capable of significantly reducing ClO_4^-, potentially to levels near the detection limit. However, instead of considering only RO, it would be prudent to also examine more productive membranes (NF and possibly UF, which operate under lower pressure requirements) that theoretically may provide sufficient rejection of ClO_4^- to be economically viable.

STRUCTURE OF THE REPORT

This report is into twelve (12) chapters. Beyond this introductory chapter (Chapter 1), Chapter 2 provides some general background information, Chapter 3 articulates study objectives and hypotheses, and Chapter 4 summarizes general methods used throughout (except for chapter-specific methods). Chapters 5 - 8 summarizes bench-scale evaluations including influence of membrane properties on perchlorate rejection (Chapter 5), probing of perchlorate rejection mechanisms (Chapter 6), quantitatively explaining perchlorate rejection through a transport model based on *hindered* diffusion (Chapter 7), and examining the effects of operational variables (e.g., recovery) on perchlorate rejection (Chapter 8). Chapter 9 addresses tradeoffs between full-stream NF versus split-stream RO treatment in terms of water quality and costs. Two special studies are highlighted in Chapter 10, use of surfactant-modified UF for enhanced perchlorate rejection, and the effects of perchlorate as an oxidizing agent on membrane flux and rejection. Chapter 11 summarizes pilot-scale tests in the form of single elements tests to examine long-term perchlorate rejection trends. The report culminates in Chapter 12, conclusions and recommendations for future research.

The applications potential provided by this report to utilities is general insight into membrane selection (RO versus NF) and operation (e.g., recovery) as a function of treatment objective (perchlorate rejection) and water quality (other anions and cations).

CHAPTER 2
BACKGROUND AND LITERATURE REVIEW

PERCHLORATE CHEMISTRY AND TRANSPORT IN NATURAL WATERS

Perchlorate is an oxidizing agent anion that originates as a contaminant in ground and surface waters from the dissolution of ammonium, magnesium, or sodium salts. Perchlorate is very mobile in aquatic systems and can persist for many decades under typical ground and surface water conditions. Under typical pH conditions found in natural waters, perchlorate exists as an anion (ClO_4^-) associated with perchloric acid ($HClO_4$), a strong acid. Perchlorate (ClO_4^-), the oxidation product of chlorate (ClO_3^-), represents the highest oxidation state (+7) of the element chlorine (Cl). Perchlorate is an anion with a molecular mass of 100 and an ionic radius of ~ 1.37×10^{-10} m (calculated from Stokes Einstein Equation based on its diffusion coefficient). The ionic radius of ClO_4^- is larger than that of OH^- (4.73×10^{-11} m) and Cl^- (1.21×10^{-10} m) and smaller than that of SO_4^{2-} (2.30×10^{-10} m) (Brandhuber, 1999).

MEMBRANE CHARACTERIZATION

Membrane Stabilization and MMCO/MWCO

Nystrom *et al.* (1995) studied fouling and retention of four different nanofiltration membranes using different model substances. In their study, the membranes were first stabilized for approximately ~ 4 h with a pressure of 10 bars at 25 °C, then the pure water flux of the membranes was measured. Stabilizing the membrane and then measuring its pure water permeability is an important initial step before making other measurements, since it provides an indicator of overall porosity/permeability among samples of a given type of membrane.

Several materials were tested to study molecular mass cut-off of porous membranes in previous studies. An uncharged model species, polyethylene glycols (PEGs), has been widely used for MMCO measurements of porous membranes. It is important to determine whether and how much membrane pore sizes may vary in the presence of co- and counter ions. Cho *et al.* (2000) reported rejection tests with an uncharged model species, polyethylene glycols (PEGs), to determine the effects of mono-valent and di-valent ions, Cl^-, Na^+, and Ca^{2+}, on membrane pore size. They found that a particular ultrafiltration (UF) membrane, whose nominal molecular mass cut-off (MMCO) was 8,000, rejected smaller molecular mass PEGs in deionized water than in NaCl or $CaCl_2$ solutions. In the study, the UF membrane apparently had MMCO value of greater than 8,000 for neutral linear macromolecules. Okazaki *et al.* (1998) performed a rejection test with several model chemicals (e.g., dodecane, squalane, and toluene) in the presence of additive agents (e.g., phenanthrene, pyrene, and triphenyl phosphate). In their study, by changing the additive agents, the MMCO of the fabricated aromatic polyimide asymmetric membranes could be adjusted in the range of 400 – 650. From their study, it is assumed that solute transport can be affected by the change of membrane pore due to the presence of organic compounds or natural organic matter, which can act as an additive agent in membrane filtration.

Zeta Potential for Membrane Surface Charge

Electrostatic exclusion is known to be one of the dominant mechanisms between negatively charged solutes (such as the perchlorate ion) and negatively charged NF and UF membranes. The surface charge or electrical potential properties of NF and UF membranes can significantly influence their separation performance and solute transport. Membrane surface charge can be estimated by zeta potential, calculated from measurements of the streaming potential. A zeta potential (ZP) is defined as the potential across the fluid shear plan at the membrane surface. Several studies were performed to measure the surface charge of clean and/or foulded membranes as a function of pH and/or ionic strength in the presence of different salts (Zhu and Nystrom, 1998, Cho *et al.,* 1998, Hong and Elimelech, 1997, Childress and Elimelech, 1996). Zhu and Nystrom compared the streaming potential of virgin, pre-cleaned (with NaOH), fouled (with protein), and cleaned (with alkaline cleaning agents) UF membranes made of polysulfone. In their study, the zeta potential of all pre-cleaned membranes became more negative. The ZP of the membranes at the four different experimental stages follows the order (negatively): virgin < fouled < pre-cleaned < cleaned.

Childress and Elimelech (1996) studied the effects of pH and various ions, Cl^-, SO_4^{2-}, Na^+, Ca^{2+}, and Mg^{2+}, on the membrane surface charge of a reverse osmosis (RO) membrane and NF membranes. For all of the membranes used in that study, the membrane surface charge was positive in the lower pH range, passed through an iso-electric point (between pH 3 and pH 5) and then became negative over the higher pH range. In addition, the surface charge of an NF membrane (NF-70) became less negative in the presence of $CaCl_2$ and Na_2SO_4 salts. It is intuitive that the measured surface charge depends on the functional groups associated with different membrane materials.

Bowen and Cao (1998) studied electrokinetic effects in pores containing electrolytes by using a numerical solution of the nonlinear Poisson-Boltzmann equation. Their study suggests that the calculation of the surface charge or electrical potential properties in terms of zeta potential requires numerical solution of the equations governing fluid flow and electrical-potential distribution in the pores to avoid substantial underestimation of the true zeta-potential.

ATR-FTIR Spectroscopy

ATR-FTIR spectra can sometimes be used to determine the membrane's functional groups. Cho *et al.* (1998) previously presented the ATR-FTIR spectra of UF (GM) and NF (ESNA) membranes. The infrared spectroscopy was also used to identify organic foulants that are readily distinguished from the absorption pattern of other compounds of an NF membrane (Her *et al.*, 2000).

MEMBRANE FILTRATION

Membrane filtration is an inherently non-equilibrium process (Mulder, 1991). Three approaches have been reviewed; the Donnan exclusion, extended Nernst-Planck, and non-equilibrium thermodynamics models; to describe perchlorate anion and other co- and counter ion transport through (and rejection by) membranes.

Hindered Diffusion and Partition in Membranes

One simple way to measure diffusion coefficients is the diaphragm (diffusion) cell. Several studies show the use of the diffusion cell to determine diffusion coefficients of solutes in porous membranes (Masselin *et al.*, 2000, Shao and Baltus, 2000, and Clark and Lucas, 1998). Masselin *et al.* (2000) determined a structural parameter, the porosity to membrane thickness ratio, by means of the diffusion experiments using aqueous solutions of neutral solutes (formamid and urea) and strong electrolytes (hydrochloric acid (HCl) and potassium chloride (KCl)) for UF and microfiltration membranes. Shao and Baltus (2000) studied the effects of solute concentration on hindered diffusion by measuring diffusion rates of dextran and polyethylene glycols (PEGs) in track-etched polycarbonate membranes. Clark and Lucas (1998) studied diffusion and partitioning of humic acid in a porous UF membrane.

In those studies, partitioning of solutes was carefully observed as an influencing parameter affecting its diffusivity when solute concentration is sufficiently large, which plays an important role influencing intermolecular interactions. Deen and Smith (1982) studied the effects of electrostatic interactions on partitioning for solutes and pores of like charge. In their study, partition coefficient increased as ionic strength increased. For partitioning of a solute, electrostatic interactions tend to be most prominent for small pore and/or low electrolyte concentrations even though steric and hydrodynamic interactions between the solute and pore will also affect the partitioning (Deen and Smith, 1982). In this study, partitioning of perchlorate was considered as the main phenomena influencing hindered diffusion with two assumptions. First, only a minimum amount of partitioning occurred between the perchlorate anion and the negatively charged membranes used in this study. Second, the potential partitioning of a perchlorate concentration of 10^{-6} M was very small due to competition between the perchlorate anion and a co-ion (e.g., Cl^- or SO_4^{2-}) having a relatively high concentration (0.002 to 0.008 M) compared to the perchlorate ion.

There are many parameters such as solute size, charge, shape, and membrane pore size, charge, and pore density, which influence hindered transport of the solutes in liquid-filled membrane pores. In our research, two parameters, pH and conductivity (ionic strength), were carefully studied to evaluate how each plays an important role in the change of membrane surface charge. An assumption was made that the membrane surface charge significantly affects hindered transport of the perchlorate ion in membrane filtration tests. Several studied have shown the effects of membrane charge on diffusion. Latex particles were used to examine the effects of electrostatic interactions on hindered diffusion in track-etched mica membranes (Malone and Anderson, 1978). Both particles and membrane pores were negatively charged. In their study, effective diffusivity of the solutes significantly declined as ionic strength was reduced, which caused an electric double layer increase, as expected, for electrostatic repulsion occurring between the negatively charged particle and the negatively charged membrane. Deen and Smith (1982) determined solute and membrane pore charge from measurements of electrophoretic mobility and streaming potential to determine the effects of charge on hindered diffusion. In their study, effective diffusion of a small solute (glucose) increased as ionic strength (KCl concentration) increased from 0.005 M to 0.1 M.

Transport Models for Charged Membranes

The non-equilibrium thermodynamic transport model and hindered transport will be employed in this study to predict perchlorate ion transport through NF and UF membranes. In addition, the theory of the extended Nernst-Planck Model combined with Donnan Equilibrium will be employed to describe perchlorate ion transport through the membranes.

Non-Equilibrium Thermodynamic Transport Model

The non-equilibrium thermodynamic approach views the membrane as a black box. Since filtration is a non-equilibrium, irreversible process, free energy is dissipated and entropy is continuously produced. Therefore, the thermodynamics of the irreversible processes can also be used to describe membrane transport based on a *three coefficient* system taking into account the interactions: solute-solvent, solute-membrane, and solvent-membrane (Kedem and Katchalsky, 1958). By determining the dissipation function, membrane performance can be interpreted thermodynamically (Mulder, 1991, Spiegler and Kedem, 1966, and Kedem and Katchalsky, 1958) even though this approach may not address the mechanisms of rejection. In fact, the model gives a 'real' description of ion transport through membranes even though the membrane is treated as a "black box" and the model provides little insight with no physico-chemical view into the physical processes involved in solute transport and solvent across a membrane. Therefore, the thermodynamic model can be widely applied, especially when the parameters used in the model are experimentally measured (or calculated with minimum assumptions).

A thermodynamic model was used by several authors to predict ion transport in porous membranes. Bauza *et al.* (1990) determined the cation transport number and the water transport number of a cellulose acetate membrane separating two solutions of mono-valent electrolytes. Pusch (1990) used a thermodynamic model to obtain a linear relationship for solute rejection as a function of volumetric flux.

Equations 2.1 and 2.2 indicate that transport across a membrane is characterized by three transport parameters, the solvent (water) permeability, the solute permeability, and reflection coefficient.

$$J_s = diffusion + convection = \omega \Delta \Pi + (1-\sigma) C_{avg} J_v \tag{2.1}$$

$$J_v = L_p(\Delta P - \sigma \Delta \Pi) \tag{2.2}$$

where J_s = solute flux, ω = molecular transport coefficient (solute permeability) = $D_{eff}\, H/lRT$, D_{eff}= observed *effective* diffusion coefficient in the membrane, H = observed partition coefficient of the electrolyte in the membrane, l = thickness of the separating layer of the membrane, R = universal gas constant, T = absolute temperature, $\Delta\Pi$ = osmotic pressure gradient, σ = molecular reflection coefficient, C_{avg} = average bulk fluid concentration between feed and permeate side, J_v = solvent flux, and L_p = filtration coefficient (solvent permeability).

Equations 2.1 and 2.2 indicate that three transport parameters, the solvent (pure water) permeability, the solute permeability, and the reflection coefficient, are important for predicting the transport across a membrane (Spiegler and Kedem, 1966, and Kedem and Katchalsky, 1958). It is assumed that the parameters shown in equations 2.1 and 2.2 are influenced by solution pH, ionic strength, and co-ions and counter-ions. More details of the theory on this thermodynamic model are described in Chapter 7.

Extended Nernst-Planck Model Combined With Donnan Equilibrium

When a dilute solution containing ions is brought in contact with a membrane possessing a fixed charge, the passage of ions possessing the same charge as the membrane (called co-ion) can be inhibited. This condition is termed Donnan Exclusion (Strathmann, 1992). The use of the extended Nernst-Planck (ENP) equation in conjunction with the Donnan condition provides the possibility of characterizing the effective membrane pore size and effective charge density to predict the separation of mixtures of electrolytes at the membrane/solution interface (Bowen and Mukhtar, 1996). The effective membrane charge depends on the bulk co-ion concentration associated with mono-valent and divalent co-ions if solution concentrations are described in equivalents (Bowen and Mukhtar, 1996). For example, Brandhuber (1999) used the combined Donnan Equilibrium/Extended Nernst-Planck model to predict As (III), a nonionic species, and As (V), an anion, rejections, as influenced by feed water pH and ionic composition, for charged NF membranes.

Structural models, such as the Hagen-Poiseuille or the Carmen-Kozeny equations are also available (Mulder, 1991). In these equations, however, ΔP is assumed to be the only potential gradient. Electrostatics, that is, the presence of an electrical potential gradient, can also influence perchlorate transport through (rejection by) a charged membrane. A model describing perchlorate transport through a charged membrane must include pressure and electrical potential terms. The extended Nernst - Planck equation, as formulated by Bowen and Mukhtar (1996), is presented as a means to evaluate membrane performance for this proposal. The equation is:

$$j_i = -D_{i,p}\frac{\mathrm{d}c_i}{\mathrm{d}x} - \frac{z_i c_i D_{i,p}}{RT}F\frac{\mathrm{d}\psi}{\mathrm{d}x} + K_{i,c}c_i J_v \qquad (2.3)$$

where j_i = flux if ion i, c_i = concentration of ion i in the membrane, z_i = valance of ion i, J_v = volumetric flux, ψ = the electrical potential gradient, F = Faraday constant, R = universal gas constant, T = absolute temperature, x = position in membrane, and $D_{i,p}$ and $K_{i,c}$ are fitted hindrance factors for diffusivity and convection respectively. The left hand side of this equation is the flux of the ion in question (perchlorate), the three terms on the right hand side represent the effects of diffusion, electrostatic potential, and convection on the ion in question.

The Nernst-Planck equation was modified to predict solute transport in terms of convection, diffusion, and electrostatic potential (Van der Horst *et al.*,1995 and Timmer *et al.*, 1993). Garba *et al.* (2000) developed a simple model based on Nernst-Plank and film theory equations to describe transport mechanism of cadmium salts through an NF membrane.

MEMBRANE OPERATION

Mass Transfer Coefficient and J_o/k Ratio

A hydrodynamic operating parameter, the ratio of the initial pure water flux (J_o) to the estimated back-diffusional mass transfer coefficient (k) through the boundary layer due to increasing concentration polarization associated with solutes being concentrated by membrane rejection near the membrane interface, can be used to facilitate comparison of results from different membranes in bench-scale cross-flow flat-sheet tests (Cho *et al.*, 1999). The J_o/k was used as a comparable hydrodynamic operating condition in that study. The mass transfer coefficient (k) can be estimated by using a diffusion coefficient (D) of 1.79×10^{-5} cm^2/s for ClO_4^- in water (Lide, 1995) and the equation for the mass transfer coefficient under laminar flow in a channel presented by Porter (1972):

$$k = 1.177\left[\frac{U_b D^2}{hL}\right]^{1/3} \tag{2.4}$$

where U_b is the average cross-flow velocity of bulk fluid (cm/s), D is the diffusion coefficient of solute in water (cm^2/s), h is the channel height (cm), and L is the channel length (cm).

Effect of J_o/k and Recovery on Perchlorate Rejection and Inorganic Scale Formation With Natural Waters

An important objective of this study is to determine the limits of membrane operation (i.e. recovery and J_o/k ratio) in treating perchlorate-contaminated water. Higher recovery and J_o/k ratio increase the product stream and reduce the rejection (concentrate) stream, perchlorate rejection is decreased and inorganic scale formation is increased.

In this study, we have investigated rejection of perhchlorate, water quality parameters, and flux decline over a various range of recoveries, and J_o/k values, analysed inorganic foulants on membrane surfaces, simultated inorganic precipitation by an equilibrium software package MINEQL, and tested inhibition of inorganic fouling with antiscalants.

MEMBRANE FLUX-DECLINE WITH NOM

Several studies have shown effects of natural organic matter (NOM) on membrane flux as a function of pH and ionic strength in the presence of anions and cations. Braghetta *et al.* (1997) used hollow membranes to compare flux declines with Suwannee River organic matter at varying pH conditions. In their study, electrostatic effects at different pH conditions were shown to compare flux declines. Hong and Elimelech (1997) performed flux-decline tests with humic acids for a TFC NF membrane in the presence of Ca^{2+} and Cl^-. In that study, the increased Ca^{2+} concentration and ionic strength reduced the negatively charge of humic acids and the membrane surface charge due to compression of the electrical double layer.

SURFACTANT MODIFIED ULTRAFILTRATION FOR ClO_4^- REMOVAL

Ultrafiltration (UF) alone is not effective in perchlorate rejection because of the pore size of UF membranes, observed rejection can be attributed to electrostatic exclusion. In this study, we have investigated how the presence of a surfactant (cationic or anionic) affects electrostatic interaction (charge repulsion) and steric (size) exclusion mechanism between a negatively charged porous UF membrane and perchlorate.

CHAPTER 3
RESEARCH OBJECTIVES AND HYPOTHESES

OBJECTIVES

The primary objectives of the research proposed were to:

1. determine the removal/rejection of perchlorate ion (ClO_4^-) by pressure-driven (ΔP) membranes, including RO, NF, and *tight* UF membranes;
2. evaluate the effects of water quality parameters, pH, ionic strength (conductivity), and co- and counter ions, on process performance;
3. study membrane operational condition such as J_o/k ratio (ratio of initial flux to back-diffusional transport), associated with the goal of maximizing system (feed water) recovery and minimizing the volume of retentate; and
4. provide/develop transport parameters/models that can be used in engineering correlations.
5. investigate the rejection of perchlorate, and determine/simulate inorganic foulants on membrane surfaces over a various range of operating conditions (J_o/k ratio or recovery).
6. investigate perchlorate transport (rejection) by a surfactant-modified UF membrane.
7. evaluate ClO_4^- removal performance in single element tests (with RO, NF, and UF) and to determine long term rejection of perchlorate in natural waters.

HYPOTHESES

The following hypotheses have been tested:

1. ClO_4^- rejection:

   a. Membrane properties enhancing ClO_4^- rejection include lower MMCO and a greater (negative) surface charge.
   b. Water quality conditions promoting ClO_4^- rejection include high pH, lower conductivity (ionic strength), and lower calcium concentration.

2. Quantitative tools for ClO_4^- rejection:

Quantitative models can be developed to predict perchlorate-rejection trends that include water composition and membrane characteristics based on a solution-diffusion model (irreversible thermodynamics) with the solute transport parameters explicated by an extended Nernst-Planck-Donnan Exclusion equation.

CHAPTER 4
EXPERIMENTAL METHODS AND MATERIALS

MEMBRANES

Perchlorate rejection measurements were evaluated with two RO, three NF, and four UF membranes. The two RO membranes, LFC-1 and TFC-ULP, and three NF membranes, ESNA, TFC-S, and MX07, are thin film composites made of aromatic polyamide (according to the manufacturers). These membranes have ionizable functional groups such as carboxylic acids and have different membrane pore sizes based on their manufacturer's nominal MMCOs. The four UF membranes, GH, GK, GM, and GN, (all from the same manufacturer) are made of the same material (purportedly a sulfonated polyethersulfone coated with an ultra-thin polyimide) and according to the manufacturer only differ in their pore sizes or nominal MMCOs. These UF membranes were selected to systematically compare the results of steric exclusion for perchlorate rejection while the material chemistry was unchanged.

The surface charge of the LFC-1, ESNA, MX07, GH, GK, GM, and GN, membranes was estimated by measuring streaming potential of the membranes. A commercial electrokinetic analyzer (EKA) measurement apparatus was employed to estimate zeta potential (ZP) of the membranes with the measured streaming potential. Pure water permeability (PWP) was determined with five different pressures for the RO, NF, and UF membranes used in this study. Table 4.1 describes the characteristics of the membranes based on zeta potential (surface charge), nominal MMCO (pore size) and material, and pure water permeability (which is related to pore size and/or pore density).

SOURCE WATERS

Seven natural waters; Colorado River water (CRW) from CA, a blend of Colorado River water and State Project water (CRW/SPW) from CA, San Gabriel Valley Ground Water (SG-GW) from CA, Lake Mead Surface Water (LM-SW) from NV, Bunker Hill Water Basin Ground Water (BHWB-GW) from CA, and Riverside Ground Water (RGW) from CA; and various synthetic water solutions were used to perform bench-scale cross-flow flat-sheet membrane tests. The Colorado River, a major drinking water source for southern California, southern Nevada, and Arizona, contains 4 to 9 μg/L of ClO_4^-, with the Las Vegas Wash of Lake Mead identified as the major contamination source. Synthetic water solutions were prepared with pure water, which was deionized with two proprietary cation-exchange beds and an anion-exchanged bed followed by 0.2 μm-filter filtration. The perchlorate anion (present at a total concentration of 100 μg/L of ClO_4^- "spiked" with $KClO_4$) was introduced to the membrane as a pure component, in binary mixtures with other salts (KCl, K_2SO_4, and $CaCl_2$), and at varying pH conditions (4, 6, 8, and 10) and solution conductivities (30, 60, and 115 mS/m). Potassium hydroxide (KOH) or hydrochloric acid (HCl) was added to pure water to adjust solution pH. Unless otherwise indicated, the natural source waters were pre-filtered with a 0.45 μm filter for all membrane tests.

Table 4.2 describes the characteristics of the feed waters used in this study. Several analyses were performed including DOC, UVA_{254}, conductivity, TDS and pH of the natural

Table 4.1
Characteristics of the membranes

Membrane	Manufacturer	Material	MMCO*	Ionizable groups†	PWP‡ (L/d-m^2-kPa)
LFC-1	Hydranautics	Polyamide TFC	n/a	yes	0.75
TFC-ULP	Fluid System	Polyamide TFC	n/a	yes	0.36
TFC-S	Fluid System	Polyamide TFC	200	yes	1.99
ESNA	Hydranautics	Polyamide TFC	200	yes	1.05
MX07	Osmonics	Polyamide TFC	400	yes	0.47
GH	Desal	Proprietary	2,500	yes	0.74
GK	Desal	Proprietary	3,500	yes	1.69
GM	Desal	Proprietary	8,000	yes	2.81
GN	Desal	Proprietary	12,000	yes	3.68

* Manufacturer's information
† Indicated by zeta potential determinations
‡ Pure water permeability

Table 4.2
Feed water characteristics

Source water	Conductivity (mS/m)	TDS (mg/L)	pH	DOC (mg/L)	UVA_{254} (1/cm)	SUVA (L/mg-m)	ClO_4^- (µg/L)
CRW	114.8	541	8.1	2.25	0.036	1.6	3.8*/100†
CRW/SPW	75.7	359	7.6	2.68	0.040	1.5	3.1*/100†
SG-GW	49.4	233	8.2	0.51	0.007	1.4	72.4*/100†
LM-SW	88.1	425	8.3	2.65	0.048	1.8	9.9*/100†
BHWB-GW	44.5	209	7.8	0.45	0.007	1.6	77.5*/100†
RGW	69.8	331	8.4	0.58	0.008	1.4	68.4*/100†
Synthetic waters	30, 60, and 115‡	141, 282, and 542	4, 6, 8, and 10	N/A	N/A	N/A	100†

N/A: not available

* Ambient ClO_4^- concentration in source waters
† Total feed ClO_4^- concentration after spiking
‡ Conductivity based on KCl, K_2SO_4, or $CaCl_2$

waters. All of the natural waters (three surface and three ground waters) used in this study represent relatively low natural organic matter (NOM) aromaticity based on their specific UVA_{254} (SUVA) value. Anions and cations in the source waters were measured to predict how mono- and divalent co- and counter ions (e.g., Cl^-, SO_4^{2-}, K^+, and Ca^{2+}) would influence perchlorate transport. As shown in Tables 4.2 and 4.3, the CRW exhibits a relatively high conductivity from the variety of both mono- and divalent ions, and therefore it is expected that significant electrostatic screening will occur.

Table 4.3
Ions in feed waters

Anions (mg/L)	CRW	CRW/SPW	SG-GW	LM-SW	BHWB-GW	RGW
Cl^-	80.5	70.3	13.9	77.4	6.05	13.8
SO_4^{2-}	248.8	229.1	36.1	222.9	34.5	89.4
NO_3^-	1.24	0.78	32.9	1.9	20.1	44.2
HCO_3^- *	79.3	42.6	180.2	126.4	203	216.0
CO_3^{2-} *	0.94	0.16	2.7	2.4	2.5	5.1
Cations (mg/L)	CRW	CRW/SPW	SG-GW	LM-SW	BHWB-GW	RGW
Na^+	216.7	81.4	22.5	74.6	13.8	40.2
Ca^{2+}	227.9	72.7	62.8	74.8	63.1	97.5
Mg^{2+}	118.9	25.8	14.6	24.8	10.7	15.3
K^+	4.8	4.2	2.4	3.9	1.9	3.4
Ba^{2+}	< 0.1	< 0.1	0.1	0.1	< 0.1	< 0.1
Other (mg/L)	CRW	CRW/SPW	SG-GW	LM-SW	BHWB-GW	RGW
SiO_2	31.8	8.6	25.7	9.1	22.1	21.6

* Estimated from alkalinity and pH measurements

MEMBRANE FILTRATION TESTING UNITS

Continuous Cross-Flow Flat-Sheet Filtration Membrane Test Apparatus

Figure 4.1 shows a schematic of the experimental system used in the bench-scale cross-flow flat-sheet tests at variable pressures. A commercial test cell was employed for these measurements. This cell permits testing of a single 9.5 cm × 14.6 cm membrane coupon under tangential feed flow conditions with a channel height of 0.142 cm. The module provides an active filtration area of approximately ~ 139 cm^2 and, without a spacer, provides for a laminar flow condition at the volumetric feed rate. The Reynolds number was ~ 35 to 78 (note: the equipment did not permit easy, independent control for constant permeate flow without some changes in both feed flow rate and retentate back pressure for some of the membranes). The J_o/k was ~ 0.5 to 0.6 by using a nominal feed flow rate ranging 100 ~ 225 mL/min and a constant permeate flow rate ranging from ~ 4.4 to 5.8 mL/min (the recovery ranged from 2.6 to 5.4%, and the cross-flow velocity ranged from 2.5 to 5.6 cm/s), which depends on the specific membrane (see Table 4.4). The experiments were performed at room temperature, 22 ± 1 °C, and the transmembrane pressure varied to obtain the constant J_o/k ranging from 0.5 to 0.6. These experiments were performed at low recoveries. The effects of recovery on perchlorate transport are discussed in Chapter 8.

Each new membrane tested was prefiltered with a pure water volume of 8 L without recirculation through the crow-flow flat-sheet unit (a feed flow rate of 200 mL/min and a permeate flow rate of 5 mL/min) to remove any materials which may be coated onto/into the membrane surface/pores. Additionally, the new membrane was then soaked in pure water for at least 24 hours. During this period, the pure water was replaced every two or three hour with

Table 4.4
Experimental operating conditions for cross-flow flat-sheet filtration

Membrane	KPa(psi)	Feed, mL/min	Perm.mL/min	J_o/k	R*, %	U†, cm/s
TFC-S	179(26)	225	5.8	0.5	2.6	5.6
ESNA	366(53)	225	5.8	0.5	2.6	5.6
MX07	614(89)	100	4.4	0.5	4.4	2.5
GH	518(75)	225	5.8	0.5	2.6	5.6
GK	228(33)	225	5.8	0.5	2.6	5.6
GM	138(20)	225	5.8	0.5	2.6	5.6
GN	104(15)	225	5.8	0.5	2.6	5.6

* R = % volumetric recovery
† U = cross-flow velocity

another new volume of pure water for membrane stabilization. Then the new membrane was filtered again overnight with pure water and the pure water permeate flow was monitored for 3 hours just before each test until an approximately constant pure water flux was obtained, then the feed waters were filtered. Unless otherwise indicated, all of the membranes used in these filtration tests, perchlorate rejection and flux-decline tests, PEG rejection tests, streaming potential measurements, and hindered diffusion coefficient measurements were prepared by following the prefiltration methods above. Pure water permeability results for the RO, NF and UF membranes evaluated at different pressures are described in Table 4.1. The experiments were performed over a pressure range of 104 kPa to 552 kPa (15 psi to 80 psi) at a room temperature of 22 °C. The pure water fluxes increased with increasing pressure; the slope indicates the permeability.

The feed solution was drawn from a 4-L reservoir, and supplied to the flat-sheet test cell. The retentate and permeate streams were returned to the reservoir after passing through the test cell. The temperature of the feed solution was maintained at ± 1.0 °C of the room temperature (~ 22 °C) by immersing the reservoir in a temperature controlled water bath. The source waters were kept at room temperature for 24 hours prior to the filtration tests to assure thermal equilibration. The retentate flow, permeate flow, and temperature were monitored over time and the feed flow was calculated by summing the retentate and permeate flows. Duplicate analytical samples were taken every 15 min during ClO_4^- rejection tests. Observed perchlorate or solute rejection, R_{obs}, based on feed concentration, was calculated by using Equation 4.1,

$$R_{obs},\% = \frac{C_f - C_p}{C_f} \times 100(\%) \tag{4.1}$$

where C_f is perchlorate or solute concentration in the feed channel and C_p is the perchlorate or solute concentration in the permeate.

Dead-End Stirred Cell Unit

PEG rejections were performed by using a dead-end stirred-cell filtration unit. Figure 4.2 shows a schematic of the experiment apparatus used in dead-end stirred-cell filtration tests (Millipore, MA) at variable pH and conductivity conditions.

MEMBRANE CHARACTERIZATION

Membrane MMCO/MWCO Determination With Polyethylene Glycols (PEGs)

PEG rejections for the ESNA, MX07, and GM membranes were performed with a range of PEGs (relative molecular mass = 200 - 10,000) by using a dead-end stirred-cell filtration unit. Total PEG concentrations varied between 5 to 6 mg/L as DOC for each of the four to five PEG rejection measurements. The PEG was introduced to the membrane as a pure component (pH 8 and conductivity < 3.0 mS/m), in binary mixtures with other salts (KCl, K_2SO_4, and $CaCl_2$), and at pH 8 and varying solution conductivity (30 mS/m, 60 mS/m, and 115 mS/m). KOH and HCl were added to adjust the solution pH. The experiments were performed at room temperature, 22 ± 1 °C, and the pressure was varied to maintain the same permeate flow rate of 0.6 mL/min in order to minimize artifacts from mass transfer at the membrane interface. Each solution was filtered and the permeate was discarded for 30 minutes to stabilize the membrane structure. Then a permeate sample of 30 mL was collected for analysis by DOC measurement. PEG rejection (by DOC analysis), R_p, based on initial feed concentration, was calculated using:

$$R_p(initial\ feed),\% = \frac{C_{if} - C_{pp}}{C_{if}} \times 100(\%) \tag{4.2}$$

where C_{if} is the PEG concentration in the initial feed fluid and C_{pp} is the PEG concentration in the permeate. These measurements were performed not strictly to estimate the MMCO of the membrane but to determine the effects of ions on rejection of a standard neutral species, a linear macromolecule, in terms of changes of membrane pore geometry (size).

Streaming Potential Measurements

Streaming potentials of the membrane surfaces were measured with an electrokinetic analyzer apparatus (EKA). The same procedures and the same EKA as those used by Chapman-Wilbert *et al.* (1999) were used to measure streaming potential. The electrolyte solution was held in a jacketed reservoir kept at constant temperature by water circulated through a temperature control bath. External pH and conductivity sensors were placed in the electrolyte reservoir. Conductivity, temperature, pressure, and streaming potential were monitored with internal sensors. For the streaming potential measurements, membrane samples were cut to fit the measurement cell and then wetted in KCl, K_2SO_4, or $CaCl_2$ solution at the desired pH and stored in a refrigerator for the soaking time specified in the experimental design. The ESNA, MX07, and GM flat-sheet membrane specimens were intensively tested with three different background solutions, KCl, K_2SO_4, and $CaCl_2$, at varying pH, 4, 6, 8, and 10.

Zeta potential was calculated from the measured streaming potential using the method described by Fairbrother and Mastin (1924), which is based on the Helmholtz-Smoluchowski relationship,

$$\zeta = \frac{\Delta U \cdot \eta \cdot L}{\Delta P \cdot \varepsilon\varepsilon_0 \cdot A \cdot R} \tag{4.3}$$

where ζ is the zeta potential (mV), $\Delta U/\Delta P$ is the change in streaming potential with pressure (mV Pa^{-1}), η is the viscosity of the solution (Pa s), L is the channel length (m), ε is the dielectric constant of the solution, ε_o is the permittivity of free space ($m^{-1}\ \Omega^{-1}$), A is the cross-sectional area (m^2), and R is the ohmic resistance of the electrolyte and channel (Ω). It is assumed that the electrolyte solution, with conductivity k, carries most of the current and is used as an approximation of $L\ A^{-1}\ R^{-1}$ for ionic solutions > 10^{-3} M. This approximation is acceptable for the membrane materials evaluated and results in the following simplification where k is the specific conductance:

$$\zeta = \frac{\Delta U \cdot \eta \cdot k}{\Delta P \cdot \varepsilon \varepsilon_0} \tag{4.4}$$

DIFFUSION CELL TESTS

Diffusion Cell

A schematic of the diffusion cell is presented in Figure 4.3. The diffusion cell constructed for this work is a two-chamber glass cell with a membrane separating the left (feed) and right (permeate) cells. The exposed membrane area in the cell is 44.77 cm^2. At the beginning of the experiment, the left cell (Cell A) was filled with a 521 mL ClO_4^- solution in the presence of variable salts at varying pH conditions, while the right cell (Cell B) was filled with a 1398 mL solution of pure water at the same pH of the feed cell with and without background conductivity adjustments. There were two purposes to perform tests both with and without conductivity adjustments in the permeate cell. The tests with conductivity adjustment (no osmotic pressure difference between the feed and permeate cells) were to simulate the condition of 100% salt rejection in cross-flow flat-sheet filtration, which represents the same salt concentration in the feed and permeate sides. In this case, perchlorate transport was not significantly affected by the osmotic pressure occurring due to other salts. However, the tests without conductivity adjustment (osmotic pressure, feed cell >> permeate cell) were to simulate the condition of osmotic pressure between the feed and permeate sides in cross-flow flat-sheet filtration. In this case, perchlorate transport can be significantly affected by the osmotic pressure that significantly influences diffusion of solutes.

Ten samples were taken from each feed and permeate cell over the experimental period (5 to 10 days) to measure perchlorate concentration. 2 mL from the feed cell and 5 mL from the permeate cell were collected at each sampling to minimize the effect of hydraulic pressure. All diffusion tests were performed at a temperature of 25 ± 1 °C and temperature was measured at each sampling. The solutions in both cells were vigorously stirred by magnetic stirrers positioned under the cell. Each cell solution was recirculated with a pump.

Model Used to Determine Effective/Hindered Diffusion (D_{eff} = Intrapore Diffusion (D_p)) Associated With Partitioning (H)

The model used in the diffusion experiment starts with a pseudo steady state application of Fick's First Law. The solution diffuses across the membrane from the feed cell to the permeate cell because of the concentration gradient. The hindered diffusion (D_{eff}) based on the Fick's law can then be calculated:

$$C_{B(t)} - C_{A(t)} = \left[C_{B(t_0)} - C_{A(t_0)}\right]\exp\left[-\frac{D_{eff}}{l}\left(\frac{1}{V_B}+\frac{1}{V_A}\right)t\right] \quad (4.5)$$

where C_B is the concentration in the fluid adjacent to the permeate surface of the membrane in diffusion cell (mol/m^3), C_A is the concentration in the fluid adjacent on the feed surface of the membrane in diffusion cell (mol/m^3), D_{eff} is the *effective* (hindered) diffusion coefficient through membrane pores (m^2/s), l is the membrane thickness (m), V_B is the volume of permeate side of the membrane in diffusion cell (m^3), V_A is the volume of feed side of the membrane in diffusion cell (m^3), and t is the time during the diffusion test period (s).

The membrane top layer (microporous layer coated with ultrathin barrier layer) was peeled off and then measured with a micrometer (Starrett T230FL), which has a measuring limit of 0.00025 cm (2.5 μm). The membrane top layer was 0.0104 cm for MX07 and 0.0101 cm for GM. For the membranes, perchlorate transport can be significantly affected by the ultrathin barrier layer because this layer has ionizable functional groups and the microporous layer has relatively large pores compared to solutes. The ultrathin barrier layer of the each membrane was estimated to be approximately (50 – 100 Å /0.005 – 0.01 μm). After each test, the membrane was rinsed several times with pure water and put into a beaker containing the same background solution of which was used for the next test and stored in the refrigerator for at least two days for membrane stabilization before being reused.

NOM CHARACTERIZATION

Dissolved Organic Carbon

Dissolved organic carbon, operationally defined by filtration through a 0.45 μm, hydrophilic, synthetic membrane filter (Millipore, Bedford, MA), was measured using an Ionic Sievers Total Organic Analyzer (Model 800, Ionic Sievers Instruments, Boulder, CO). A standard reagent, potassium hydrogen phthalate (KHP), was used to prepare an organic carbon calibration curve.

The Ionic Sievers DOC analyzer utilizes an ultraviolet/persulfate combination to oxidize total carbon to CO_2, which is detected using a CO_2 selective conductivity cell. Organic carbon is calculated by taking the difference between the total carbon and inorganic carbon, using an internal acidification (H_3PO_4) step. The Ionic Sievers analyzer is internally calibrated. However, the KHP standard was used periodically to confirm the accuracy of the instrument. The measured and predicted standards were consistently within 1% of one another.

The filter used for the 0.45 μm prefiltration was rinsed with a sufficient volume of pure water to remove the leachable carbon inherent to the filter membrane. All natural feed-waters were 0.45 μm-filtered, providing an operational definition of dissolved NOM.

UV Absorbance

UV absorbance was measured at a 254 nm wavelength with a Shimadzu UV-160A UV-Visible Spectrophotometer (Shimadzu, Colombia, MD), using a 1 cm path length quartz cell. For the UVA_{254} measurements, all samples were also filtered through a 0.45 μm filter prior to analysis. UVA_{254} values are used in determining the humic or aromatic content of NOM.

Specific UVA (SUVA), which is UVA_{254} divided by DOC, was used as an index of humic/aromatic character (Edzwald *et al.*, 1985).

ARSENIC (III); AN UNCHARGED MODEL SPECIES

As (III) was used as a model uncharged species to study a size surrogate for ClO_4^- for which incremental electrostatic repulsion occurs between anionic species and a negatively charged membrane. The As (III) exists as uncharged H_3AsO_3 at a pH < 9.13 (under $pK_{a1} = 9.13$), with a molecular mass of 126, compared to ClO_4^- which is a charged species at all experimental pH conditions ($pK_a = -1.6$) and has a molecular mass of 100. Several As (III) rejection tests were performed at varying pH and conductivity conditions. As (III) was measured by ICP (ARL Fusion Technology 3410). Observation of arsenic emission by the monochromator was performed at the primary arsenic detection wavelength of 193.79 nm.

ANALYTICAL METHODS AND QA/QC

For all analyses performed, quality assurance and quality control (QA/QC) was a priority; therefore, analysis of replicate samples and a spike addition of a known concentration of the analyte into experimental samples were performed routinely. Whenever possible and necessary, experiments were performed in either duplicate or triplicate to ensure that observed differences between systems was repeatable and statistically relevant.

Perchlorate Measurement Summary

The analytical method employed to measure perchlorate was a Dionex DX300 Ion Chromatograph (IC). All standards and solutions were made with 18 mega-ohms pure water (dissolved organic carbon, anion, and cation free water). Five concentration standards of 1, 5, 20, 50, and 100 μg/L were used in the calibration. A ClO_4^- calibration curve was constructed with the pure (> 99%) standards, with the criterion for acceptance being a calibration curve with a linear coefficient of determination (r^2) of ≥ 0.95. The minimum reporting limit (MRL) is 2 μg/L while the method has a minimum detection limit (MDL) of 1.4 μg/L. Uncertainty has been determined through replicate analyses with the resultant data used to calculate a mean and a standard deviation (SD) (see Table 4.5).

Table 4.5
Standard for minimum detection limit of perchlorate (n = 11)

ClO_4^- (μg/L)	Peak intensity ($*10^6$)	S.D.*	C.V.† (%)
1	0.18	0.02	11.51
5	1.14	0.05	4.81
20	5.10	0.25	4.85
50	12.54	0.27	2.14
100	25.78	0.46	1.78

* S.D. = standard deviation
† C.V. = coefficient of variance

The correlation between standard concentration and intensity was excellent throughout the range of perchlorate concentrations by using the Dionex IonPac AS11 analytical column, as shown in Figure 4.4. This column was used because it did not require the addition of either p-cyanophenol or methanol organic modifiers in the mobile phase (Harrington and Shen, 1998). Another study also proved the benefit of the Dionex IonPac AS11 analytical column: the column showed a higher laboratory control standard (LCS) mean recovery at 20 μg/L and a better performance in terms of control limits for LCS than the Dionex IonPac AS5 analytical column (Eaton *et al.*, 1998).

Zeta Potential Calculation

Zeta potential was calculated from the measured streaming potential as described earlier. Table 4.6 summarizes the ZP values of the GM membrane at varying pH and conductivity conditions. Two duplicate measurements for the ZP were performed such that standard deviation was mostly within 1 mV.

Hindered Diffusion/Partitioning Measurement

Hindered diffusion and partitioning were measured by the diffusion cell as described earlier. Table 4.7 summarizes the hindered diffusion/partitioning ($D_{eff}H$) through the GM membrane at varying pH conditions in the presence of KCl. Two duplicate measurements for the hindered diffusion/partitioning were performed such that standard deviation was mostly within 2.15×10^{-11} cm^2/s (the coefficient of variation was less than 7% over the range of hindered diffusion/partitioning values).

Other Ion Measurements Summary

Co-ions, Cl^- and SO_4^{2-}, were also measured by using the same IC with an AS14-4 column. A 0.5 N $NaHCO_3$/0.5 N $NaCO_3$ eluent with a 50 mM H_2SO_4 regenerant was used with

Table 4.6
Standard deviations of zeta potential measurement for the GM membrane at varying pH and conductivities

pH at conductivity 30 mS/m	ZP w/ KCl (mV)	S.D.* (mV)	ZP w/ K_2SO_4 (mV)	S.D.* (mV)	ZP w/ $CaCl_2$ (mV)	S.D.* (mV)
4	-4.50	0.42	-5.85	1.63	-3.20	0.14
6	-8.85	0.78	-8.95	0.49	-4.65	0.07
8	-9.70	0.42	-10.20	0.57	-5.65	0.49
10	-10.65	0.07	-10.80	0.28	-6.70	0.28
Conductivity at pH 8 (mS/m)	**ZP w/ KCl (mV)**	**S.D.* (mV)**	**ZP w/ K_2SO_4 (mV)**	**S.D.* (mV)**	**ZP w/ $CaCl_2$ (mV)**	**S.D.* (mV)**
30	-9.70	0.42	-10.20	0.57	-5.65	0.49
60	-7.20	1.27	-6.50	0.71	-4.95	0.35
115	-3.40	1.41	-5.90	1.27	-3.35	0.49

* S.D. = standard deviation

Table 4.7
Standard deviation and coefficient of variance of hindered diffusion/partitioning of the GM membrane at varying pH conditions

pH at conductivity 30 mS/m	$D_{eff}H$ (cm^2/s)	S.D.* (cm^2/s)	C.V.† (%)
4	1.52E-09	2.15E-11	1.42
6	1.11E-09	7.77E-11	7.00
8	5.59E-10	6.38E-12	1.14
10	3.71E-10	4.92E-12	1.33

* S.D. = standard deviation
† C.V. = coefficient of variance

Table 4.8
Summary of overall analytical methods

Determination	Equipment	Reference method
pH	Orion 250	SM 4500H^+ B
Conductivity	Orion 195	SM 2510 B
UVA_{254}	Shmadzu UV160U	SM 5910 B
TOC/DOC	Sievers 800 TOC Analyzer	SM 5310 C
Chloride	Dionex DX300 IC	SM 4110 B
Sulfate	Dionex DX300 IC	SM 4110 B
Sodium	ICP-ES	SM 3120 B
Potassium	ICP-ES	SM 3120 B
Calcium	ICP-ES	SM 3120 B
Magnesium	ICP-ES	SM 3120 B

an injection volume of 250 μL. Counter-ions, K^+ and Ca^{2+}, were measured using a Liberty-Series II Inductivity Coupled Plasma Emission Spectroscopy (ICP-ES). A high purity standard solution with 1000 μg/μL potassium and calcium was used to make various standard solutions.

Fouled Membrane Characterization

Fouled membrane characterization encompassed X-ray diffraction (XRD) and scanning electron microscopy (SEM, DS-130, AKASHI) to determine precipitative inorganic fouling (e.g. CaCO3), and FTIR (Nicolect 752) analysis to investigate the change of membrane surface chemistry by inorganic fouling. Neutron activation analysis (NAA, GMX-25195, ORPEC) has recently been employed to quantitatively estimate the total concentration of each element (e.g. Ca^{2+}, Ba^{2+}) on the membrane surface.

Summary of Overall Analytical Methods

Overall analytical methods are described in Table 4.8. Table 4.8 summarizes the analytical determination, equipment used, and the appropriated standard method. Procedures outlined in *Standard Methods* (SM) (APHA, AWWA, and WEF, 1998) were followed.

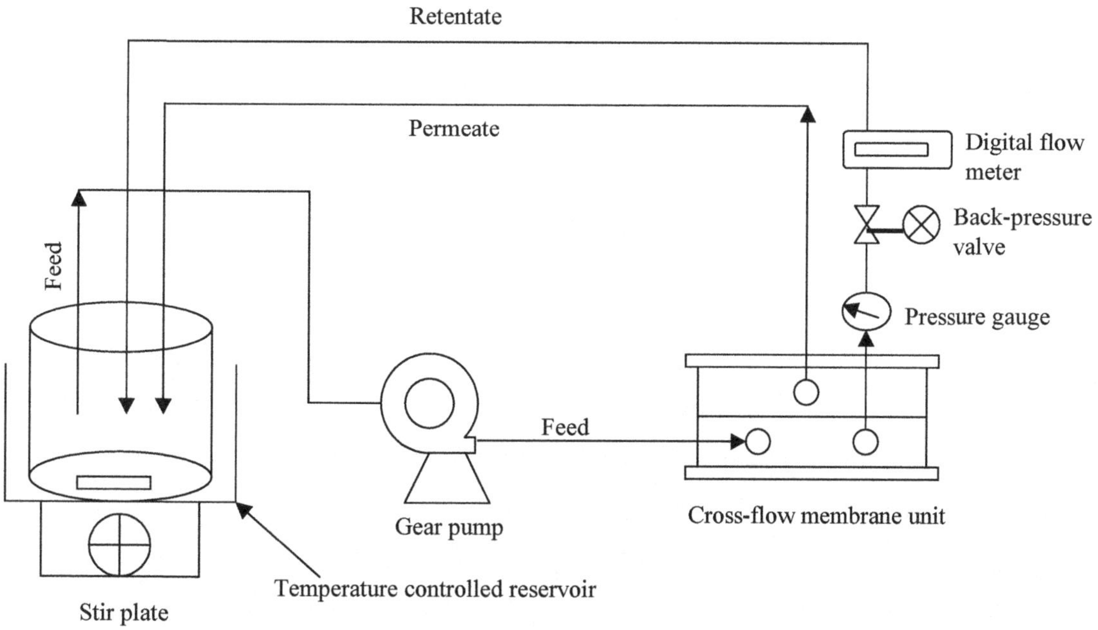

Figure 4.1 Schematic of cross-flow flat-sheet membrane filtration apparatus

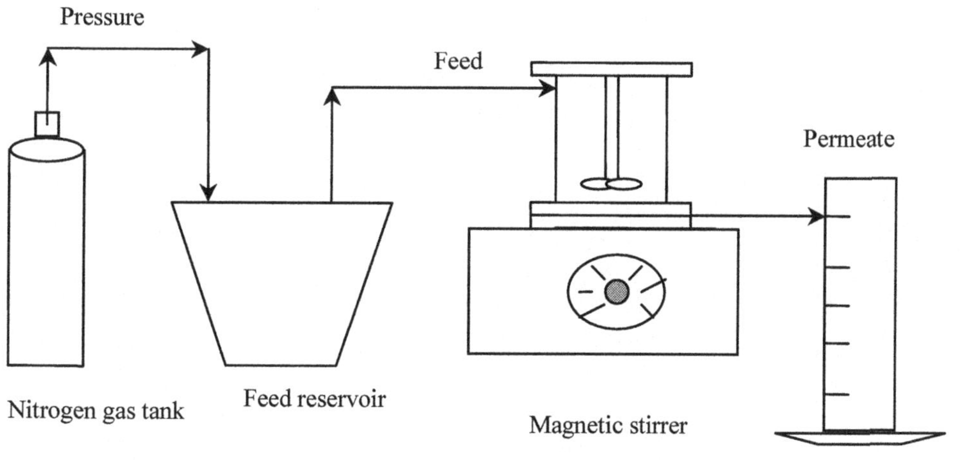

Figure 4.2 Schematic of dead-end stirred-cell filtration unit

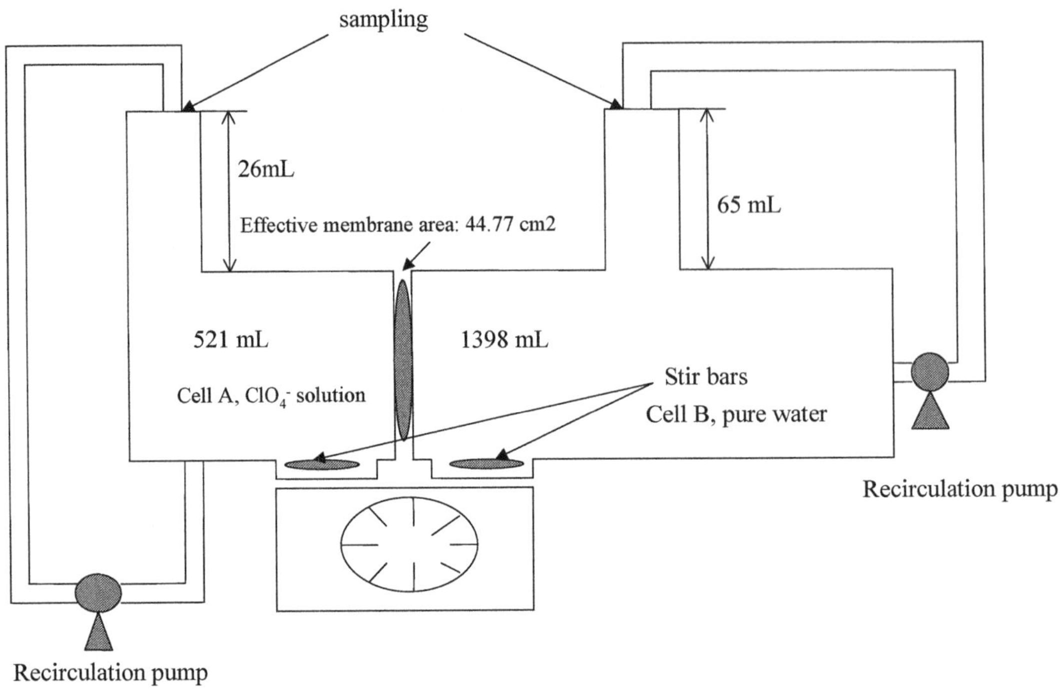

Figure 4.3 Schematic of diffusion cell apparatus

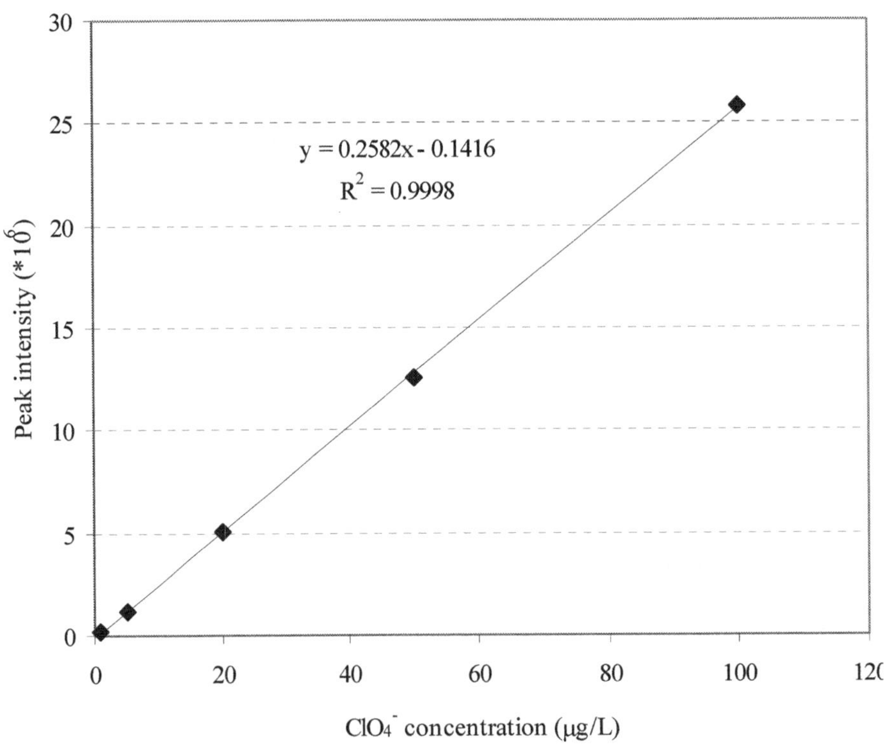

Figure 4.4 Typical standard curve for perchlorate analysis by IC

CHAPTER 5
BASIC UNDERSTANDING OF MEMBRANE PORE SIZE AND CHARGE AND THEIR INFLUENCE ON PERCHLORATE TRANSPORT

MEMBRANE STABILIZATION AND PERMEABILITY

The membranes shown in Table 4.1 were first stabilized for at least 24 hours in pure water, and the pure water flux of the membranes was monitored with a pressure of 414 kPa (60 psi) at 22 °C (see Figure 5.1). A two- to four-hour membrane setting period was observed, after which flux reached a constant value.

Pure water permeabilities at different pressures obtained from the experiments performed with a cross-flow flat-sheet filtration unit are shown for the RO, NF, and UF membranes in Table 4.1. The experiments were performed with 104 to 552 pKa (15 to 80 psi) of pressure at a room temperature. The pure water flux increased with increasing pressure; the slope is indicative of the pure water permeability (PWP) or mass transfer coefficient for water. The pure water permeability of the membranes (especially porous membranes, NF and UF) can be used to estimate membrane pore geometry (size and number) of the membranes, which can influence solute transport through the membranes. The TFC-S and ESNA membranes have a greater PWP than the MX07 membrane, even though the TFC-S and ESNA membranes ostensibly have a smaller pore size than the MX07 membrane (based on their nominal MMCOs). It is therefore reasonable to assume that the TFC-S and ESNA membranes either have greater pore density or a thinner pore layer (or some combination of both) than the MX07 membrane. The PWPs of the UF membranes follow the order of their nominal MMCOs, GH < GK < GM < GN (see Table 4.1) as expected.

ATR-FTIR SPECTROSCOPY

ATR-FTIR spectra of the clean ESNA, MX07, and GM membranes are shown in Figure 5.2. Infrared spectroscopy has been widely used to identify certain organic compounds or moieties that are readily distinguished from the adsorption pattern of other compounds. All the membranes show almost the same spectral pattern with slightly different intensities even though the ESNA and MX07 membranes are comprised of an aromatic polyamide TFC, and the GM membrane is possibly comprised of sulfonated polyethersulfone (PES) with an ultrathin (~ 0.005 – 0.01 μm) top layer (possibly made of polyimide). In general, a TFC membrane is made of three parts, top layer, microporous support, and support (reinforcing fabric). However, the top layer FTIR peaks cannot be well detected since the top layer is too thin to obtain distinct FTIR peaks, because the evanescent wave passes through the ultrathin top layer and primarily loses energy thorough the microporous support. This is a limitation of FTIR spectra measurement for membranes having an ultrathin top layer and microporous support; thus, ionizable functional groups that may be dominant on the top layer of the membranes may not be well detected. Therefore, ZP measurement can be useful to confirm that ionizable functional groups exist at a membrane surface. The ZP results that follow can support the assumption that the ESNA, MX07, and GM membranes have ionizable acidic functional groups on the top layer due to the negative membrane surface charge, even though their FTIR spectra do not directly show ionizable functional groups.

As shown in Figure 5.2, the characteristic FTIR peaks of polysulfone are seen at 1,592 cm^{-1} and 1,110 cm^{-1} (aromatic double bonded carbons), at 1,016 cm^{-1} (ether), at 1,492 cm^{-1} (methyl), and at 1,151 cm^{-1} and 694 cm^{-1} (sulfone) (Bellamy, 1975). Thus, it can be assumed that all at the membranes are based on polysulfone microporous layers.

MMCO WITH PEG

PEG Rejection Measurements

Figure 5.3 presents the PEG-rejection trends of the ESNA, MX07, and GM membranes at pH 8 and varying solution conductivities, as established with KCl, K_2SO_4, and $CaCl_2$. The objective of these PEG rejection measurements was to evaluate the effects of mono- and divalent ions on the change of membrane pore size based on the rejection of PEGs, neutral linear macromolecules. All of the membranes show similar trends, that is, the PEG rejection increases with increasing conductivity; a possible explanation is compaction of the electrical double layer associated with pore charge and thus a material relaxation that reduces pore size. The results also show that PEG rejection by the membranes follows the order KCl < K_2SO_4 < $CaCl_2$, due to ion binding (such as with Ca^{2+}) and increasing ions ($Cl^- < K^+ < SO_4^{2-}$) at the interface of the membranes. A supposition can be inferred that ClO_4^- steric rejection may increase (similar to the PEG rejection) due to size exclusion associated with a decrease in membrane pore size, in the presence of mono- and divalent co- and counter ions.

Effect of Ions on Flux

Figure 5.4 shows the solvent flux of the membranes in the presence of anions and cations. The relatively small pore size membranes (ESNA and MX07) show a decrease of the water flux with increasing conductivity (ionic strength). However, the relatively large pore size membrane (GM) shows an increase of the water flux with increasing conductivity (ionic strength). Generally water flux decreases with increasing ionic strength due to membrane compaction. However, the water flux of the GM membrane increased with increasing conductivity using KCl, K_2SO_4, and $CaCl_2$. The increase of the water flux for the GM membrane may be caused by the decreasing membrane surface charge, which can influence the increase in the viscosity of the permeate fluid near the membrane pore wall (electroviscous effects). This originates through interactions between the permeate fluid and the ions in the diffuse layer near the pore walls, such as hydrogen bonds between the solvated water and permeate. As a result, an additional resistance to flow can be imparted on water as it moves through the membrane pores.

ZETA POTENTIAL DETERMINATIONS: EFFECTS OF pH AND CONDUCTIVITY

ZP results obtained at varying pH levels and conductivities are shown in Figures 5.5 and 5.6. These measurements identify the qualitative expectations for the electrostatic contribution to perchlorate rejection from charge repulsion between the perchlorate anion and the negatively charged membrane under varying pH conditions and salts. Previous studies have revealed that pH and mono- and divalent co- and counter ions affect the surface charge of RO and NF membranes (Childress and Elimelech, 1996 and Hong and Elimelech, 1997). The surface charge of a membrane is significantly affected by increasing pH. All of the membranes (ESNA, MX07, and GM) acquired more negative charge when pH increased at constant conductivity = 30 mS/m, regardless of the salt type. As can be seen in Figure 5.5, for different type of salts at constant pH and solution conductivity, the ZP of the membranes follows the order (negatively), $CaCl_2$ < KCl < K_2SO_4. However, the membranes exhibit less negative charge at relatively higher conductivity with the salts.

Two mechanisms can be employed to explain these results based on previous studies (Mallevialle et al., 1996, Bauza C. R. et al., 1990, Bellamy L. J., 1975, Booth et al., 2000, and Bowen and Cao, 1998) which relate to the Ca^{2+} binding as an outer-sphere complex and an inner-sphere ion pair for mono- and divalent co-ions, Cl^- < SO_4^{2-}. The ZP of the membranes would thus be decreased by $CaCl_2$ due to complexation and ion binding. However, an assumption is made that for the co-ion, Cl^-, bonding/adsorption is also reduced with increasing conductivity (ionic strength) due to a decrease of the electrical double layer. We can also apply this mechanism to a divalent co-ion, like SO_4^{2-}. SO_4^{2-} bonding/adsorption can significantly decrease even with increasing K_2SO_4 concentration due to a decrease in the electrical double layer since the SO_4^{2-} bonding/adsorption occurs mostly as inner-sphere, which is related to the electrical double layer. Overall, independent of the type of mono- and divalent co- and counter ions, the ZP of the membranes become less negative almost linearly with increasing solution conductivity at pH = 8. Our results and interpretations are consistent with the general observations of previous studies. Previous studies explain the effects of pH and co- and counter ions on the ZP of RO, NF, and UF membranes with following phenomena.

1. In aqueous solutions, since anions are less hydrated than cations, they can more closely approach the membrane surface (Childress and Elimelech, 1996).
2. Preferential adsorption of anions has been used to explain surface charge behavior of several non-ionogenic surfaces (i.e. surfaces without ionizable functional groups) (Childress and Elimelech, 1996).
3. Since thin film composite membranes are made by the interfacial polymerization reaction of a monomeric polyamine with a polyfunctional acyl halide, anions be expected on the membrane surface (Elimelech *et al.*, 1994).
4. In surface complex formation, the ion may form an inner-sphere complex (chemical bond), and an outer-sphere ion pair, or be in the diffuse swarm of the electrical double layer (Stumm, 1992).
5. If electrical attraction is the only bonding mechanisms, the Ca^{2+} would form an outer-sphere complex. However, if co-ion bonding or some combination of co-ion and ionic bonding occurs, inner-sphere complexes are formed (Childress and Elimelech, 1996).

SUMMARY OF RESULTS

The TFC-S and ESNA membranes have a greater PWP than the MX07 membrane even though the TFC-S and ESNA membranes have a smaller pore size than the MX07 membrane (based on their nominal MMCOs). It is reasonable to assume that the TFC-S and ESNA membranes have a greater pore density than the MX07 membrane. The PWPs of the UF membranes follows the order, GH < GK < GM < GN, since the membranes are made of the same material and only have different membrane pore sizes based on their MMCOs.

The results obtained from FTIR measurements provide direct insight into the microporous support and support materials, which are made of mostly polysulfone and polyethersulfone. The results support the conclusion that membrane functional groups exists mostly on the top layer or ultrathin top layer.

PEG rejection by the membranes follows the order KCl < K_2SO_4 < $CaCl_2$, due to ion binding (such as with Ca^{2+}) and increasing ions (Cl^- < K^+ < SO_4^{2-}) at the interface of the membranes. The ion binding and concentration polarization at the membrane surface influence the electrical double layer. These results can be explained by electrostatic interaction between a negatively charged membrane and the anions, Cl^- and SO_4^{2-}, increasingly introduced to the membrane surface.

The relatively small pore size membranes (ESNA and MX07) show a decrease of the water flux with increasing conductivity (ionic strength). However, the relatively large pore size membrane (GM) shows an increase of the water flux with increasing conductivity (ionic strength). Generally, water flux decreases with increasing ionic strength presumably influencing membrane pore size becoming smaller for relatively small pore size membranes. The water flux of the GM membrane increased with increasing conductivity using KCl, K_2SO_4, and $CaCl_2$.

All of the membranes (ESNA, MX07, and GM) acquire more negative charge when pH increased regardless of the salt at the same conductivity (30 mS/m). The ZP values of the membranes became less negative with increasing conductivity regardless of salt. At the same pH and solution conductivity, the ZP of the membranes follows the order (negatively) in the presence of different type of salts: $CaCl_2$ < KCl < K_2SO_4.

RELATION OF THIS CHAPTER TO RESEARCH HYPOTHESES

The results presented in this chapter addressed two hypotheses; (i) membrane properties enhancing ClO_4^- rejection include lower MMCO and a greater (negative) surface charge (hypothesis 3.2.1.a) and (ii) water quality conditions promoting ClO_4^- rejection include high pH, lower conductivity (ionic strength), and lower calcium concentration (hypothesis 3.2.1.b).

Several qualitative predictions for perchlorate ion transport through the membranes can be made from the membrane characterizations with PEG rejection and streaming potential measurements at varying pH and solution conductivities (with the different type of salts):

1. Perchlorate rejection by the membranes may increase with increasing conductivity due to decreasing membrane pore size. This contributes to size exclusion, particularly for the relatively smaller pore size membranes (ESNA and MX07), supporting hypothesis 3.2.1.a that lower MMCO enhances perchlorate rejection.
2. Perchlorate rejection may increase (due to electrostatic repulsion) with increasing pH due to the increasingly negative membrane surface charge (based on ZP). This contributes to electrostatic exclusion by the membranes, supporting hypothesis 3.2.1.a that greater (negative) surface charge, at relatively high pH, enhances perchlorate rejection.
3. Perchlorate rejection may decrease (due to electrostatic repulsion) with increasing solution conductivity due to a less negative membrane surface charge (based on ZP), supporting hypothesis 3.2.1.b that higher conductivity reduces perchlorate rejection.

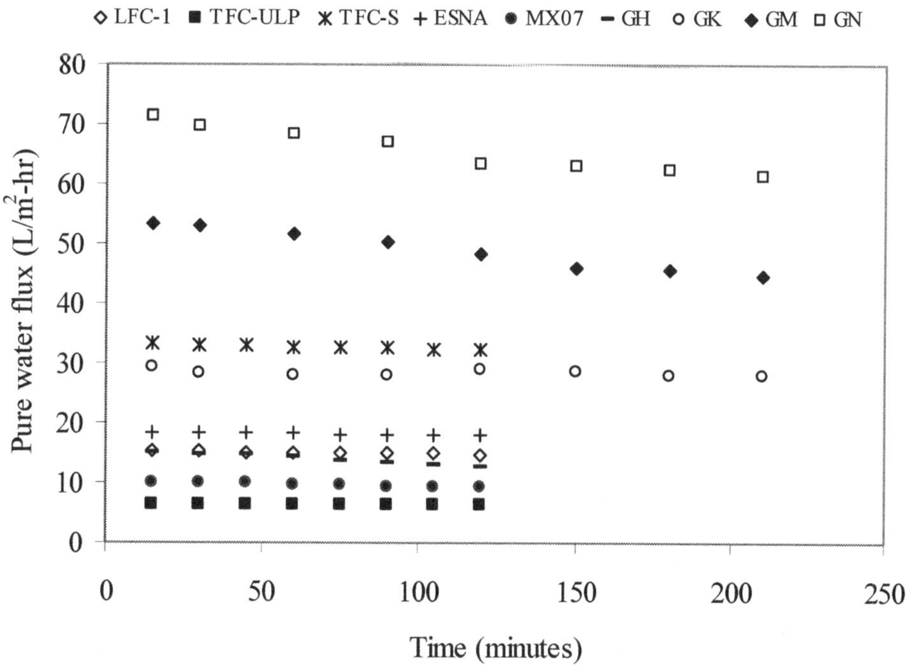

Figure 5.1 Pure water flux during stabilization of the RO, NF, and UF membranes

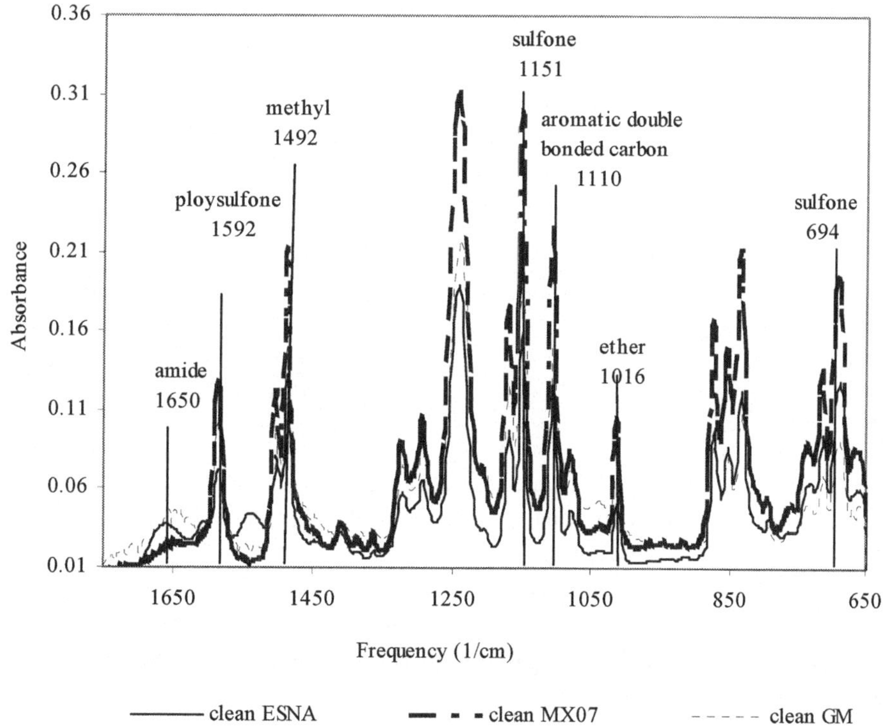

Figure 5.2 Comparison of FTIR spectra of the clean NF and UF membranes

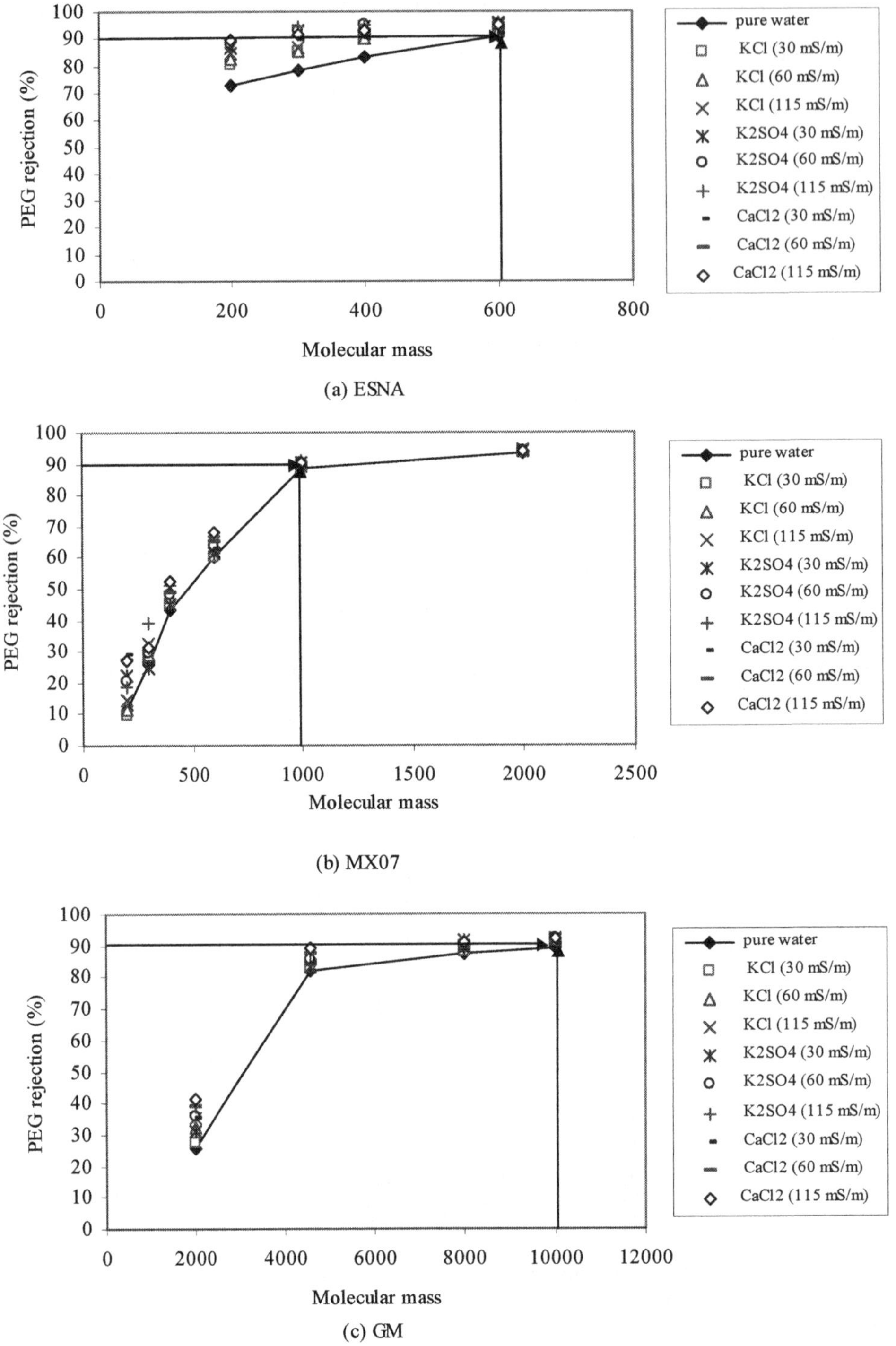

Figure 5.3 Effect of ions on membrane pore size based on PEG rejection for the NF and UF membranes

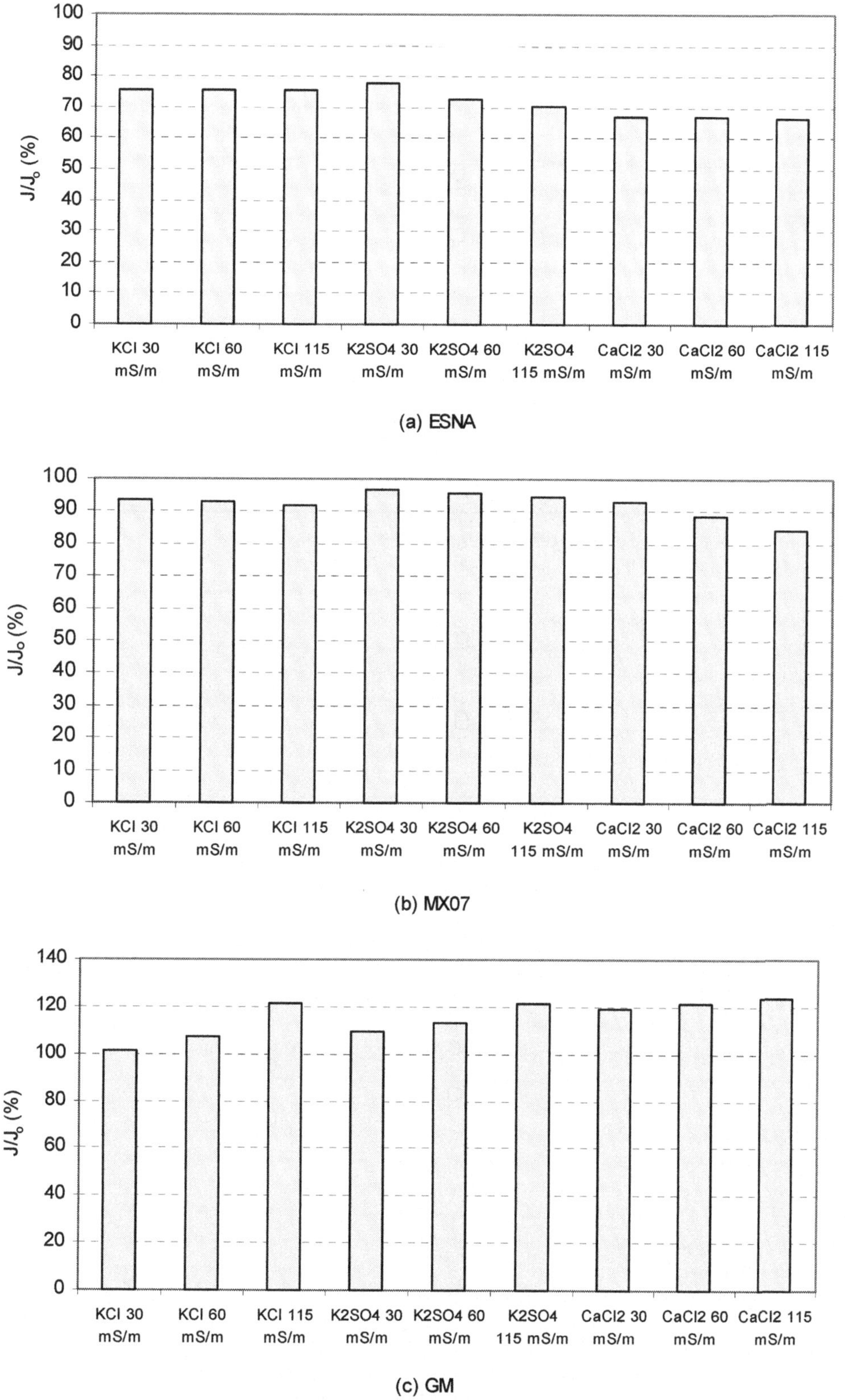

Figure 5.4 Effect of ions on flux for the NF and UF membranes

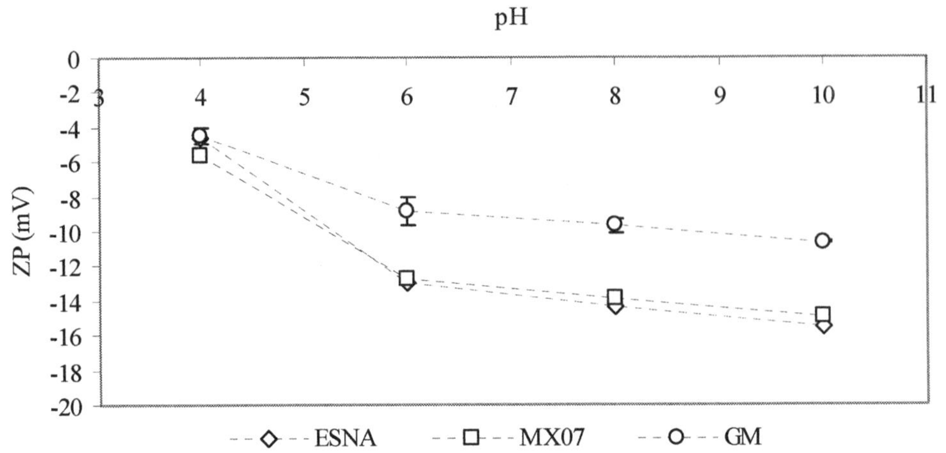

(a) Conductivity 30 mS/m KCl

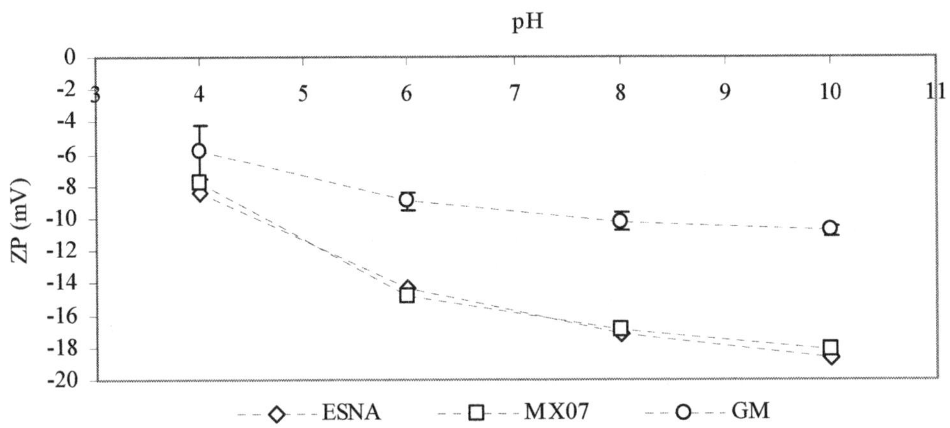

(b) Conductivity 30 mS/m K_2SO_4

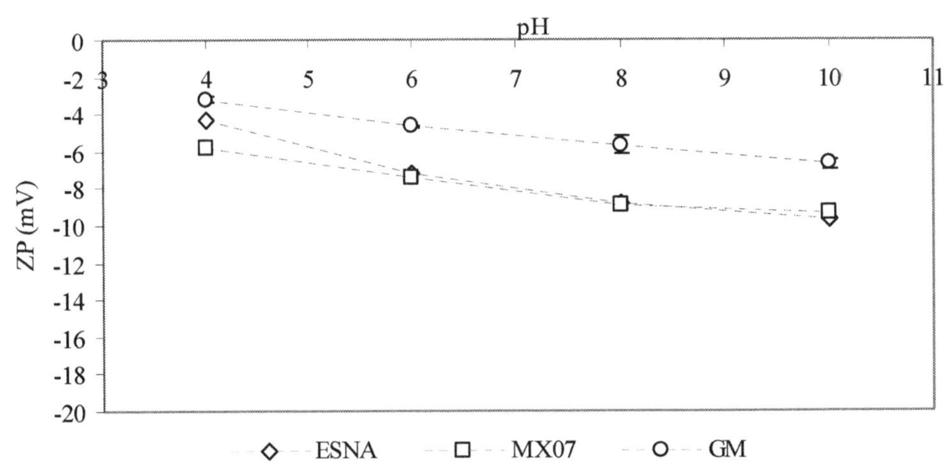

(c) Conductivity 30 mS/m $CaCl_2$

Figure 5.5 Effect of pH on zeta potential for the NF and UF membranes

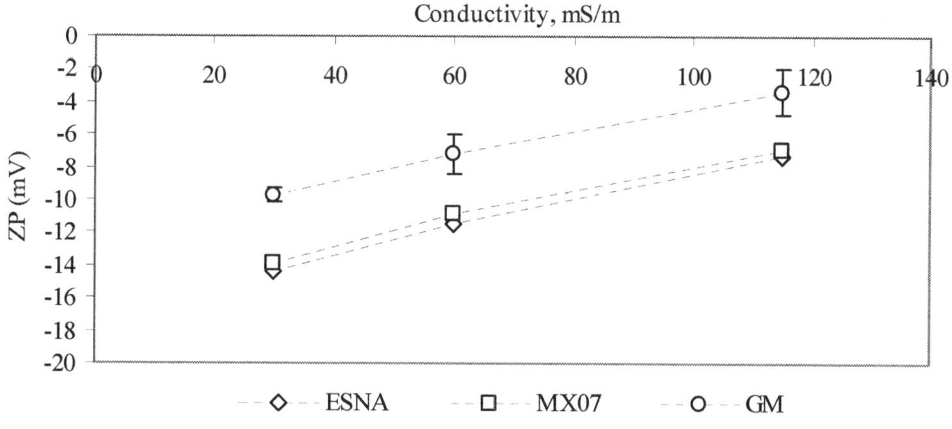

(a) pH 8 (KCl)

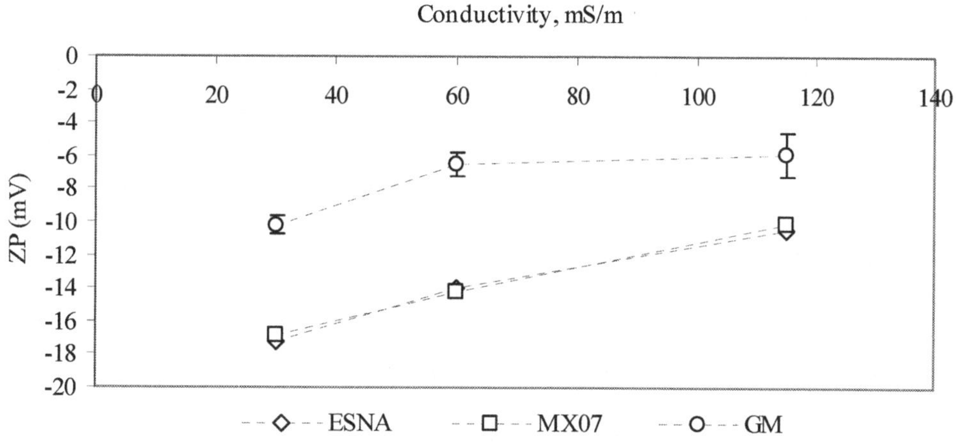

(b) pH 8 (K_2SO_4)

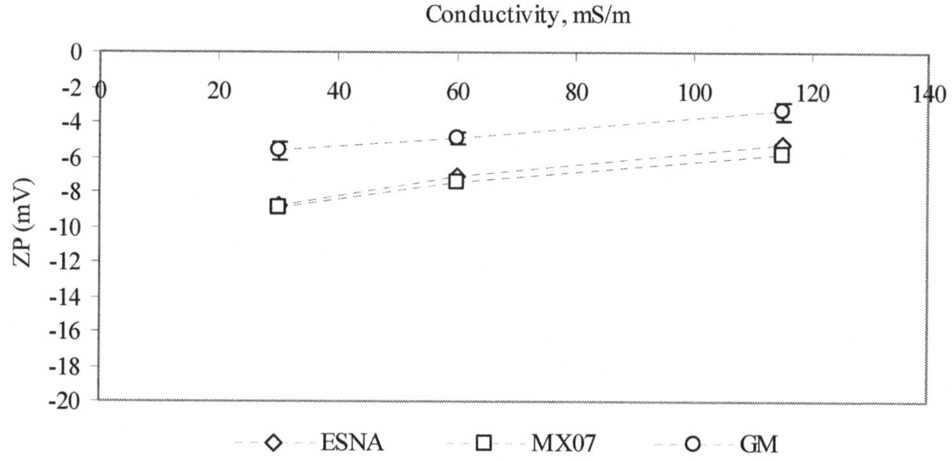

(c) pH 8 ($CaCl_2$)

Figure 5.6 Effect of conductivity on zeta potential for the NF and UF membranes

CHAPTER 6
SYSTEMATIC BENCH-SCALE ASSESSMENT OF PERCHLORATE TRANSPORT MECHANISMS BY NF AND UF MEMBRANES

PERCHLORATE REJECTION AND FLUX-DECLINE WITH SYNTHETIC WATERS AND CRW

Two NF membranes, ESNA and MX07, and one UF membrane, GM, were tested for perchlorate rejection and flux-decline in cross-flow tests with synthetic waters and CRW. As the experimental conditions for this study are described in Chapter 4 and Table 4.4, the J_o/k ratio was kept approximately constant at 0.5 by employing a feed flow rate of ~ 100 to 225 mL/min, a constant permeate flow rate of ~ 4.4 to 5.8 mL/min, a constant recovery of ~ 2.6 to 4.4%, and constant cross-flow velocity of ~ 2.5 to 5.6 cm/s for the membranes as a comparable hydrodynamic operating condition in this study.

The characteristics of the CRW used as a feed water are described in Tables 4.2 and 4.3, and show that the CRW exhibits a relatively high conductivity based on mono- and divalent co- and counter ions, which can significantly change membrane physical and chemical properties (e.g., pore size and surface charge) influencing perchlorate transport. A synthetic water (pure water) was employed at pH 8.0 and a conductivity < 3.0 mS/m in the absence of NOM. The synthetic water (with a concentration of 100 μg/L of ClO_4^- provided by $KClO_4$) was tested for perchlorate rejection by the ESNA, MX07, and GM membranes to determine how perchlorate transport occurs when co- and counter ions and NOM affect membrane pore size and surface charge.

Perchlorate rejection by the ESNA, MX07, and GM membranes was monitored over time with pure water and spiked CRW, with 100 μg/L spiked in pure water, and 96 μg/L spiked into CRW ClO_4^- since the CRW with a background concentration of 4 μg/L. Flux decline trends were also monitored at the same time. All of the membranes showed relatively high (75% to 90%) perchlorate rejection with pure water. Also, the ESNA membrane consistently provided approximately 80% perchlorate rejection over time with CRW, higher than that observed for either the MX07 (25%) or the GM (< 5%) membranes (see Figure 6.1). In this work, perchlorate rejection did not appear to be affected by flux decline during short-term bench-scale filtration tests in the presence of NOM in CRW (see Figure 6.2). However, the objective of this study is not only to determine the effects of flux decline or fouling on perchlorate rejection, but also to elucidate the fundamental ClO_4^- rejection mechanisms, steric/size exclusion and electrostatic exclusion, with the results obtained with the synthetic water and CRW for the various membranes.

CRW was selected as a model feed water for this study since CRW contaminated with perchlorate has been widely used as a drinking water source. Since CRW has relatively high levels of anions and cations, it can be used as a representative feed water to study perchlorate transport through the negatively charged membranes. From the results, both steric/size exclusion and electrostatic exclusion significantly contribute to perchlorate rejection from the synthetic waters for the relatively small pore size, negatively charged ESNA membrane.

Only electrostatic repulsion can be a dominant mechanism for perchlorate rejection from the synthetic waters for the relatively large pore size MX07 and GM membranes. Perchlorate rejection is significantly reduced when filtering CRW compared to relatively low ionic-composition synthetic waters for the MX07 and GM membranes. The ions in the CRW cause the

ZP of those membranes to significantly decrease; thus, electrostatic repulsion decreases between the perchlorate ion and the MX07 and GM membranes. This is consistent with the expectations from the membrane characterization measurements.

PERCHLORATE REJECTION AT VARYING pH AND CONDUCTIVITY

Two NF membranes, ESNA and MX07, and four UF membrane, (GH, GK, GM, and GN), were tested for perchlorate rejection under varying pH and conductivity conditions with synthetic waters in the absence of NOM. The operating conditions used in the cross-flow flat-sheet filtration are described in Chapter 4 and Table 4.4.

The effects on perchlorate rejection by the ESNA, MX07, GH, GK, GM, and GN membranes from changes in solution pH and conductivity were determined with the synthetic waters having three different salts, KCl, K_2SO_4, and $CaCl_2$ (see Figures 6.3 and 6.4). Perchlorate rejection increased with increasing pH, regardless of salt type, at constant conductivity = 30 mS/m for all the membranes. This is likely attributable to co-ion bonding/adsorption occurring at the inner-sphere between OH^- and the membrane surface (Elimelech and O'Melia, 1990). Perchlorate rejection followed the order, $CaCl_2 < K_2SO_4 < KCl$, at constant pH and conductivity conditions for all of the membranes tested. In terms of electrostatic interaction, the Ca^{2+} binding causes the membrane surface charge to significantly decrease, thus perchlorate rejection is much lower in a solution having $CaCl_2$ than in those with KCl and K_2SO_4. Perchlorate rejection is greater with the KCl solution than with the K_2SO_4 solution even though the apparent ZP is greater with the K_2SO_4 solution than with the KCl solution as shown in Figures 5.5 and 5.6. The results are opposite to the prediction based on electrostatic interaction with measured ZP results. We explain these results by increasing concentration polarization (CP) with the salts, KCl and K_2SO_4, at the interface of the membranes. The K_2SO_4 concentration increases more than KCl concentration over time at the membrane interface (GH, GK, GM, and GN) during filtrations (see Figures 6.5 and 6.6). An assumption is made that ZP at the membrane interface, where solute concentration increases, can be different than what is obtained at the bulk concentration. Thus, since there is more concentration polarization with increasing K_2SO_4 concentration, the electrical double layer is being reduced more with increasing K_2SO_4 than with increasing KCl. Thus, an *effective ZP* at the interface of the membrane during filtration may be more negative when filtering the KCl solution than with the K_2SO_4 solution. Perchlorate rejection follows the order, GN < GM < GK < GH < MX07 < ESNA, at varying pH conditions. These results are consistent with the size exclusion mechanism based on their nominal MMCOs, ESNA < MX07 < GH < GK < GM < GN.

Perchlorate rejection decreases with increasing solution conductivity (see Figure 6.4) for all of the membranes except the ESNA membrane which is hardly affected. The results are explained by a reduction in electrostatic repulsion and an increase in both the hindered diffusion and the partitioning (between the perchlorate anion and the membrane) with increasing conductivity. In terms of electrostatic repulsion, the electrical double layer is reduced with increasing solution conductivity provided by the KCl and K_2SO_4 salts, and thus the ZP of the membrane becomes less negative due to less Cl^- or SO_4^{2-} bonding/adsorption at the inner-sphere of the membrane. However, the Ca^{2+} binding is the most efficient at screening the co-ion bonding/adsorption. This reduction in the electrical double layer may thus favor partitioning and also increase the hindered diffusion within the membranes, which affects perchlorate transport through the membranes.

EFFECTS OF CO- AND COUNTER ION REJECTION ON PERCHLORATE TRANSPORT WITH SYNTHETIC WATERS: EFFECTS OF pH AND CONDUCTIVITY

The concentrations of each co- and counter-ion species, K^+, Ca^{2+}, Cl^- and SO_4^{2-}, as well as ClO_4^- in the feed and permeate during filtration were measured to determine how the target solute's composition at the interface of the membranes may change. These measurements were done for filtration with the GH, GK, GM, and GN membranes (see Figures 6.5 through 6.8). The measurements were performed at varying pH and solution conductivity with three different salts, KCl, K_2SO_4, and $CaCl_2$. The experimental conditions, pressure, feed flow rate, permeate flow rate, J_o/k ratio, recovery, and cross-flow velocity, are as described in Table 4.4. The theory of the Donnan exclusion and the extended Nernst-Planck model is employed to evaluate the results obtained from the co- and counter ion rejection by the membranes. As mentioned earlier for Donnan exclusion, more specifically for Cl^- and SO_4^{2-} rejection, when a dilute solution containing anionic chloride and sulfate is brought in contact with a membrane possessing a fixed negative charge, the rejection of chloride and sulfate may be greater than if the membrane were completely uncharged. Also, it was mentioned earlier that Ca^{2+} binding screens the co-ion bonding/adsorption. When the charge screening by Ca^{2+} occurs for the negatively charged membranes, anion rejection is significantly reduced. Earlier it was discussed that the theory of the extended Nernst-Planck model was employed to generally explain how effective charge density is used to predict the separation of mixtures of electrolytes at the membrane/solution interface.

As shown in Figures 6.5 through 6.8, different amounts (2% to 90%) of K^+, Ca^{2+}, Cl^-, and SO_4^{2-} are rejected by the membranes depending on pH and conductivity. The increase of ion concentration at the membrane interface should affect the Donnan exclusion and charge density, which can significantly influence perchlorate transport. It is anticipated that perchlorate rejection is reduced because of the effect on the electrical double layer (a reduction) due to increasing concentration of the solutes, K^+, Cl^- and SO_4^{2-}. For example, SO_4^{2-} is rejected more than the mono-valent co-ion, Cl^-, in the presence of K^+ at varying pH and conductivity for the membranes. Also K^+ is rejected to a greater degree in the presence of SO_4^{2-} than in the presence of Cl^- (due to the requirement of electroneutrality). As previously observed, because of its increased rejection, K_2SO_4 affects the charge density at the membrane/solution interface more than KCl. These ion rejection results are consistent with the trends anticipated from the ZP results shown in Figures 5.5 and 5.6 for the GM membrane. For example, the influence of conductivity (or concentration) on the ZP provides a self-consistent framework for the observations of the magnitude of electrostatic repulsion that different ionic solutes would experience at the interface of the membrane.

EFFECTS OF ZETA POTENTIAL ON ARSENIC (III) REJECTION: EFFECTS OF pH AND CONDUCTIVITY

A model uncharged species, As (III), was used as a size-surrogate for ClO_4^- to verify that electrostatic interaction significantly occurs between anionic species and a negatively charged membrane, and that steric/size exclusion mechanism is more dominant for the ESNA membrane than for the MX07 membrane. The experiments were performed at the same conditions, pH and conductivity with KCl salt, as perchlorate rejection tests. The As (III) rejection only slightly varies at pH 4, 6, and 8 and 30 mS/m, 60 mS/m, and 115 mS/m since the As (III) exists primarily as an uncharged species H_3AsO_3 below pH 9.13 (see Figures 6.9 and 6.10). However, As (III) rejection significantly increases at pH 10 when As (III) is anionic ($H_2AsO_3^-$). In comparing these results to ClO_4^- rejection at the same pH and conductivity conditions, anionic ClO_4^- over pH 4 to 8 is rejected to a greater degree than As (III) by the negatively charged membranes. The steric/size exclusion mechanism is dominant for rejection of the uncharged species, As (III), until it becomes negatively charged at pH > 9.13, where the electrostatic exclusion mechanism also plays a role.

SUMMARY OF RESULTS AND RELATION OF THIS CHAPTER TO RESEARCH HYPOTHESES

All of the membranes show similar and relatively high (75% to 90%) perchlorate rejections with pure water. Also, the ESNA membrane consistently provided approximately 80% perchlorate rejection over time with CRW, higher than that observed for the MX07 (25%) and GM (< 5%) membranes. Perchlorate rejection did not appear to be affected by flux decline during short-term bench-scale filtration tests in the presence of NOM in CRW. The results support hypothesis 3.2.1. The relatively small pore size membrane, ESNA, based on its MMCO showed greater perchlorate rejection than the relatively large pore size membranes, MX07 and GM, with CRW. Ions in CRW significantly affected the membrane surface charge (becoming less negative). Therefore, less electrostatic repulsion between the perchlorate anion and the membranes resulted in a decrease of perchlorate rejection with CRW. However, the MX07 and GM membranes showed similar perchlorate rejection compared to the ESNA membrane with a synthetic water (pure water) even though they have different membrane pore sizes. This was because of the unaffected surface charge of the MX07 and GM membranes, due to a minimum amount of ions in the synthetic water, providing enough electric charge to reject the perchlorate.

Perchlorate rejection increased with increasing pH at the same conductivity (30 mS/m) for all of the membranes, regardless of salt type. Perchlorate rejection followed the order, $CaCl_2$ < K_2SO_4 < KCl, at the same pH and conductivity conditions for the membranes. Perchlorate rejection followed the order, GN < GM < GK < GH < MX07 < ESNA, at varying pH conditions. These results support hypothesis 3.2.1.a that greater (negative) surface charge due to increasing pH enhances perchlorate rejection and lower MMCO also enhances perchlorate rejection.

Different amounts (2% to 90%) of K^+, Ca^{2+}, Cl^-, and SO_4^{2-} are rejected by the membranes depending on pH and conductivity. SO_4^{2-} is rejected more than the mono-valent co-ion, Cl^-; in the presence of K^+ at varying pH and conductivity for the membranes. K^+ is rejected more in the presence of SO_4^{2-} than in the presence of Cl^-. These ion rejection results are also related to the ZP results for the GM membrane. These results support hypothesis 3.2.1.b that higher ionic strength, especially with Ca^{2+}, can reduce perchlorate rejection because an increase of CP in the presence of Ca^{2+} at the membrane surface resulted in membrane surface charge becoming significantly less negative.

As (III) rejection only slightly varied at pH 4, 6, and 8 and 30 mS/m, 60 mS/m, and 115 mS/m for the membranes since the As (III) exists primarily as an uncharged species below pH 9.13 based on its pK_{a1} of 9.13. However, As (III) rejection significantly increased at pH 10, where As (III) is anionic. These results support an existing theory that significant electrostatic repulsion occurs between an anion and a negatively charged membrane.

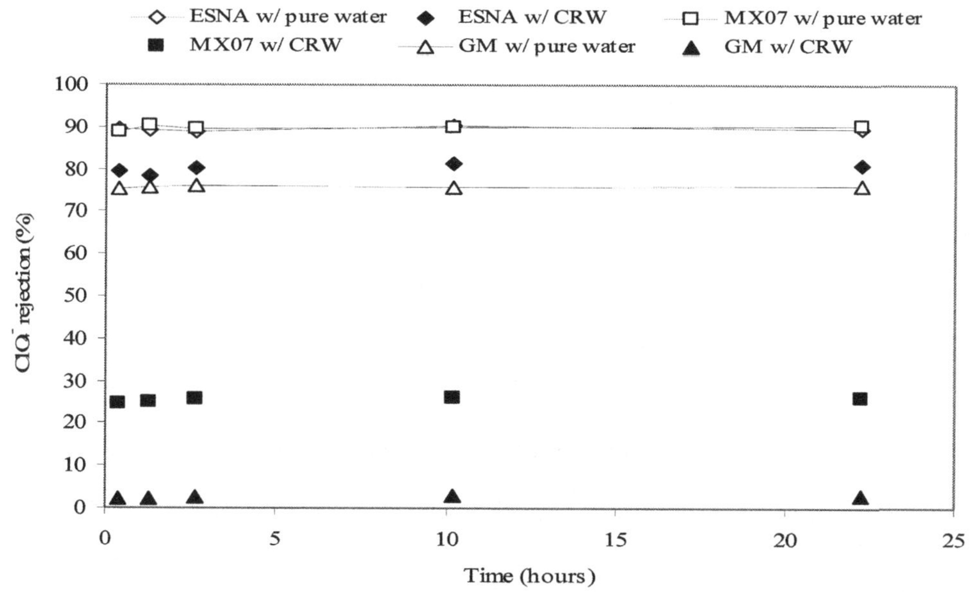

Figure 6.1 Comparison of perchlorate rejection with synthetic (pure) water and CRW for the NF and UF membranes

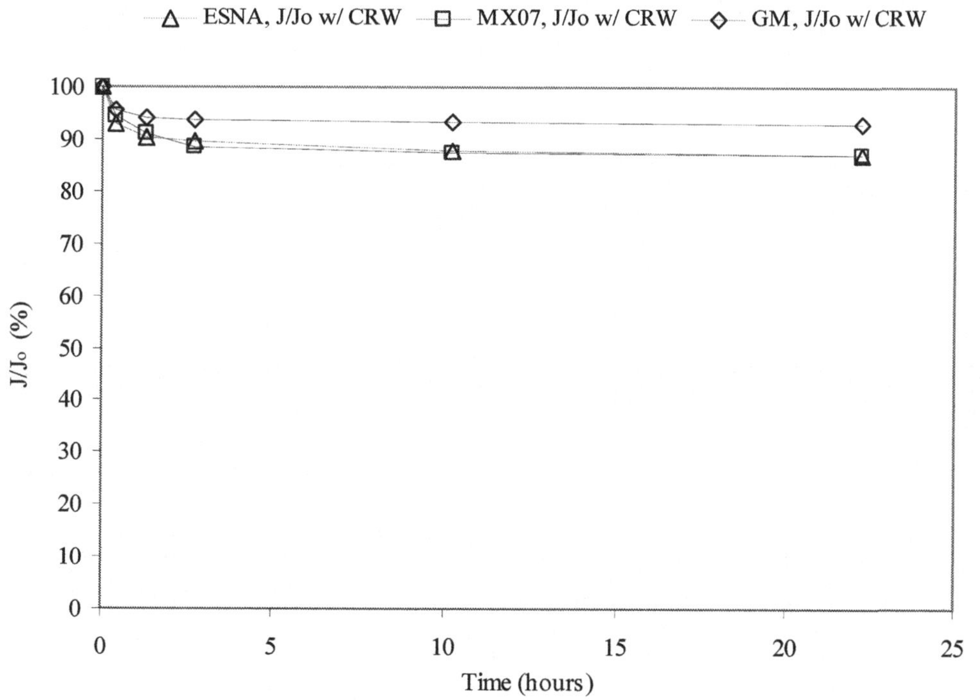

Figure 6.2 Comparison of flux-decline trends with CRW for the NF and UF membranes

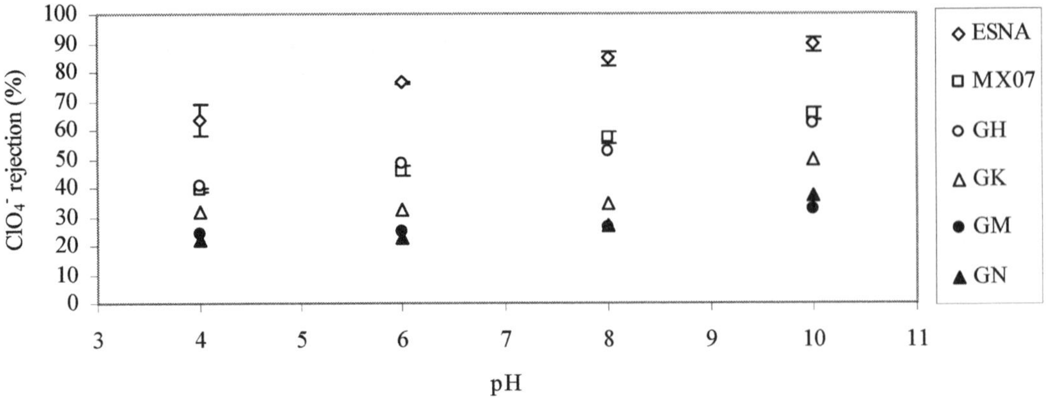

(a) Conductivity 30 mS/m KCl

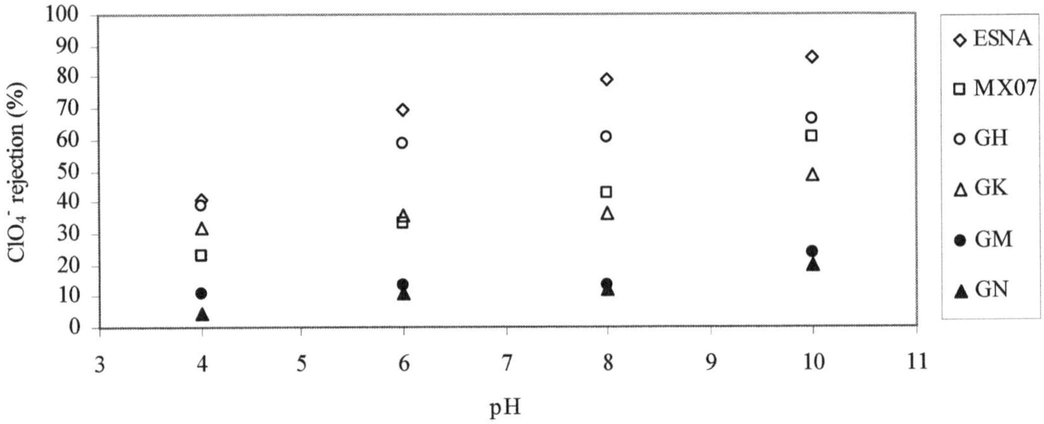

(b) Conductivity 30 mS/m K_2SO_4

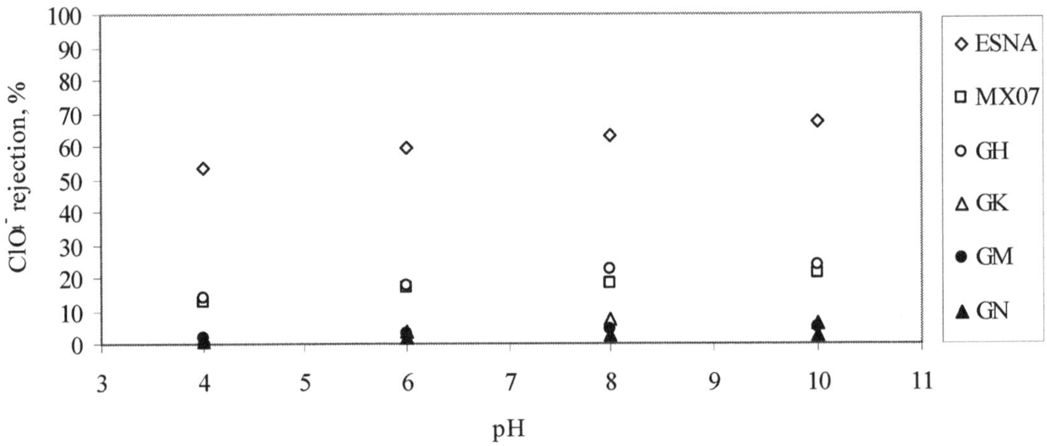

(c) Conductivity 30 mS/m $CaCl_2$

Figure 6.3 Effect of pH on perchlorate rejection for the NF and UF membranes with variable salts, KCl, K_2SO_4, and $CaCl_2$

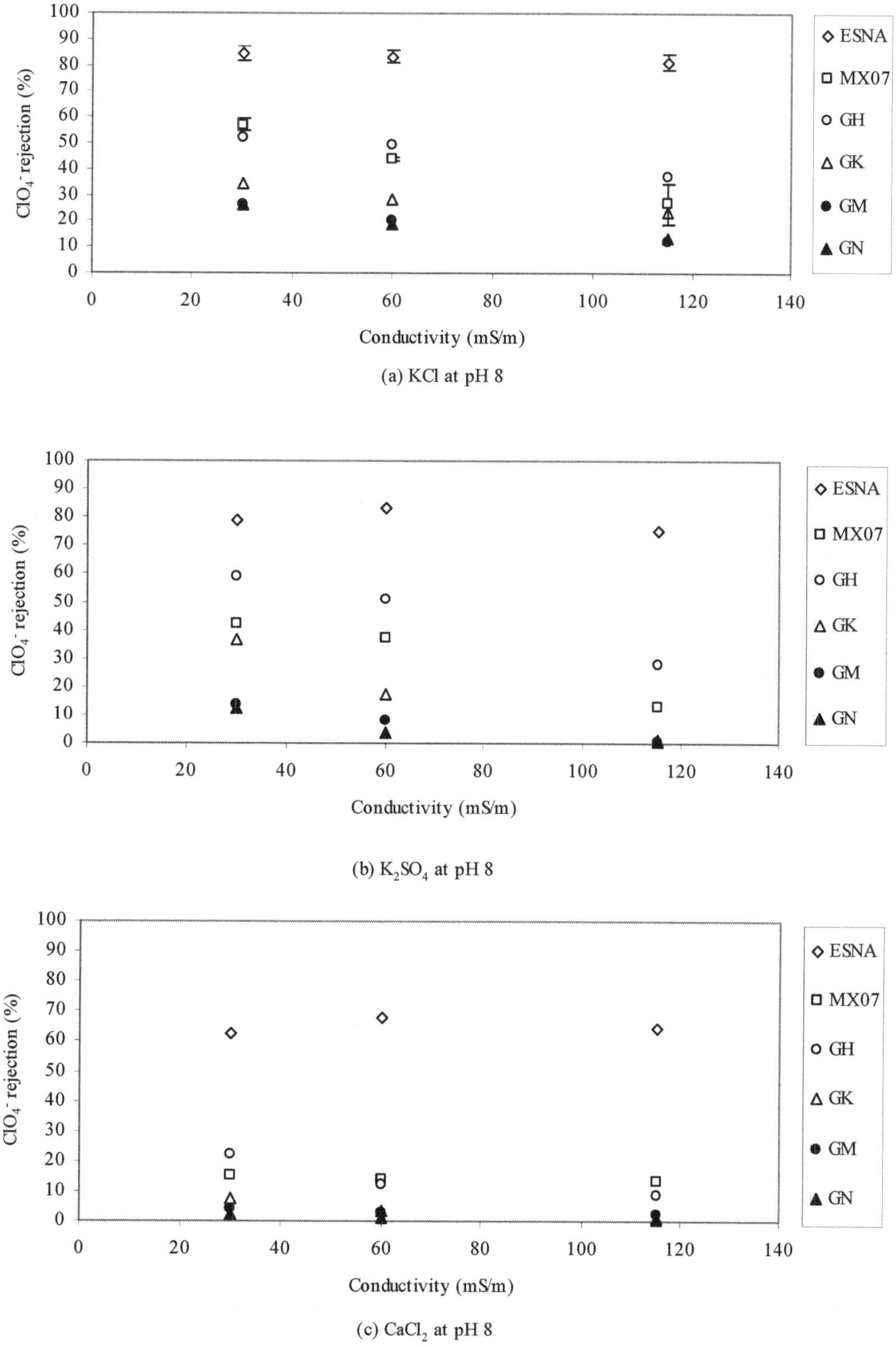

(a) KCl at pH 8

(b) K_2SO_4 at pH 8

(c) $CaCl_2$ at pH 8

Figure 6.4 Effect of conductivity on perchlorate rejection for the NF and UF membranes with variable salts, KCl, K_2SO_4, and $CaCl_2$

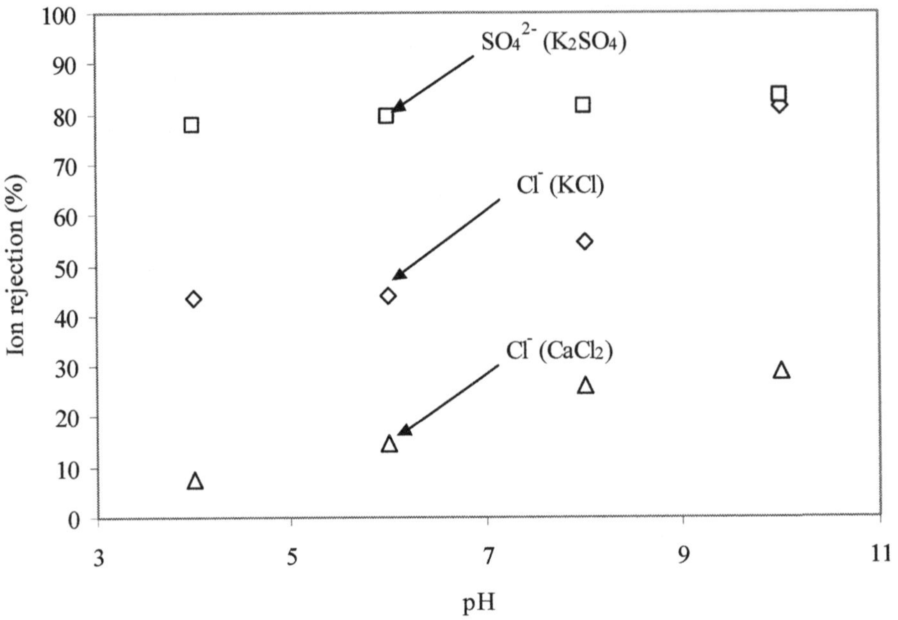

a. GH (MMCO = 2,500)

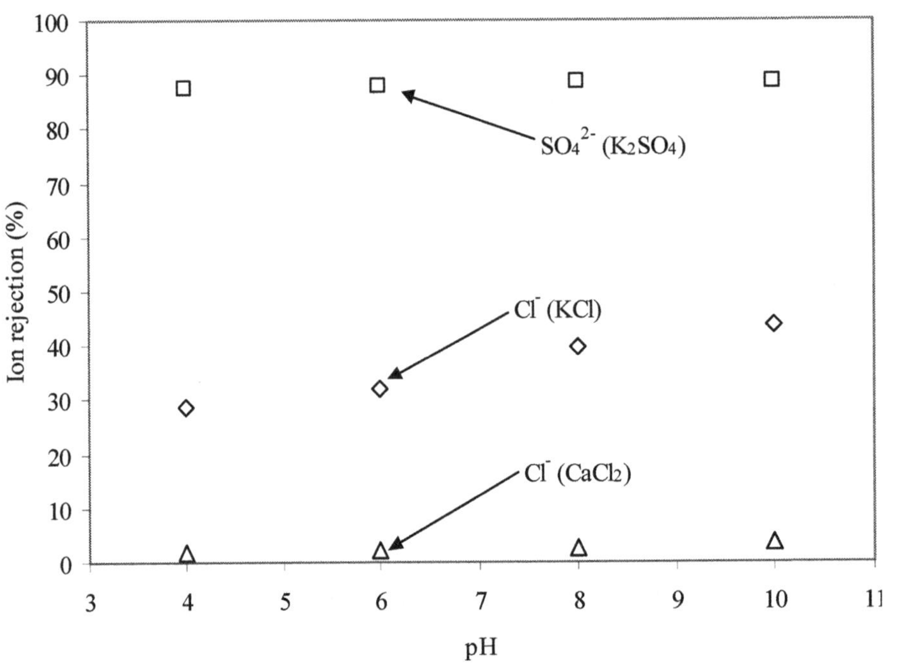

(b) GK (MMCO = 3,500)

(continued)

Figure 6.5 Comparison of mono- and divalent co-ion rejection as a function of pH at conductivity 30 mS/m for the UF membranes

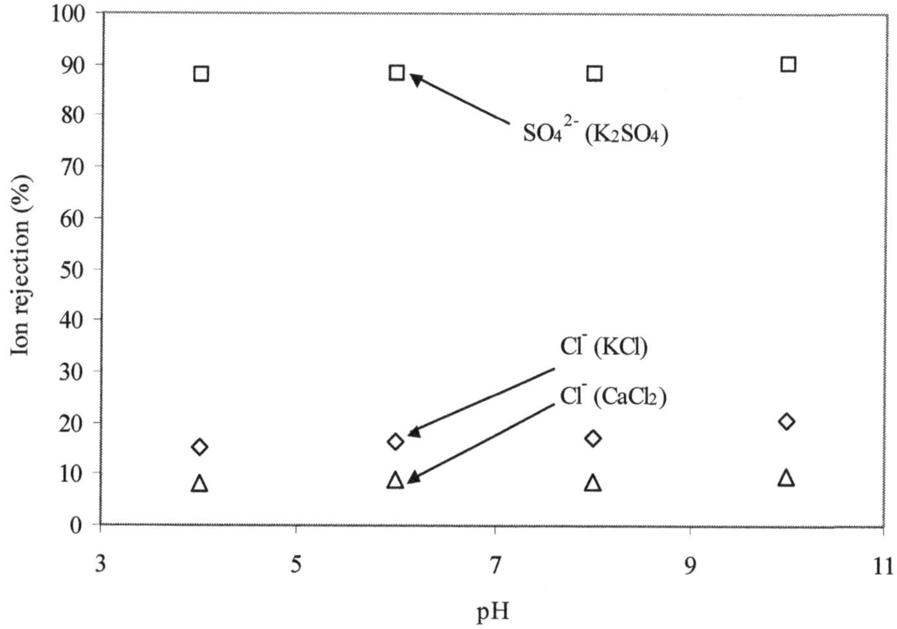

(c) GM (MMCO = 8,000)

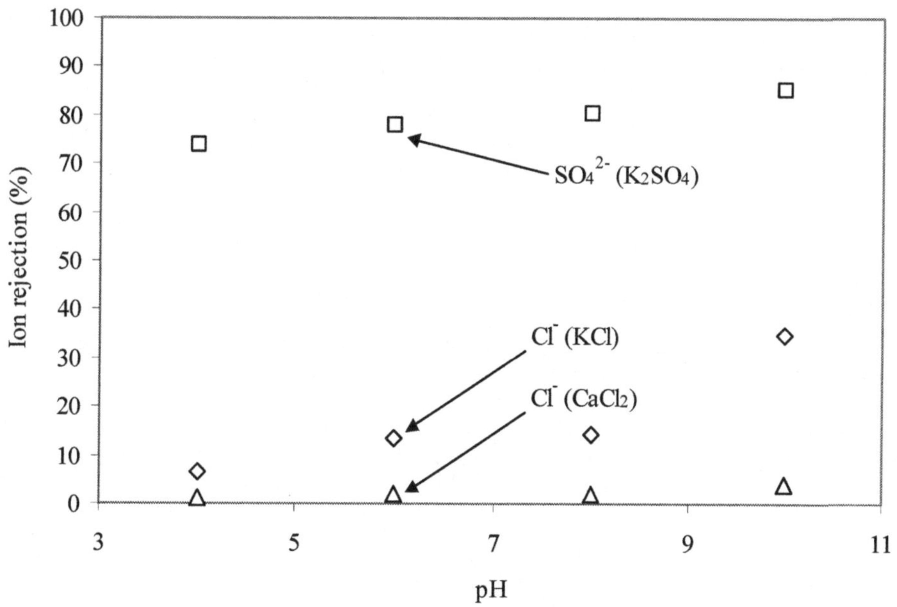

(d) GN (MMCO = 12,000)

Figure 6.5 (Continued)

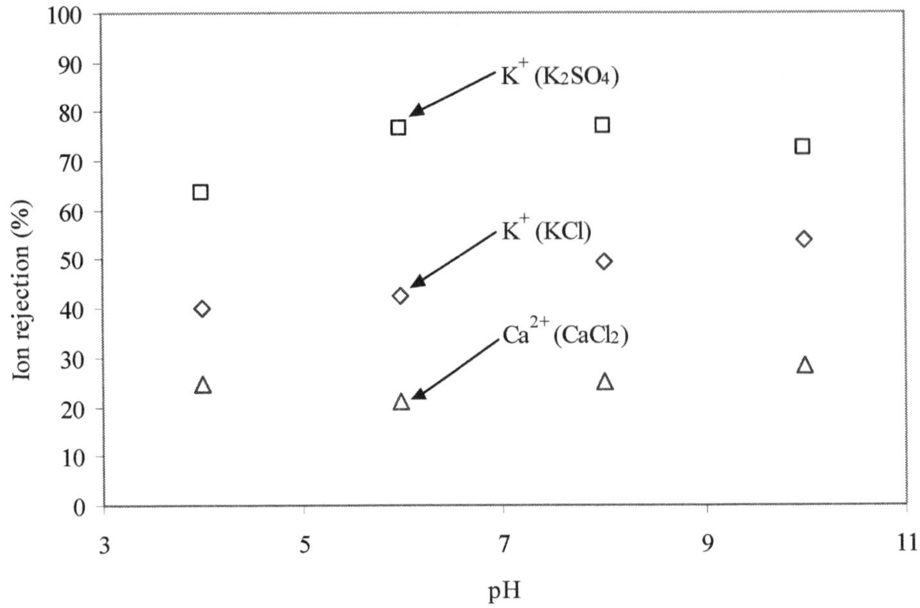

(a) GH (MMCO = 2,500)

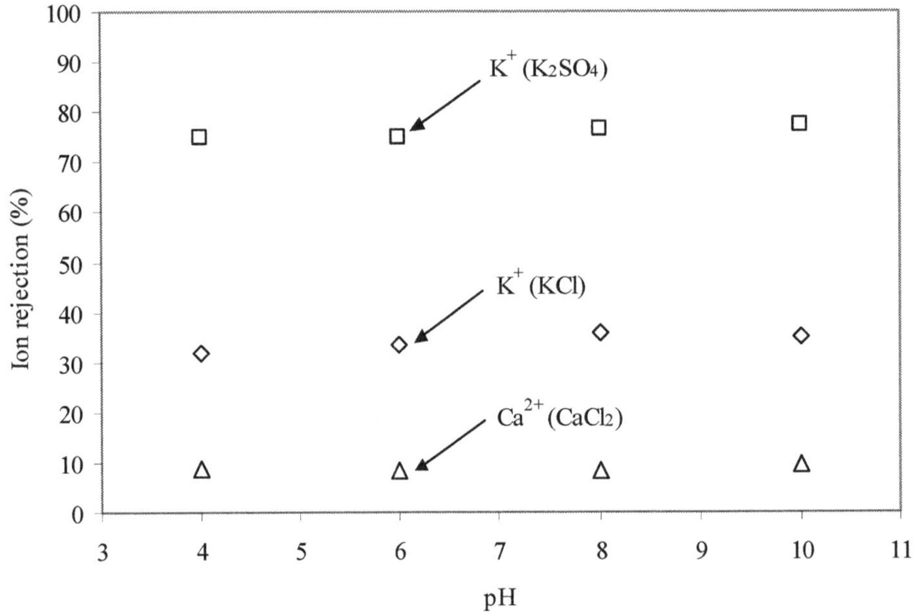

(b) GK (MMCO = 3,500)

(continued)

Figure 6.6 Comparison of mono- and divalent counter ion rejection as a function of pH at conductivity 30 mS/m for the UF membranes

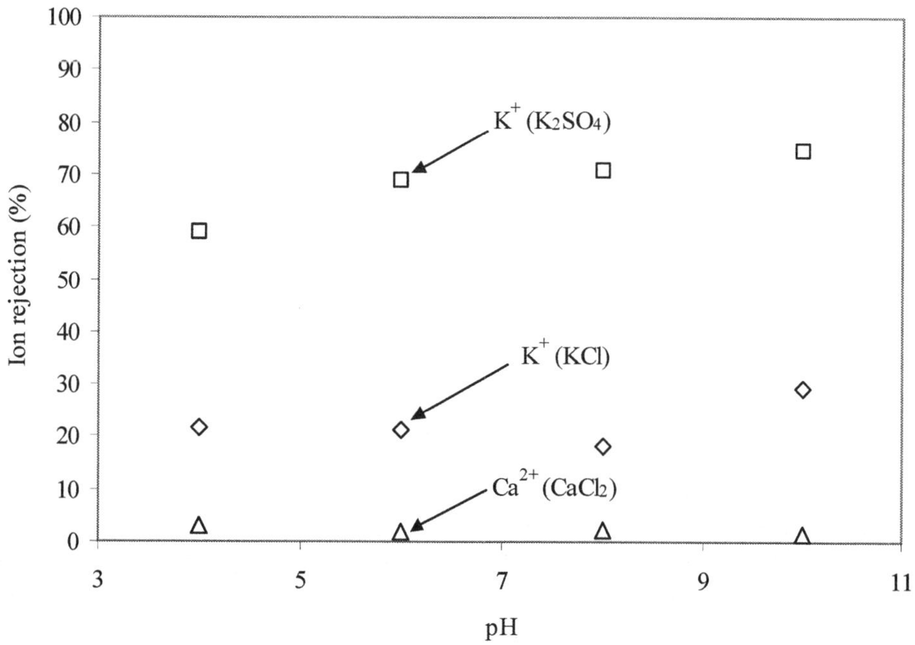

(d) GN (MMCO = 12,000)

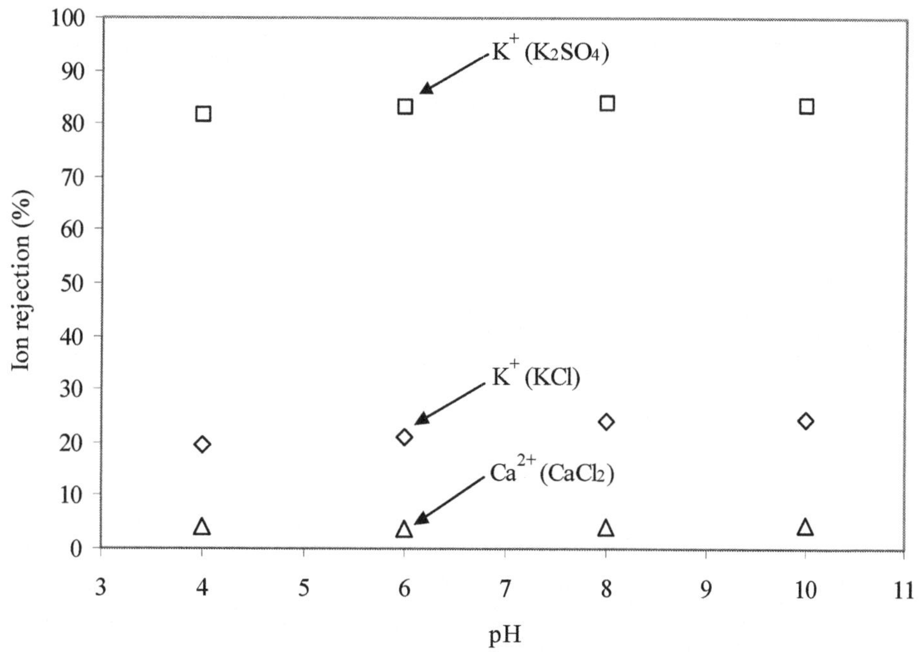

(c) GM (MMCO = 8,000)

Figure 6.6 (Continued)

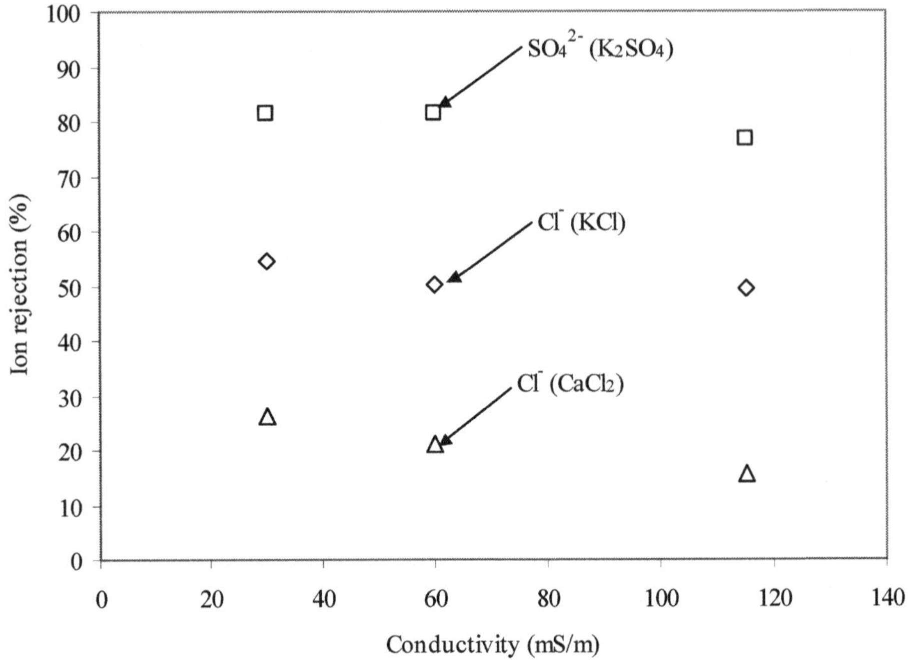

(a) GH (MMCO = 2,500)

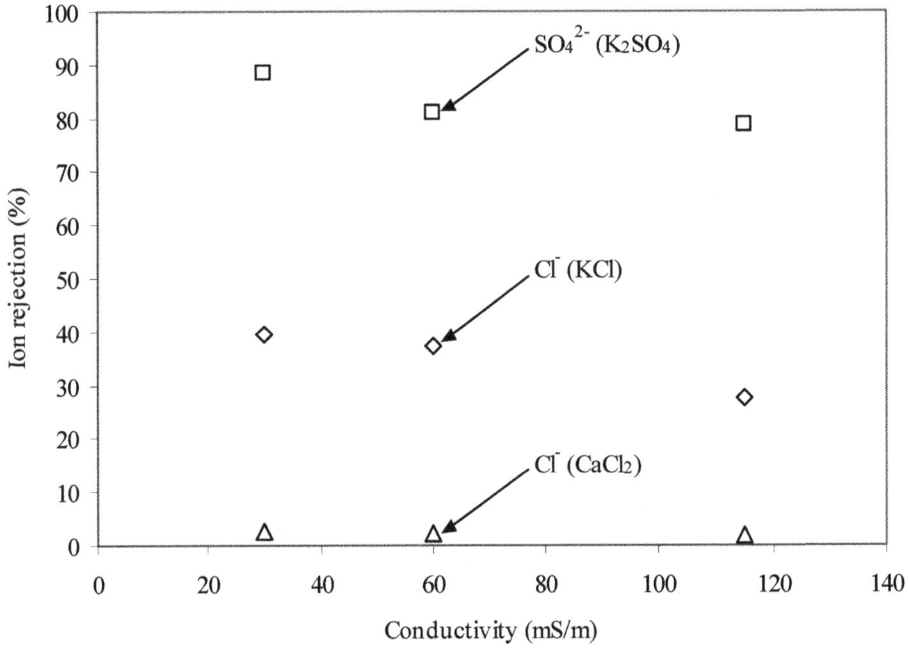

(b) GK (MMCO = 3,500)

(continued)

Figure 6.7 Comparison of mono- and divalent co-ion rejection as a function of conductivity at pH 8 for the UF membranes

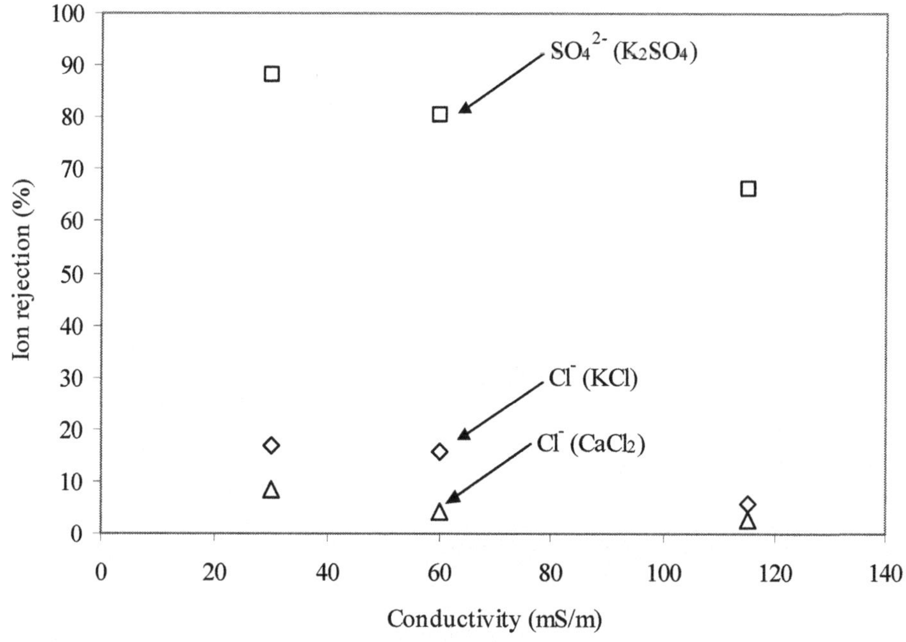

(c) GM (MMCO = 8,000)

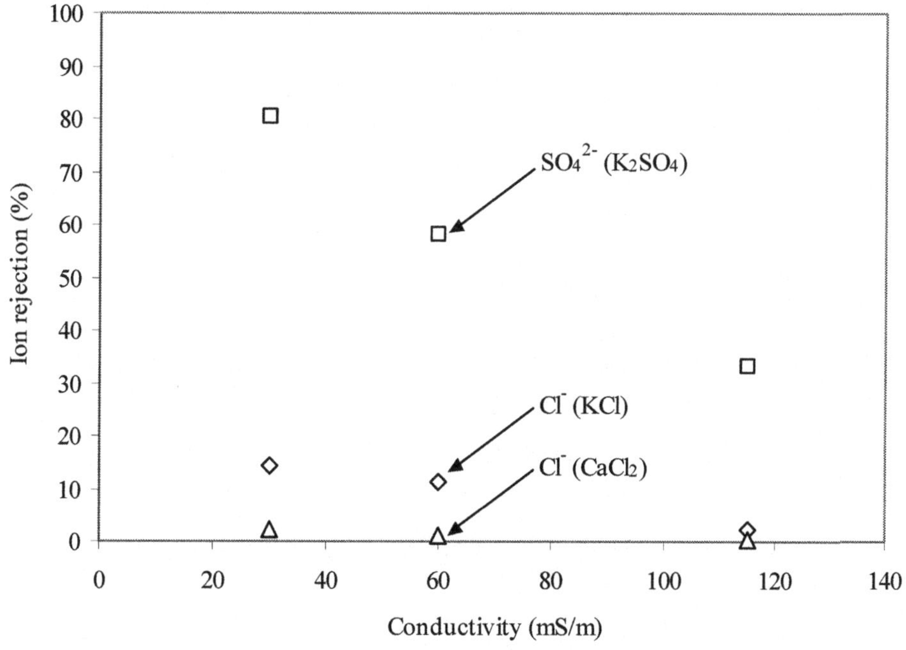

(d) GN (MMCO = 12,000)

Figure 6.7 (Continued)

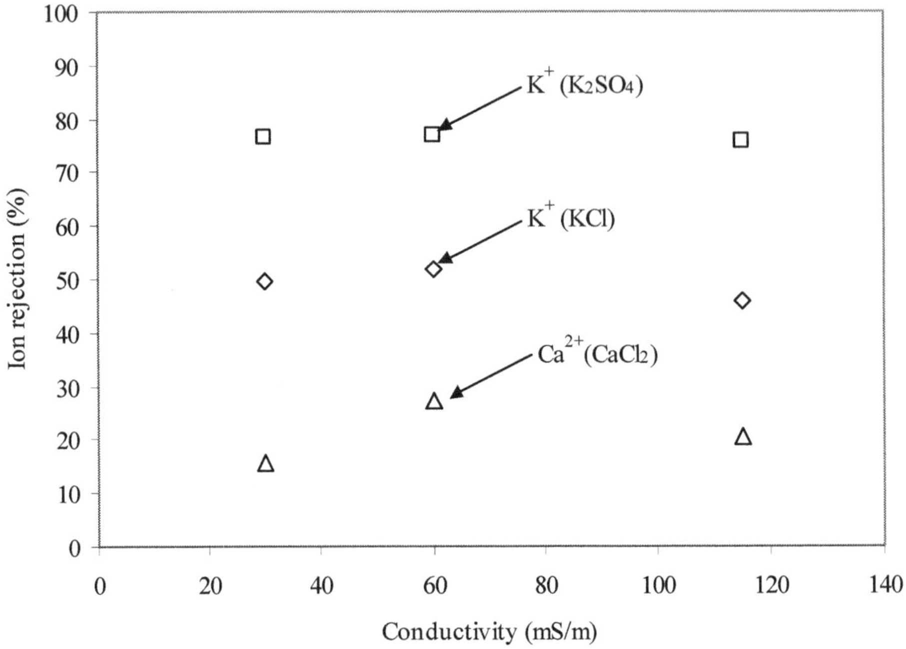

(a) GH (MMCO = 2,500)

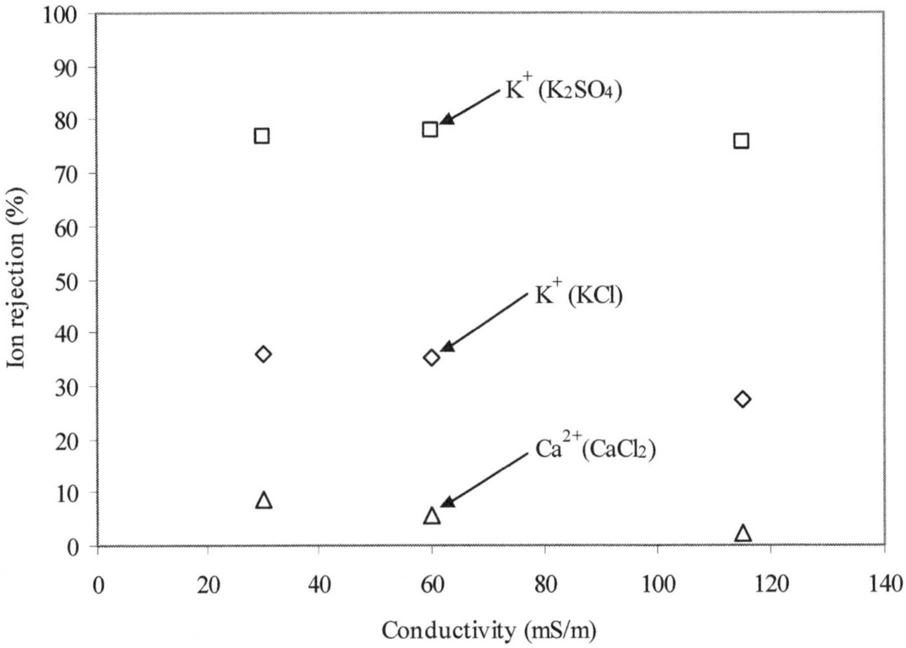

(b) GK (MMCO = 3,500)

(continued)

Figure 6.8 Comparison of mono- and divalent counter ion rejection as a function of conductivity at pH 8 for the UF membranes

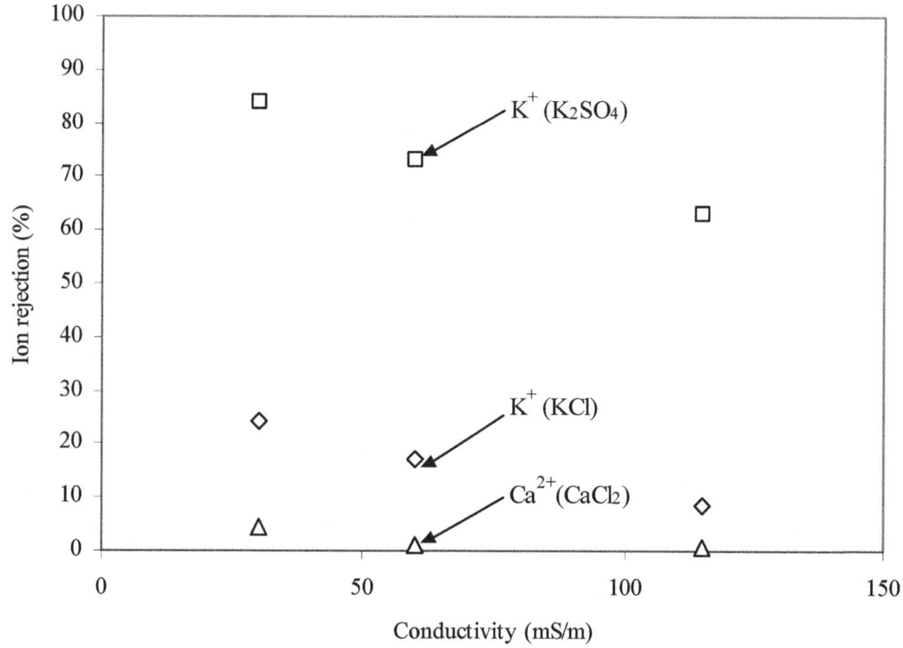

(c) GM (MMCO = 8,000)

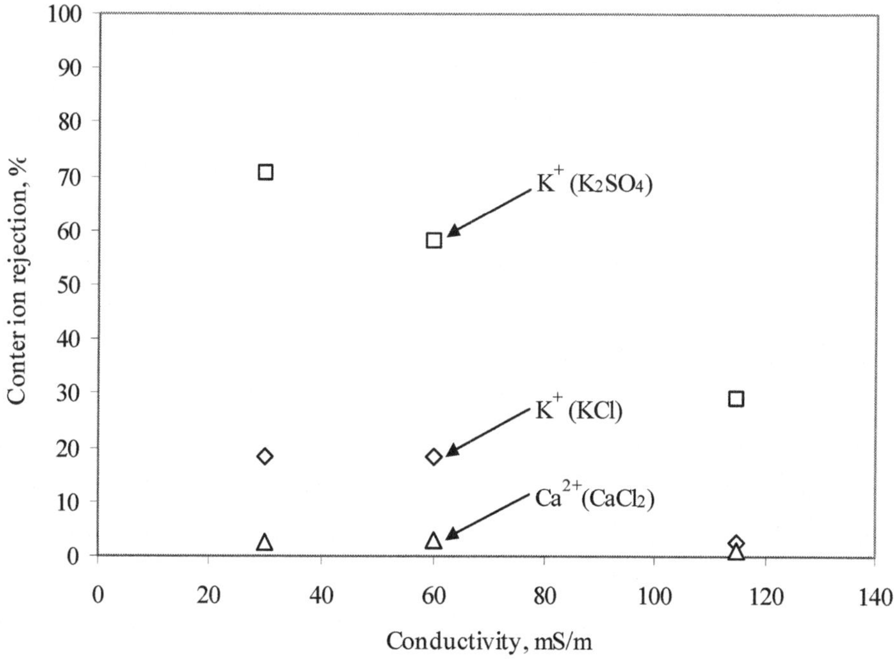

(d) GN (MMCO = 12,000)

Figure 6.8 (Continued)

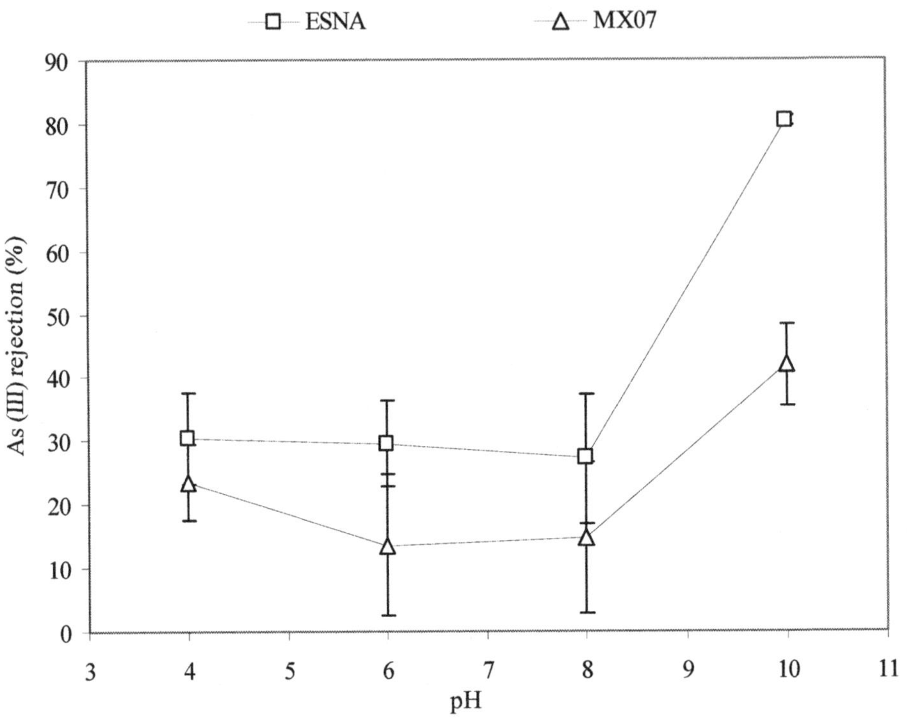

Figure 6.9 Effect of pH on As (III) rejection at 30 mS/m KCl for the NF membranes

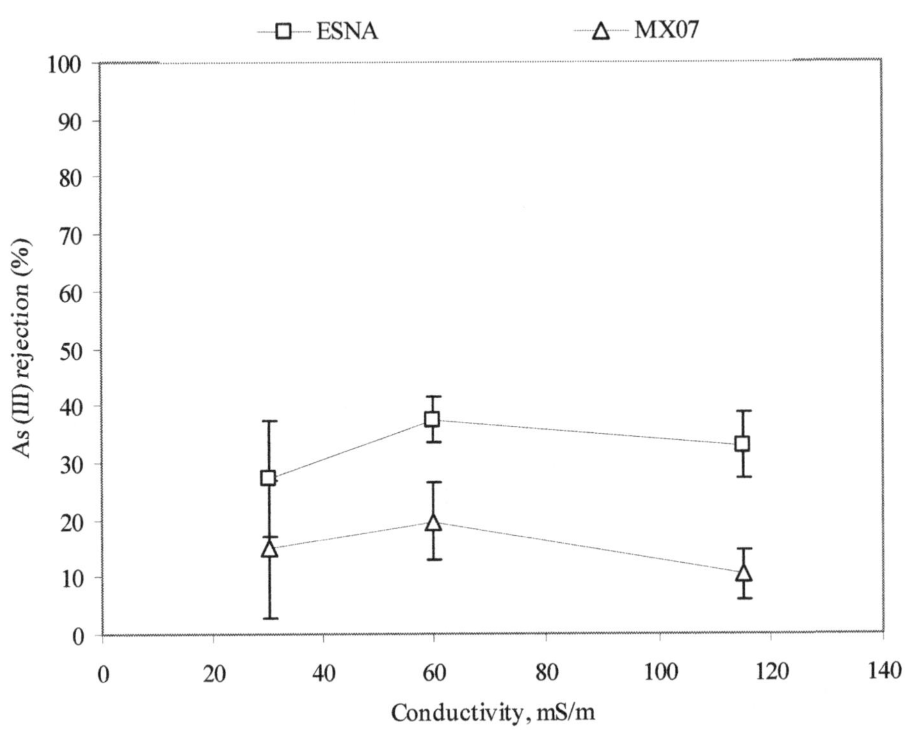

Figure 6.10 Effect of conductivity (KCl) on As (III) rejection at pH 8 for the NF membranes

CHAPTER 7
MEASUREMENT OF HINDERED DIFFUSION/PARTITIONING AND ESTIMATION OF PERCHLORATE TRANSPORT THROUGH NEGATIVELY CHARGED NF AND UF MEMBRANES

HINDERED DIFFUSION AND PARTITIONING THROUGH MEMBRANE PORES

It has been known that the diffusion coefficient of a solute through membrane pores, called the *effective* diffusion coefficient (D_{eff}), is usually found to be less than its corresponding value in water. This transport phenomenon is called hindered diffusion. Hindered diffusion (D_{eff}) plays an important role for target solute (in this case ClO_4^-) transport in membrane filtration affected by diffusion and convection forces. A non-equilibrium thermodynamic model was initially proposed by Kedem and Katchalsky (1958) and modified by Spiegler and Kedem (1966) to predict solute transport. The effective diffusion coefficient associated with partitioning is an important parameter for the non-equilibrium thermodynamic model when diffusion is dominant in membrane filtration. This is especially the case for a relatively small pore membranes.

Hindered Diffusion and Partitioning in Diffusion Cell

Diffusion experiments with synthetic water involved the use of one NF membrane (MX07) and one UF membrane (GM) at varying pH and conductivity conditions. Experimental methods for this study are described in Chapter 4. Figure 7.1-a presents the change of perchlorate concentration in Cells A and B over time for a negatively charged porous membrane (GM) when 100 μg/L (10^{-6} M) of ClO_4^- was spiked into a 0.0022 M KCl solution at pH 8 in Cell A. The ClO_4^- concentration in Cell A linearly decreased from 100 μg/L to 86 μg/L during a 250 hour period. The ClO_4^- concentration in Cell B linearly increased up to 7 μg/L during the experiment period.

The use of Equation 4.5 is shown in Figure 7.1-b, portraying a trend of D_pH values calculated with the data presented in Figure 7.1-a. The D_pH was consistent at approximately 2.5×10^{-10} cm^2/s after 140 hours. Therefore, a steady value of the D_pH was obtained with an average of the data after stabilization of hindered diffusion and partitioning at each given pH and conductivity conditions. In this study, D_pH is being used as a composite parameter that was directly measured by diffusion cell. D_p and H were not determined individually by the diffusion cell.

Prediction of Hindered Diffusion and Partitioning Based on ZP

ZP results obtained at varying pH levels and conductivities are shown in Figures 7.2 and 7.3. The ZP results were adopted from Chapter 5 to demonstrate the effect of electrostatic repulsion on hindered diffusion and partitioning of perchlorate between the perchlorate anion and the negatively charged membranes.

The membrane surface charge is significantly affected by the changes in pH. All of the membranes (MX07 and GM) acquire more negative charge when pH increases regardless of the salt type at the same conductivity of 30 mS/m (see Figure 7.2). As shown in Figure 7.2, at the same pH and solution conductivity, the ZP of the membranes follows the order (negatively) in the presence of different type of salts, $CaCl_2$ < KCl < K_2SO_4. However, the membranes exhibit less negative charge at relatively higher conductivities (see Figure 7.3). In particular, the ZP of the membranes may be significantly affected by increasing solution conductivity, as provided by $CaCl_2$. The ZP of the membranes became less negative with increasing divalent cation, Ca^{2+}, due to ion binding.

Two assumptions are made with the ZP results as a function of pH and conductivity. One assumption is that the hindered diffusion of perchlorate associated with partitioning decreases as pH increase. This is due to more electrostatic repulsion occurring between the perchlorate anion and the negatively charged membrane at relatively higher pH values. The other assumption is that the hindered diffusion of perchlorate associated with partitioning increases as conductivity increases. This is due to less electrostatic repulsion occurring between the perchlorate anion and the negatively charged membrane, which is in turn caused by a decrease of the electrical double layer at a relatively high conductivity.

Effect of pH and Conductivity on Hindered Diffusion and Partitioning

Effect of pH

The D_pH values of perchlorate under varying pH conditions (4, 6, 8, and 10) are shown in the presence co- and counter ions (e.g., K^+, Ca^{2+}, Cl^-, and SO_4^{2-}) for the GM membrane (see Figure 7.4). The D_pH value of perchlorate significantly decreased as pH increased regardless of the salt type at the same conductivity of 30 mS/m. As shown in Figure 7.4, at the same pH and solution conductivity, the D_pH of perchlorate follows the order in the presence of different type of salts, $CaCl_2 > KCl > K_2SO_4$. The mechanism of electrostatic repulsion is employed to explain these results. As shown in Figure 7.2, the surface charge of the GM membrane becomes more negative as pH increases, thus an increase of electrostatic repulsion occurs between the perchlorate anion and the negatively charged GM membrane. The D_pH of perchlorate with a background solution of $CaCl_2$ is larger than background solutions of KCl and K_2SO_4. This is likely due to Ca^{2+} binding as an outer-sphere complex (Childress and Elimelech, 1996), reducing the negative surface charge of the GM membrane. In addition, the D_pH of perchlorate with a background solution of K_2SO_4 is smaller than that of a KCl based solution because of an inner-sphere ion adsorption (Stumm, 1992), $Cl^- < SO_4^{2-}$. Thus, greater electrostatic repulsion can occur in the presence of SO_4^{2-} than in the presence of Cl^-.

Figure 7.5 shows the effect of pH on D_pH of perchlorate for the GM membrane with and without background conductivity adjustment in the presence of a KCl background solution. As expected, the D_pH of perchlorate in the diffusion tests is larger without conductivity adjustment in Cell B than with conductivity adjustment in Cell B. This is likely due to the osmotic pressure occurring between the feed and permeate cells, significantly affecting perchlorate diffusion. As mentioned in Chapter 4, the osmotic pressure difference without conductivity adjustment (in Cell B) is larger than that the case with conductivity adjustment in Cell B.

Figure 7.6 shows the effect of pH on the D_pH of perchlorate for the NF (MX07) and UF (GM) membranes with a background solution of KCl. The MX07 membrane also shows similar D_pH trends compared to the GM membrane as a function of pH. The D_pH of perchlorate decreased as pH increased for the MX07 membrane. However, the MX07 membrane shows a smaller D_pH than the GM membrane because the MX07 membrane has a smaller membrane pore size than the GM membrane based on their nominal MMCOs. In addition, the MX07 membrane shows more negative charge than the GM membrane as shown in Figure 7.3, thus greater electrostatic repulsion can occur with the MX07 membrane than with the GM membrane.

Effect of Conductivity

The D_pH values of perchlorate at varying conductivity conditions (30, 60, and 115 mS/m) are shown for the GM membrane at pH 8 in Figure 7.7. The D_pH of perchlorate significantly increased as conductivity increased regardless of the salt type. According to the ZP of the GM membrane shown in Figure 7.3, an expected, a decrease of electrostatic repulsion as conductivity increased resulted in an increase of the D_pH of perchlorate for the GM membrane. Figure 7.8 shows the effects of conductivity on the D_pH of perchlorate for the GM membrane with and without conductivity adjustment in Cell B at pH 8. As expected, the D_pH of perchlorate in the diffusion tests is larger without conductivity adjustment in Cell B than with conductivity adjustment in Cell B. In addition, the MX07 membrane, which has a smaller membrane pore size and higher membrane negative surface charge than the GM membrane, shows smaller D_pH of perchlorate than the GM membrane as conductivity increases (see Figure 7.9).

Table 7.1 summarizes the overall hindered diffusion results associated with partitioning of perchlorate under various pH and conductivity conditions. Some of these results are not shown in the previous figures. The D_pH of perchlorate is also greater without conductivity adjustment in Cell B than with conductivity adjustment in Cell B for the background solutions of K_2SO_4 and $CaCl_2$ at pH 8.

ESTIMATION OF PERCHLORATE TRANSPORT BY THERMODYNAMIC MODEL

The transport of the perchlorate anion through NF and UF membranes (MX07 and GM) was predicted based on the non-equilibrium thermodynamic model. To compare the predicted perchlorate flux to the measured perchlorate flux, the measured flux was obtained from the cross-flow flat-sheet filtration tests in the presence of three electrolytes, KCl, K_2SO_4, and $CaCl_2$, under varying pH conditions (4, 6, 8, and 10). These varying conditions were used to determine their effects on perchlorate transport significantly influenced by diffusion and convection.

The model was characterized by five transport parameters: molecular transport coefficient (ω), osmotic pressure gradient ($\Delta\Pi$), molecular reflection coefficient (σ), average bulk fluid interfacial concentration between feed and permeate side (C_{avg}), and solvent flux (J_v). Each parameter used for the model was experimentally measured and/or calculated with minimum assumptions. For this model, the hindered diffusion and partitioning of perchlorate through the membrane pores was measured by the diffusion cell under varying pH and conductivity conditions as shown earlier.

Perchlorate was always substantially rejected by RO, NF, and UF membranes in this study. The mechanism is always solution diffusion, but the solubility and diffusivity values are determined by steric/size exclusion and electrostatic exclusion to varying degrees. RO can always be used to remove perchlorate, but it exhibits little selectivity among other desirable salts, requires high operating pressure, and is the most prone to fouling. For membrane processes that operate at lower pressures such as NF or UF membranes, selectivity based on size and/or charge may be effective for perchlorate removal. In this section, quantitative models are addressed to predict perchlorate transport through NF and UF membranes based on a solution-diffusion model (irreversible thermodynamics) with the measured and calculated solute transport parameters.

Table 7.1
Hindered diffusion associated with partitioning (D_pH) of perchlorate for the UF and NF membranes under varying pH and conductivity conditions

Membrane/ pH-Conductivity	Electrolyte (Salt Type)						
	KCl* (cm^2/s)	S.D.† (KCl)	KCl‡ (cm^2/s)	K_2SO_4* (cm^2/s)	K_2SO_4‡ (cm^2/s)	$CaCl_2$* (cm^2/s)	$CaCl_2$‡ (cm^2/s)
GM pH at 30 mS/m							
4	1.52E-09	2.15E-11	3.24E-10	9.41E-10		1.55E-09	
6	1.11E-09	7.77E-11	2.71E-10	7.41E-10		1.37E-09	
8	5.59E-10	6.38E-12	2.45E-10	4.71E-10	6.43E-10	9.40E-10	1.04E-10
10	3.71E-10	4.92E-12	2.31E-10	3.44E-10		7.45E-10	
Conductivity (mS/m) at pH 8							
30	5.59E-10		2.45E-10	4.71E-10		9.40E-10	
60	8.60E-10		4.03E-10	7.54E-10		1.09E-09	
115	1.08E-09		5.32E-10	8.57E-10		1.25E-09	
MX07 pH at 30 mS/m							
4	1.33E-09						
6	5.55E-10						
8	3.83E-10						
10	2.77E-10						
Conductivity (mS/m) at pH 8							
30	3.83E-10						
60	7.25E-10						
115	8.66E-10						

* Without conductivity adjustment in Cell B
† S.D.: Standard deviation
‡ With conductivity adjustment in Cell B
$D_{ClO_4^-}$ in water = $1.79 \times 10^{-5} cm^2$ (Lide, 1995)

Theoretical Background

The non-equilibrium thermodynamic equation takes into account two driving forces: convection and diffusion, combining the observations about molecular transport, osmotic pressure, molecular reflection, average bulk fluid, and solvent transport to describe solute transport through a membrane:

$$J_s = diffusion + convection = \omega\Delta\Pi + (1-\sigma)C_{avg}J_v \qquad (7.1)$$

where J_s is the solute flux (mol/m^2-s) occurring due to diffusion and convection, ω is the molecular transport coefficient (solute permeability) (mol/m^2-s-Pa), $\Delta\Pi$ is the osmotic pressure gradient (Pa), σ is the molecular reflection coefficient (dimensionless), C_{avg} is the average bulk fluid interfacial concentration between feed and permeate side (mol/m^3), and J_v is the solvent flux (m^3/m^2-s).

ω is a function of the effective diffusion coefficient (D_{eff}) associated with the partition coefficient (H), which is measured by the diffusion cell, and membrane thickness (l). It is very important to directly measure the hindered diffusion and partitioning of a target solute (in this case ClO_4^-) by the diffusion cell because the D_pH is significantly affected by ionic radius, ion charge, membrane surface charge, membrane pore size, membrane pore density, and membrane thickness. Therefore, to determine the D_pH parameter is to obtain an indirect physico-chemical view for solute transport through the membrane. The molecular transport coefficient can be given by:

$$\omega = \frac{D_p H(or\ D_{eff})}{l}\frac{1}{RT} \tag{7.2}$$

where D_p is the intrapore diffusion coefficient in membrane pores (cm^2/s), D_{eff} is the effective diffusion coefficient through membrane pores (cm^2/s), H is partition coefficient of the electrolyte at the membrane interfaces (dimensionless), l is the thickness of the separating layer of the membrane (m), R is the gas constant (J/mole-K), and T is the absolute temperature (K).

The determination of the D_pH in the diffusion experiment starts with a pseudo steady state application of Fick's First Law. The D_pH can then be calculated from Equation 7.3:

$$C_{B(t)} - C_{A(t)} = \left[C_{B(t_0)} - C_{A(t_0)}\right]\exp\left[-\frac{D_p \cdot H}{l}\left(\frac{1}{V_B}+\frac{1}{V_A}\right)t\right] \tag{7.3}$$

where C_B is concentration in the fluid adjacent to the permeate surface of the membrane in diffusion cell (mol/m^3), C_A is concentration in the fluid adjacent on the feed surface of the membrane in diffusion cell (mol/m^3), V_B is volume of permeate side of the membrane in diffusion cell (m^3), V_A is volume of feed side of the membrane in diffusion cell (m^3), and t is time during the diffusion test period (s).

The osmotic pressure influencing solute transport by diffusion arises when two solutions of different concentration (or a pure solvent and a solution) are separated by a semi-permeable membrane (i.e., one which is permeable to the solvent but impermeable to the solute). The osmotic pressure gradient, $\Delta\Pi$, is calculated by:

$$\Delta\Pi = -RT(\sum C_s^{feed} - \sum C_s^{perm}) \tag{7.4}$$

where C_s^{feed} is the total solute concentration in the feed side (mol/m^3) and C_s^{perm} is the total solute concentration in the permeate side (mol/m^3).

The reflection coefficient, σ, represents a measure of the selectivity of a membrane and usually has a value between 0 and 1. Solute transport by convection does not occur as σ is equal to one. There is no selectivity when σ is equal to zero. Experimentally determined parameters, J_s

(measured), C_{avg}, and J_v under varying pH and conductivity conditions, were used to determine the reflection coefficient by nonlinear estimation with a three dimensional plot J_s versus, C_{avg}, versus J_v. The equation was solved using STATISTICA (5.5 version, StatSoft, Inc.).

The average bulk fluid interfacial concentration, C_{avg}, between the feed at the membrane surface and the permeate is the mean logarithmic concentration and is given by:

$$C_{avg} = \frac{(C_m - C_p)}{\ln(\frac{C_m}{C_p})} \tag{7.5}$$

where C_m is the feed concentration of the membrane surface (mol/m^3) and C_p is the permeate concentration (mol/m^3). The increase in concentration of solute at the membrane surface influences their transport through the membrane. The solute feed concentration at the membrane surface is calculated from the Film Theory Model of concentration polarization (Mulder, 1991) and given by:

$$\ln\left(\frac{C_m - C_p}{C_b - C_p}\right) = \frac{J \cdot \delta}{D} \tag{7.6}$$

where C_b is the feed (bulk) concentration (mol/m^3), J is the pure water flux (m/s), δ is the thickness of the boundary layer (m), and D is the diffusion coefficient of solute in water (m^2/s). δ is the boundary layer where the solute concentration increases and reaches a maximum value at the membrane surface. δ is calculated by:

$$\delta = \frac{D}{k} \tag{7.7}$$

where k is the boundary layer back-diffusion mass transport coefficient (m/s). The back-diffusional mass transfer coefficient at the membrane interface is estimated, using a diffusion coefficient (D) of 1.79×10^{-5} cm^2/s for ClO_4^- in water (Lide, 1995). The back-diffusional mass transfer occurs where solute concentration increases at the membrane surface. Porter (1972) developed an equation to calculate the mass transfer coefficient for laminar flow in a channel:

$$k = 1.177\left[\frac{U_b D^2}{hL}\right]^{1/3} \tag{7.8}$$

where U_b is the average cross-flow velocity of bulk fluid (m/s), h is the channel height (m), and L is the channel length (m).

The solvent flux, J_v, is experimentally measured and is described as:

$$J_v = L_p(\Delta P - \sigma\Pi) \tag{7.9}$$

where L_p is the filtration coefficient (solvent permeability) (m^3/m^2-s-Pa) and ΔP is the differential pressure (Pa).

Real/intrinsic perchlorate or solute rejection, R_{real}, based on the concentration at the membrane surface, is calculated by using Equation 7.10:

$$R_{real}, \% = \frac{C_m - C_p}{C_m} \times 100\,(\%) \tag{7.10}$$

Determination of Model Parameters

The fitted parameters, ω, $\Delta\Pi$, σ, C_{avg}, and J_v, measured and/or calculated at varying pH conditions and solution conductivities are shown in Tables 7.2 through 7.5. The real rejection of perchlorate was calculated from Equation 7.10. Perchlorate real rejection increased with increasing pH at constant conductivity (30 mS/m) for the GM and MX07 membranes regardless of salt type (see Table 7.2). However, perchlorate real rejection decreases with increasing conductivity at pH 8 (see Table 7.3). These results can be explained by electrostatic interaction (repulsion), which is significantly influenced by changes in pH and ionic strength.

The molecular transport coefficient (ω) of perchlorate, a function of effective diffusion, partitioning, and membrane thickness, decreases with increasing pH and increases with increasing conductivity. As shown in Table 7.2, the effective diffusion increasing pH and conductivity more than in the presence of KCl or $CaCl_2$ solution (see Tables 7.2 and 7.3).

The reflection coefficient of perchlorate for the membranes varies with background salt. The reflection coefficient of perchlorate follows the same order as the conductivity condition for the GM membrane, $CaCl_2 < K_2SO_4 < KCl$. The MX07 membrane, which is more impermeable to the solute than the GM membrane, shows a larger reflection coefficient for perchlorate than the GM membrane in the presence of a KCl solution (see Tables 7.2 and 7.3).

The average bulk fluid interfacial concentration (C_{avg}) of perchlorate, which was calculated with C_m, C_b, and C_p, is described in Tables 7.2 and 7.3. C_b and C_p values were obtained using a bench-scale cross-flow flat-sheet membrane filtration unit. Solvent flux was also monitored over time under the pH and conductivity conditions. Solvent flux decreases with increasing pH and increases with increasing conductivity for the relatively large pore size GM membrane. However, the relatively small pore size MX07 membrane showed decreasing solvent flux with increasing pH and conductivity (see Tables 7.2 and 7.3). The solvent flux changes for the relatively large pore size membranes may be caused by changes in membrane surface charge as a function of pH and conductivity. These can influence the viscosity of the permeate fluid near the membrane pore wall (electroviscous effects). This originates through interactions between the permeate fluid and the ions in the diffuse layer near the pore walls, such as hydrogen bonds between the solvated water and permeate. A decrease in solvent flux with increasing ionic strength for the relatively small pore size MX07 membrane is derived from concentration polarization, because an increase in ionic strength results in a decrease in the membrane pore size.

The real rejection of Cl^- and SO_4^{2-} by the GM membrane at varying pH and conductivity conditions is described in Tables 7.4 and 7.5. SO_4^{2-}, a divalent anion, in the presence of K^+ was rejected much greater than Cl^-, a mono-valent anion, in the presence of K^+ or Ca^{2+}. The real rejection of Cl^- and SO_4^{2-} also increases with increasing pH and decreases with increasing conductivity. The reflection coefficient of Cl^- and SO_4^{2-} for the membrane is described in Tables 7.4 and 7.5. The reflection coefficient of SO_4^{2-} is much larger than that of Cl^-.

Table 7.2
Perchlorate real rejection as a function of pH (conductivity = 30 mS/m) and model parameters for the MX07 and GM membranes

GM	*pH*	R_{real} (ClO_4^-)	ω *(mol-m/L-day-Pa)*	$\Delta\Pi$ *(Pa)*	σ	C_{avg} *(mol/m³)*	J_v *(m/day)*	D_pH (ClO_4^-) *(cm²/s)*
KCl	4	0.34	-5.43E-11	-913	0.19	9.41.E-07	0.565	1.52E-09
	6	0.35	-3.96E-11	-1,019	0.19	9.36.E-07	0.552	1.11E-09
	8	0.36	-2.00E-11	-1,098	0.19	9.30.E-07	0.543	5.59E-10
	10	0.42	-1.33E-11	-1,286	0.19	8.95.E-07	0.53	3.71E-10
K_2SO_4	4	0.16	-3.36E-11	-3,925	0.10	9.77.E-07	0.563	9.41E-10
	6	0.20	-2.65E-11	-4,603	0.10	9.69.E-07	0.549	7.41E-10
	8	0.20	-1.68E-11	-4,453	0.10	9.68.E-07	0.542	4.71E-10
	10	0.33	-1.23E-11	-4,742	0.10	9.33.E-07	0.525	3.44E-10
$CaCl_2$	4	0.03	-5.54E-11	-317	0.03	9.96.E-07	0.581	1.55E-09
	6	0.05	-4.89E-11	-364	0.03	9.94.E-07	0.572	1.37E-09
	8	0.06	-3.36E-11	-354	0.03	9.92.E-07	0.546	9.40E-10
	10	0.08	-2.66E-11	-379	0.03	9.89.E-07	0.54	7.45E-10
MXO7	4	0.52	-4.75E-11	-2,338	0.37	8.99.E-07	0.471	1.33E-09
KCl	6	0.58	-1.98E-11	-3,973	0.37	8.67.E-07	0.452	5.55E-10
	8	0.68	-1.37E-11	-4,434	0.37	8.02.E-07	0.425	3.83E-10
	10	0.75	-9.89E-12	-4,610	0.37	7.44.E-07	0.410	2.77E-10

Table 7.3
Perchlorate real rejection as a function of conductivity (pH = 8) and model parameters for the MX07 and GM membranes

GM	Conductivity *(mS/m)*	R_{real} (ClO_4^-)	ω *(mol-m/L-day-Pa)*	$\Delta\Pi$ *(Pa)*	σ	C_{avg} *(mol/m³)*	J_v *(m/day)*	$D_pH(ClO_4^-)$ *(cm²/s)*
KCl	30	0.36	2.00E 11	-1,098	0.19	8.90.E-07	0.543	5.59E-10
	60	0.30	-3.07E-11	-1,704	0.19	9.80.E-07	0.566	8.60E-10
	115	0.19	-3.86E-11	-1,414	0.19	9.80.E-07	0.600	1.08E-09
K_2SO_4	30	0.20	-1.68E-11	-4,453	0.10	9.30.E-07	0.542	4.71E-10
	60	0.12	-2.69E-11	-8,136	0.10	1.00.E-06	0.556	7.54E-10
	115	0.01	-3.06E-11	-13,193	0.10	1.00.E-06	0.576	8.57E-10
$CaCl_2$	30	0.06	-3.36E-11	-354	0.03	9.90.E-07	0.546	9.40E-10
	60	0.05	-3.89E-11	-271	0.03	1.00.E-06	0.586	1.09E-09
	115	0.04	-4.46E-11	-340	0.03	9.00.E-07	0.592	1.25E-09
MX07	30	0.68	-1.37E-11	-4,434	0.37	8.00.E-07	0.425	3.83E-10
KCl	60	0.57	-2.59E-11	-8,082	0.37	8.80.E-07	0.403	7.25E-10
	115	0.39	-3.09E-11	-15,065	0.37	9.50.E-07	0.371	8.66E-10

Table 7.4
Cl^- and SO_4^{2-} real rejection as a function of pH (conductivity = 30 mS/m) and fitting parameters for thermodynamic model with the GM membrane in the presence of K^+ or Ca^{2+}

GM	*pH*	R_{real}	σ	C_{avg} *(mol/m³)*	J_v *(m/day)*
Cl^-	4	0.24	0.11	0.0021	0.568
(KCl)	6	0.25	0.11	0.0022	0.555
	8	0.26	0.11	0.0021	0.546
	10	0.29	0.11	0.0022	0.534
SO_4^{2-}	4	0.91	0.85	0.0009	0.577
(K_2SO_4)	6	0.91	0.85	0.0011	0.565
	8	0.91	0.85	0.0010	0.558
	10	0.92	0.85	0.0010	0.541
Cl^-	4	0.13	0.11	0.0021	0.583
($CaCl_2$)	6	0.14	0.11	0.0023	0.573
	8	0.14	0.11	0.0022	0.548
	10	0.15	0.11	0.0021	0.541

Table 7.5
Cl^- and SO_4^{2-} real rejection as a function of conductivity (pH = 8) and fitting parameters for thermodynamic model with the GM membrane in the presence of K^+ or Ca^{2+}

GM	Conductivity *(mS/m)*	R_{real}	σ	C_{avg} *(mol/m³)*	J_v *(m/day)*
Cl^-	30	0.26	0.11	0.0021	0.546
(KCl)	60	0.24	0.11	0.0041	0.572
	115	0.10	0.11	0.0079	0.605
SO_4^{2-}	30	0.91	0.85	0.0010	0.558
(K_2SO_4)	60	0.85	0.85	0.0025	0.586
	115	0.72	0.85	0.0057	0.625
Cl^-	30	0.14	0.11	0.0022	0.548
($CaCl_2$)	60	0.07	0.11	0.0042	0.588
	115	0.05	0.11	0.0075	0.593

Comparison of Real and Predicted Rejections in Mixed Systems

The anion rejection of a mixed ClO_4^- and Cl^-/SO_4^{2-} system in the presence of a counter ion (K^+/Ca^{2+}) is plotted as a function of pH and conductivity for the GM membrane in Figures 7.10 and 7.11. The model parameters for the thermodynamic model are described in Tables 7.2 through 7.5. For the predicted perchlorate rejection, both diffusion and convection transport forces were considered. However, for the predicted Cl^- and SO_4^{2-} rejection, an assumption was made that convection is a dominant driving force for the relatively large pore size GM membrane because hindered diffusion and partitioning of Cl^- and SO_4^{2-} were not measured. Cl^- and SO_4^{2-} transport by diffusion was neglected for this membrane. Both real and predicted rejection of a mixed system show consistently similar trends in that rejection increases with increasing pH and decreases with increasing conductivity. Very good agreement between predicted and measured data has been obtained.

Both real and predicted rejection of SO_4^{2} is greater in the presence of K^+ than Cl^- or ClO_4^- in the presence of K^+ or Ca^{2+} regardless of pH and conductivity conditions (see Figures 7.10 and 7.11). This is because the divalent anion, SO_4^{2}, is significantly rejected by the negatively charged membrane through electrostatic exclusion. In this case, the membrane is permeable to the solvent but partially permeable to SO_4^{2}, so the solvent exhibits a higher free energy than that of the solution. Thus, the solvent tends to flow from higher free energy to lower free energy associated with the osmotic pressure difference.

Determination of Predicted Perchlorate Transport by Diffusion and Convection

The transport of perchlorate by diffusion and convection was determined by the thermodynamic model for the GM and MX07 membranes at varying pH and conductivity conditions. As described in Table 7.6, the predicted perchlorate flux by both diffusion and convection decreases with increasing pH and increases with increasing conductivity. The predicted perchlorate flux by convection is much greater than that by diffusion regardless of pH and conductivity conditions. These results are consistent with the theory that solute flux is dominated by convection for a relatively large pore size membrane. In particular, anion (in this case ClO_4^-) transport by diffusion is significantly affected when a membrane is negatively charged because hindered diffusion of perchlorate through the negatively charged membrane pores is much lower than that in water.

Convection-dominant perchlorate flux follows the order for the GM membrane at the same pH and conductivity, $KCl < K_2SO_4 < CaCl_2$, as described in Table 7.6. This trend can be explained by electrostatic repulsion, which is affected by the Ca^{2+} binding significantly decreasing the membrane surface charge. Therefore, perchlorate flux increases in the presence of Ca^{2+}. Another phenomenon is that the K_2SO_4 concentration increases at the membrane surface more than the KCl concentration over time. Chapter 5 describes that zeta potential at the membrane interface, where solute concentration increases, becomes less negative with increasing K_2SO_4 concentration more than with KCl concentration. This is because the electrical double layer is being reduced with increasing K_2SO_4 more than with increasing KCl. Anion (e.g., SO_4^{2-} and Cl^-) adsorption mostly occurs in the electrical double layer. Therefore, perchlorate flux also increases in the presence of SO_4^{2-} more than in the presence of Cl^-.

The MX07 membrane shows smaller predicted perchlorate flux by convection than the GM membrane in the presence of K^+ and Cl^- because the MX07 membrane has a smaller membrane pore size than the GM membrane, respectively. However, the MX07 membrane shows greater predicted perchlorate flux by diffusion than the GM membrane in the presence of K^+ and Cl^-. These results support the theory that solute transport becomes dominated by diffusion for a relatively small pore size membrane, (i.e., tight NF membrane), or non-porous membrane (i.e., reverse osmosis membrane).

Table 7.6
Comparison of predicted perchlorate flux by diffusion and convection by thermodynamic model for the MXO7 and GM membranes

Membrane/ pH-Conductivity		*KCl*	*KCl*	K_2SO_4	K_2SO_4	$CaCl_2$	$CaCl_2$
GM							
pH at 30 mS/m	*Flux unit*	J_{diff}	J_{conv}	J_{diff}	J_{conv}	J_{diff}	J_{conv}
4	mol/m^2-$sec*10^9$	0.06	4.99	0.15	5.73	0.02	6.5
6	mol/m^2-$sec*10^9$	0.05	4.84	0.14	5.54	0.02	6.38
8	mol/m^2-$sec*10^9$	0.03	4.73	0.09	5.47	0.01	6.08
10	mol/m^2-$sec*10^9$	0.02	4.45	0.07	5.1	0.01	5.99
Conductivity at pH 8	*Flux unit*	J_{diff}	J_{conv}	J_{diff}	J_{conv}	J_{diff}	J_{conv}
30 mS/m	mol/m^2-$sec*10^9$	0.03	4.73	0.09	5.47	0.01	6.08
60 mS/m	mol/m^2-$sec*10^9$	0.06	5.05	0.25	5.69	0.01	6.54
115 mS/m	mol/m^2-$sec*10^9$	0.06	5.49	0.47	5.99	0.02	6.61
MXO7		*KCl*	*KCl*				
pH at 30 mS/m	*Flux unit*	J_{diff}	J_{conv}				
4	mol/m^2-$sec*10^9$	0.13	3.09				
6	mol/m^2-$sec*10^9$	0.09	2.86				
8	mol/m^2-$sec*10^9$	0.07	2.48				
10	mol/m^2-$sec*10^9$	0.05	2.22				
Conductivity at pH 8	*Flux unit*	J_{diff}	J_{conv}				
30 mS/m	mol/m^2-$sec*10^9$	0.07	2.48				
60 mS/m	mol/m^2-$sec*10^9$	0.24	2.58				
115 mS/m	mol/m^2-$sec*10^9$	0.54	2.56				

Calculated and Measured Perchlorate Flux

The measured and predicted (calculated) perchlorate flux is plotted as a function of pH and conductivity in Figures 7.12 and 7.13. As shown in Figures 7.12 and 7.13, both measured and predicted perchlorate flux decreases with increasing pH and increases with increasing conductivity for the GM and MX07 membranes. There is good agreement between the predicted and experimental data even though the thermodynamic model slightly underpredicts perchlorate flux at all pH and conductivity conditions. The MX07 membrane shows a smaller measured and predicted perchlorate flux than the GM membrane because the MX07 membrane has a smaller membrane pore size than the GM membrane based on their MMCOs. Figure 7.14 shows correlation between measured and predicted perchlorate flux obtained as a function of pH and conductivity. The correlation results show good perchlorate flux prediction compared to measured perchlorate flux with a linear coefficient of determination (r^2) of ≥ 0.88 (see Figure 7.14).

SUMMARY OF RESULTS AND RELATION OF THIS CHAPTER TO RESEARCH HYPOTHESES

The D_pH value of perchlorate significantly decreased as pH increased, regardless of the salt type at the same conductivity of 30 mS/m. At the same pH and solution conductivity, the D_pH of perchlorate follows the order, $CaCl_2 > KCl > K_2SO_4$, in the presence of different types of salts. The D_pH of perchlorate significantly increased as conductivity increased regardless of the salt type. These results support hypothesis 3.2.1.a that greater (negative) surface charge due to increasing pH and decreasing conductivity enhances perchlorate rejection by decreasing diffusion-dominant perchlorate transport. These results also support that lower calcium concentration enhances perchlorate rejection.

The MX07 membrane shows smaller D_pH than the GM membrane because the MX07 membrane has a smaller membrane pore size than the GM membrane based on their nominal MMCOs. This result supports hypothesis 3.2.1.a that smaller membrane pore size and greater (negative) membrane surface charge enhance perchlorate rejection because the MX07 membrane also shows more negative charge than the GM membrane.

Among the five main fitted model parameters, the molecular transport of perchlorate as a function of hindered diffusion and partitioning decreased with increasing pH and increased with increasing conductivity for the GM and MX07 membranes. Membrane osmotic pressure obtained from each test increased with increasing pH and conductivity. The reflection coefficient of perchlorate in the presence of each salt follows the order, $CaCl_2 < K_2SO_4 < KCl$, at the same conductivity condition for the GM membrane. The average bulk concentration for both of the membranes decreased with increasing pH and increased with increasing conductivity for the membranes. Solvent flux decreased with increasing pH and increased with increasing conductivity for the GM membrane. However, the solvent flux decreased with increasing pH and conductivity for the MX07 membrane. These model parameters, experimentally measured and calculated, support the thermodynamic model being directly developed to quantitatively predict perchlorate transport with minimum assumptions.

The predicted perchlorate flux by both diffusion and convection decreases with increasing pH and increases with increasing conductivity. The predicted perchlorate flux by convection is much greater than that by diffusion regardless of pH and conductivity conditions. These results are consistent with the theory that solute flux is dominated by convection for a relatively large pore size membrane. Both measured and predicted perchlorate flux for the GM and MX07 membranes decreases with increasing pH and increases with increasing conductivity. The correlation results show good perchlorate flux prediction compared to measured perchlorate flux. These results support hypothesis 3.2.1.a that the quantitative model developed in this study can show a good prediction of perchlorate rejection trends, which are significantly influenced by water composition and membrane characteristics.

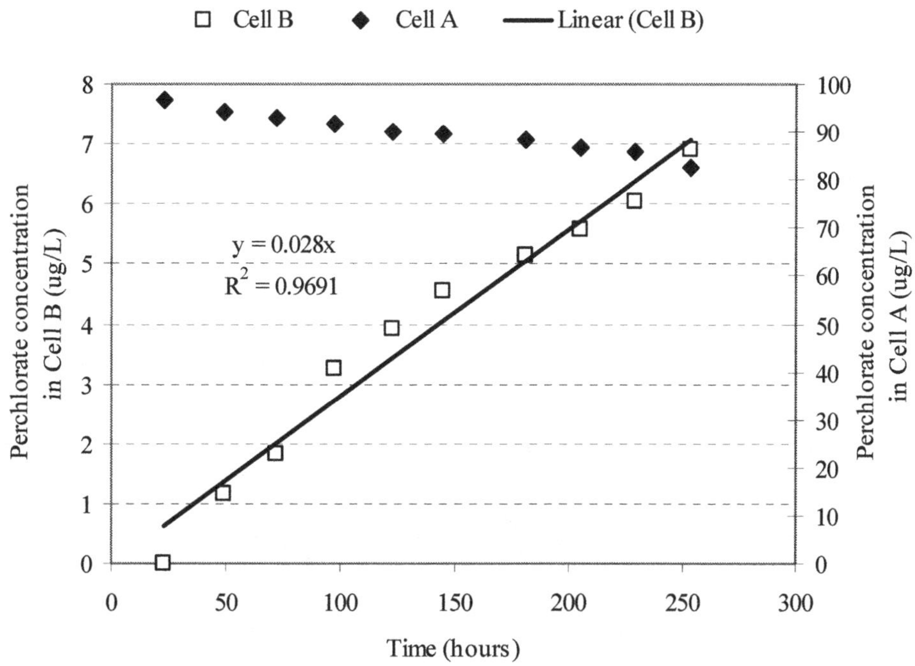

a. Concentration of perchlorate in Cells A and B over time

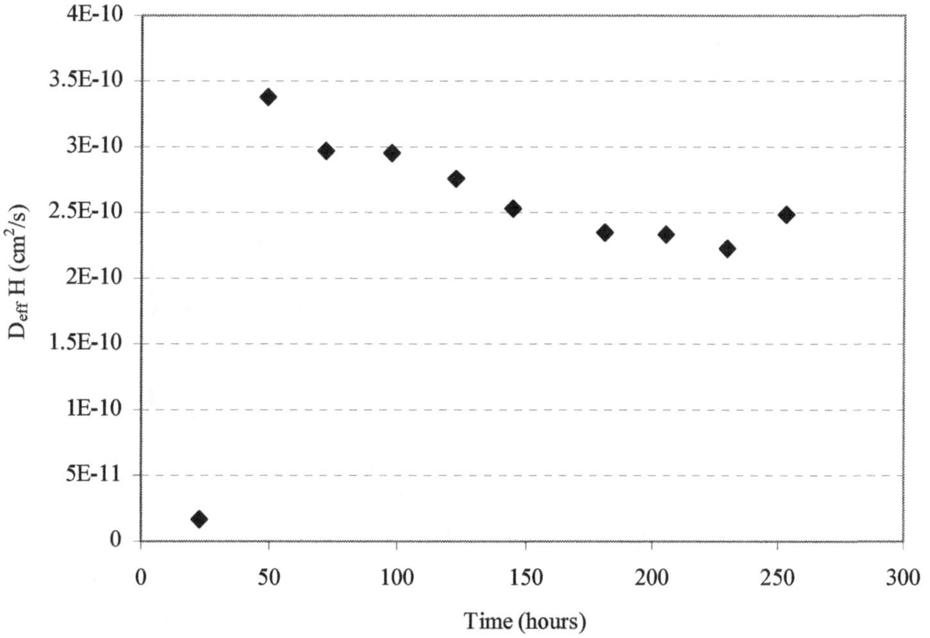

b. Effective diffusion coefficient associated with partitioning of perchlorate over time

Figure 7.1 Change of perchlorate concentrations in Cells A and B (a) and hindered diffusion and partitioning of perchlorate through the GM membrane over time (b): KCl 30 mS/m at pH 8 in Cells A and B

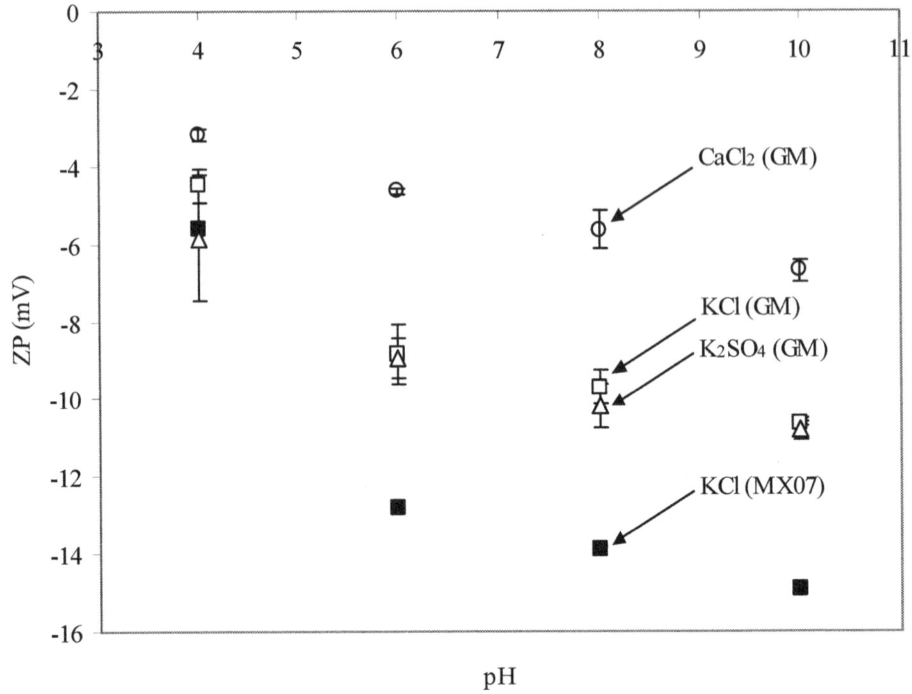

Figure 7.2 Effect of pH on ZP for the NF and UF membranes

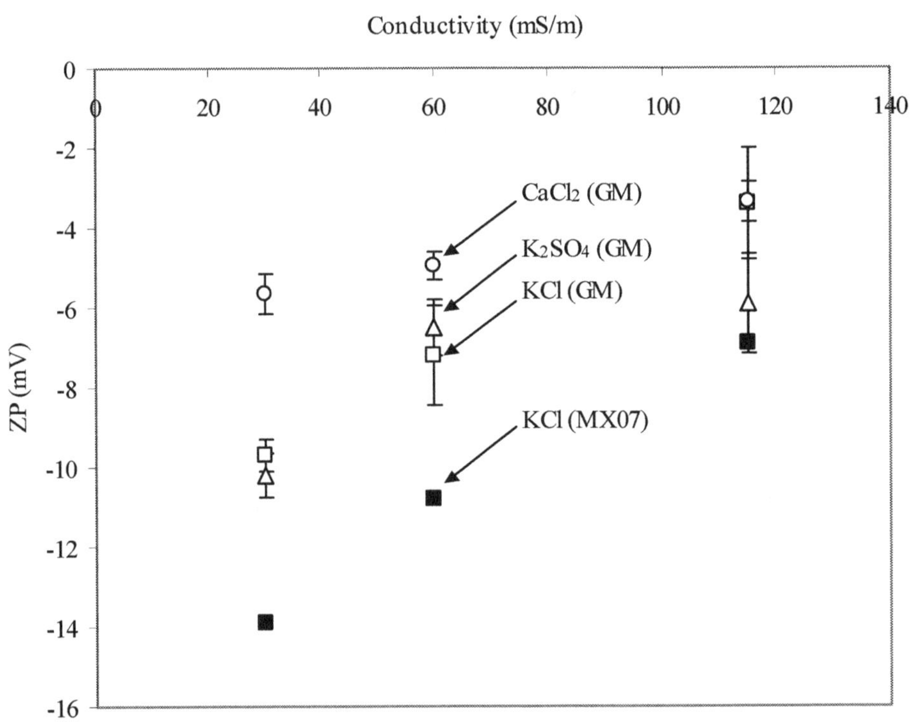

Figure 7.3 Effect of conductivity on ZP for the MX07 and GM membranes

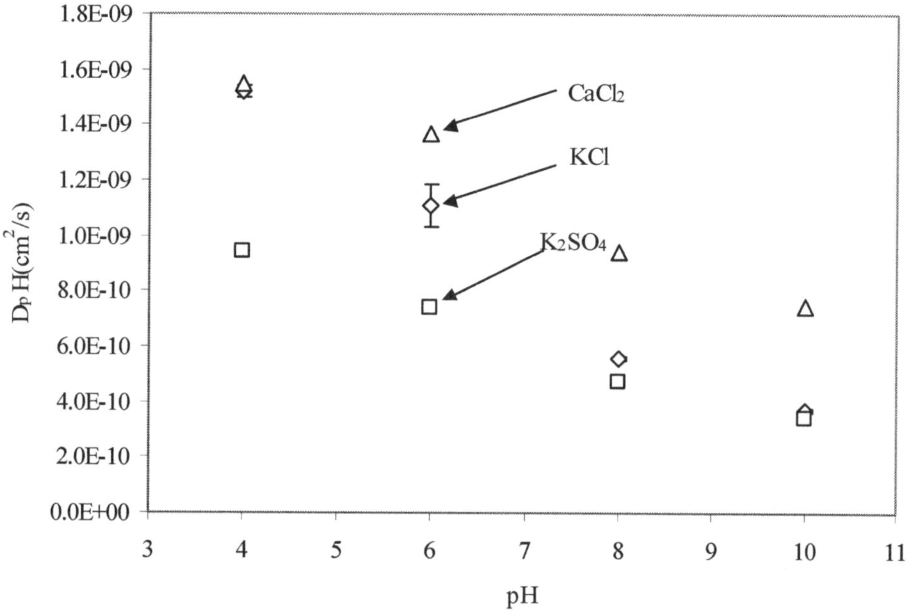

At conductivity 30 mS/m

Figure 7.4 Effect of pH on hindered diffusion and partitioning of perchlorate through the GM membrane in the presence of various salts: without conductivity adjustment in Cell B

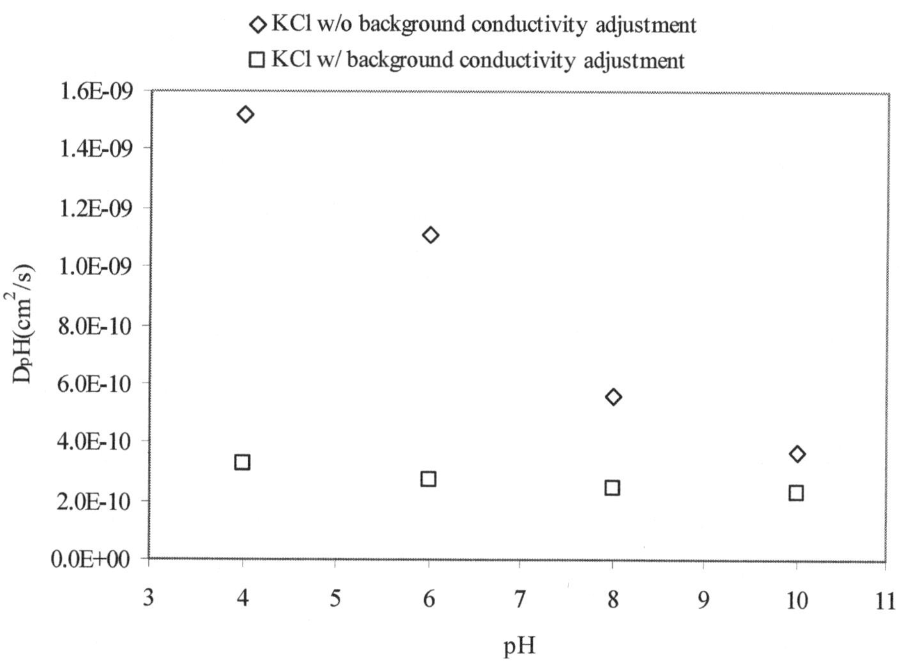

At conductivity 30 mS/m

Figure 7.5 Effect of pH on hindered diffusion and partitioning of perchlorate through the GM membrane with/without conductivity adjustment (in Cell B) in the presence of K^+ and Cl^-

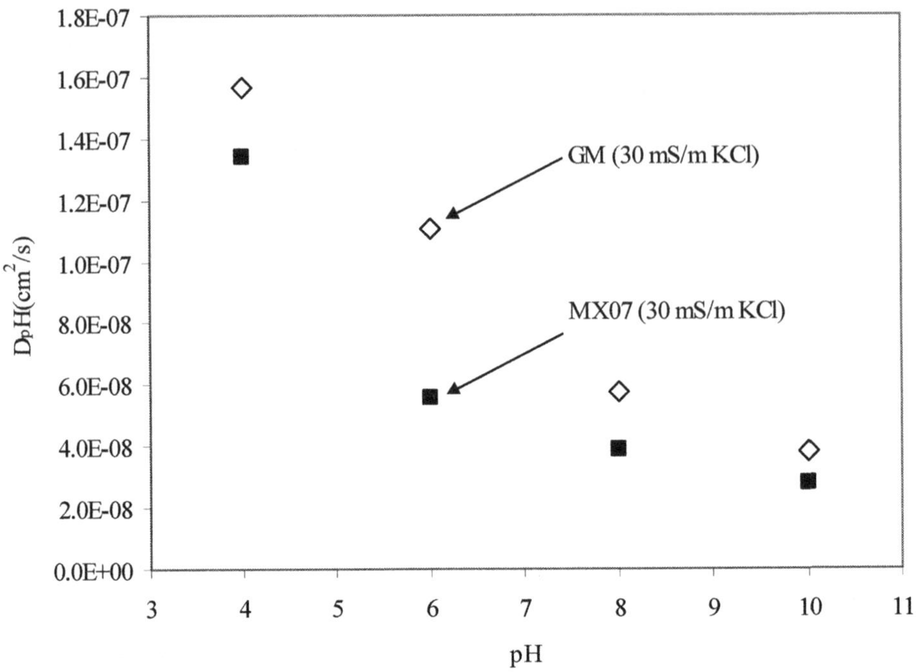

Figure 7.6 Effect of pH on hindered diffusion and partitioning of perchlorate through the MX07 and GM membranes in the presence of K^+ and Cl^-: without conductivity adjustment in Cell B

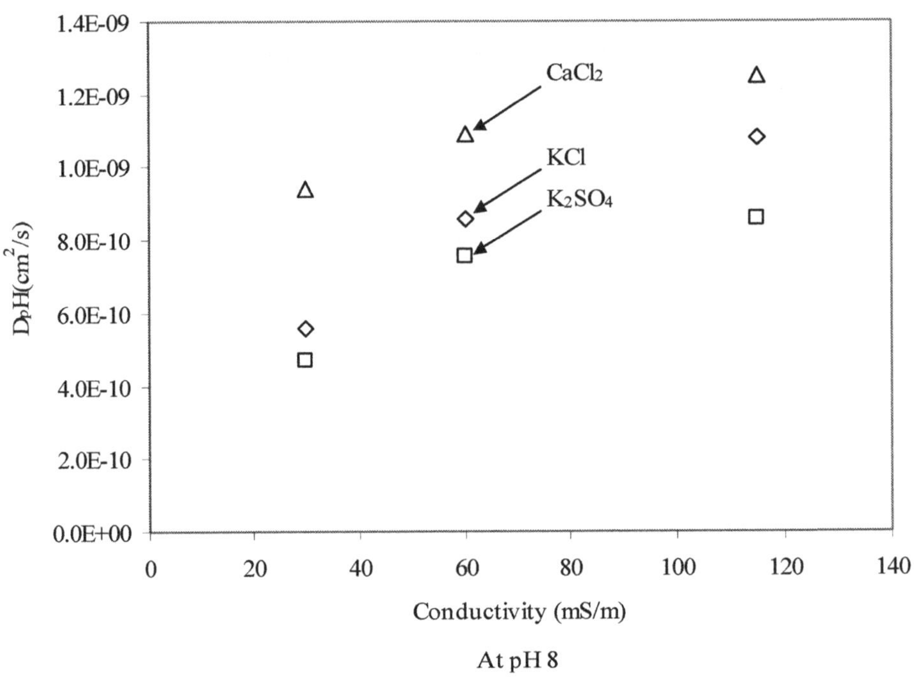

Figure 7.7 Effect of conductivity on hindered diffusion and partitioning of perchlorate through the GM membrane in the presence of various salts: without conductivity adjustment in Cell B

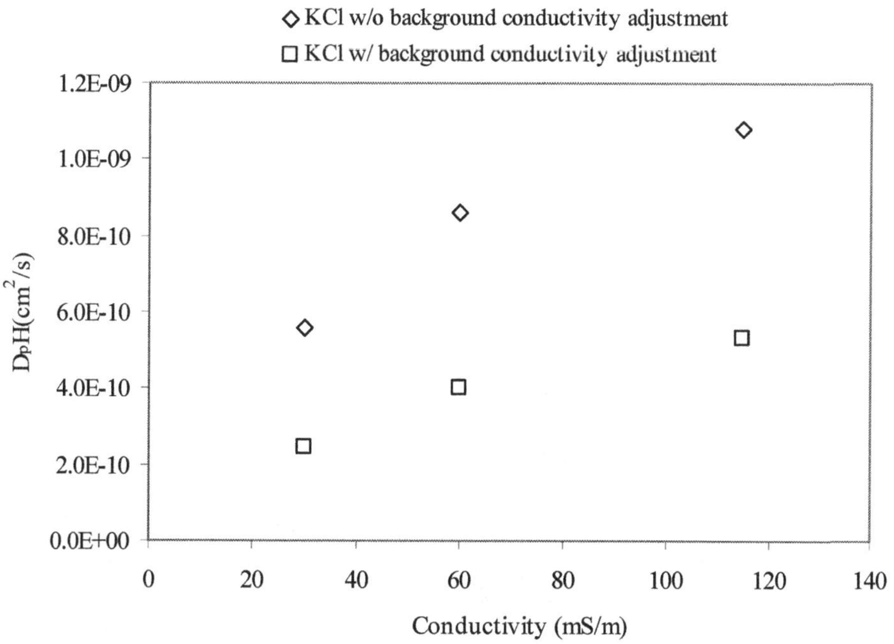

At pH 8

Figure 7.8 Effect of conductivity on hindered diffusion and partitioning of perchlorate through the GM membrane with/without conductivity adjustment in the presence of K^+ and Cl^-

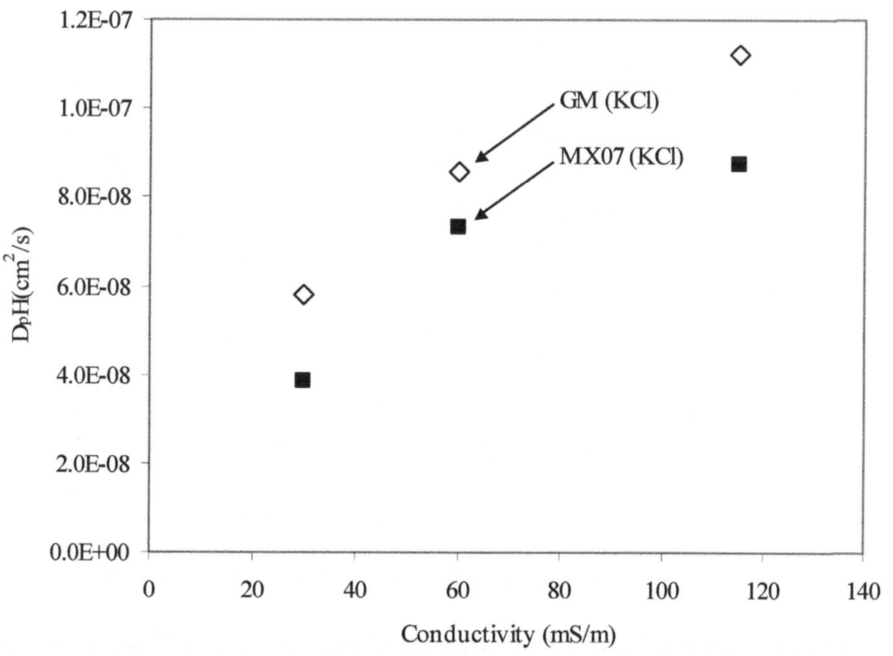

Figure 7.9 Effect of conductivity on hindered diffusion and partitioning of perchlorate through the MX07 and GM membranes in the presence of K^+ and Cl^-: without conductivity adjustment in Cell B

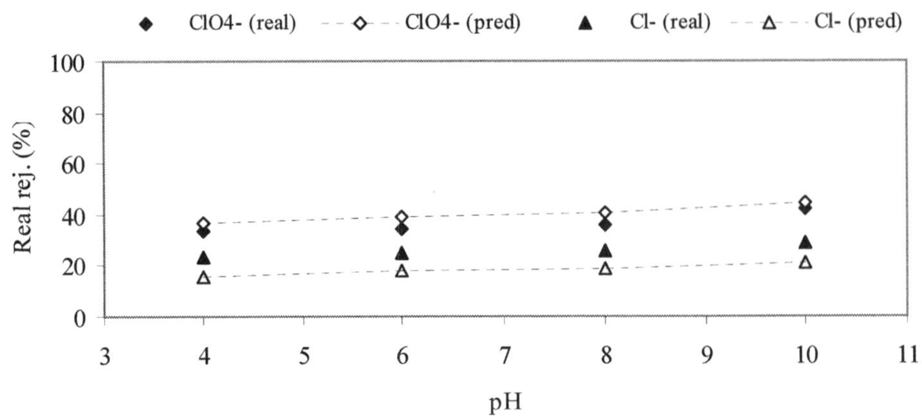

a. ClO_4^- (10^{-6} M) and Cl^- (0.002 M) in the presence of K^+

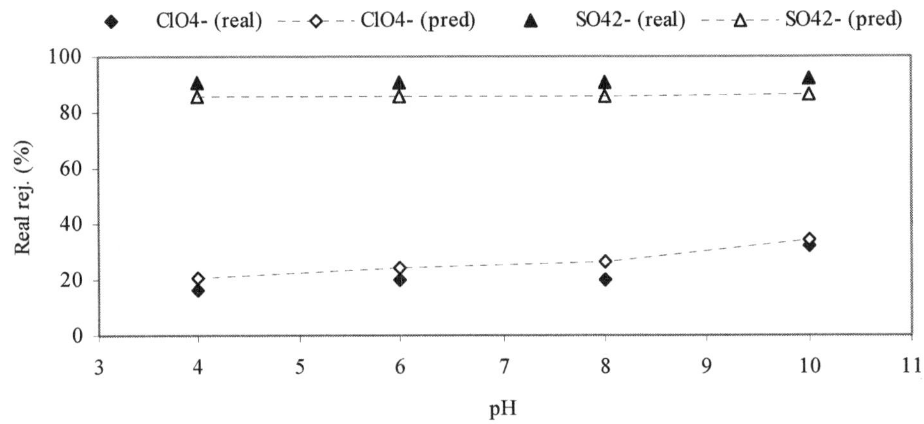

b. ClO_4^- (10^{-6} M) and SO_4^{2-} (0.002 M) in the presence of K^+

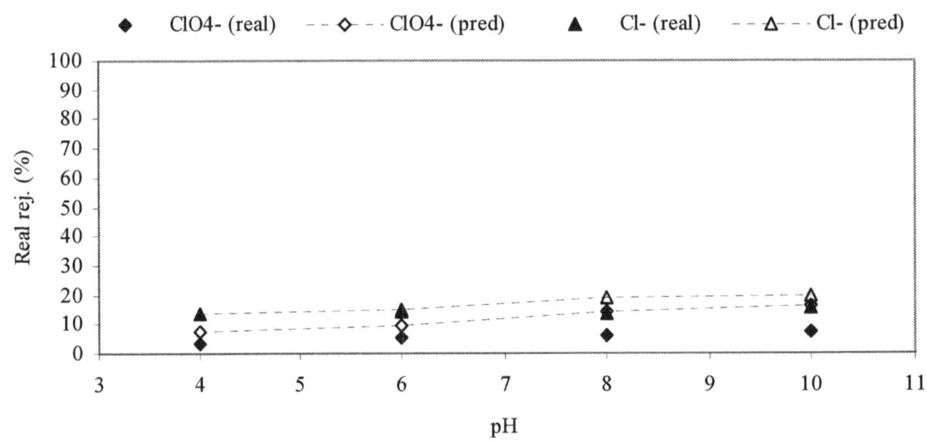

c. ClO_4^- (10^{-6} M) and Cl^- (0.002 M) in the presence of Ca^{2+}

Figure 7.10 Rejection of a mixed ClO_4^- and Cl^-/ SO_4^{2-} system by the GM membrane as a function of pH at conductivity 30 mS/m with KCl

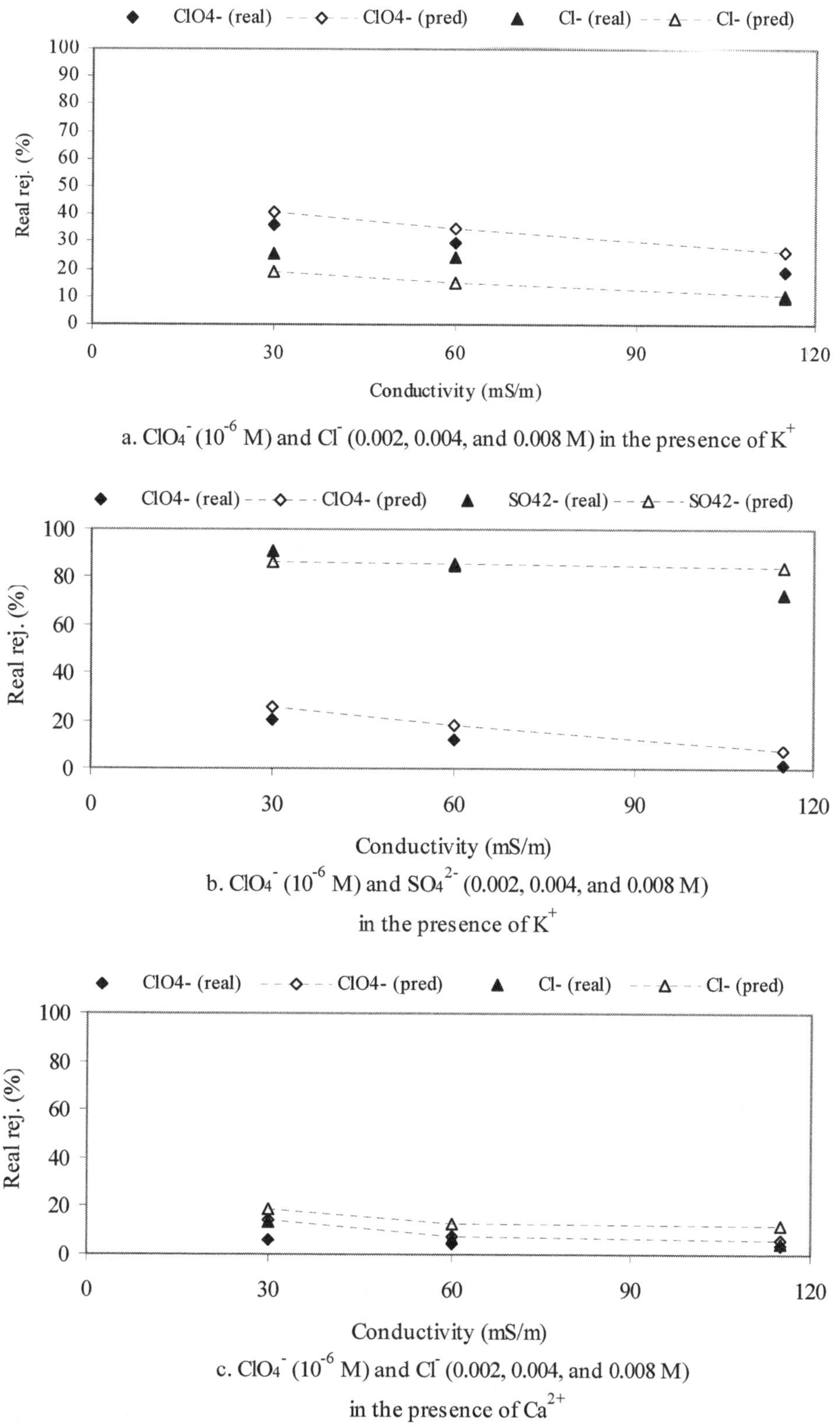

Figure 7.11 Rejection of a mixed ClO_4^- and Cl^-/ SO_4^{2-} system by the GM membrane as a function of conductivity at pH 8

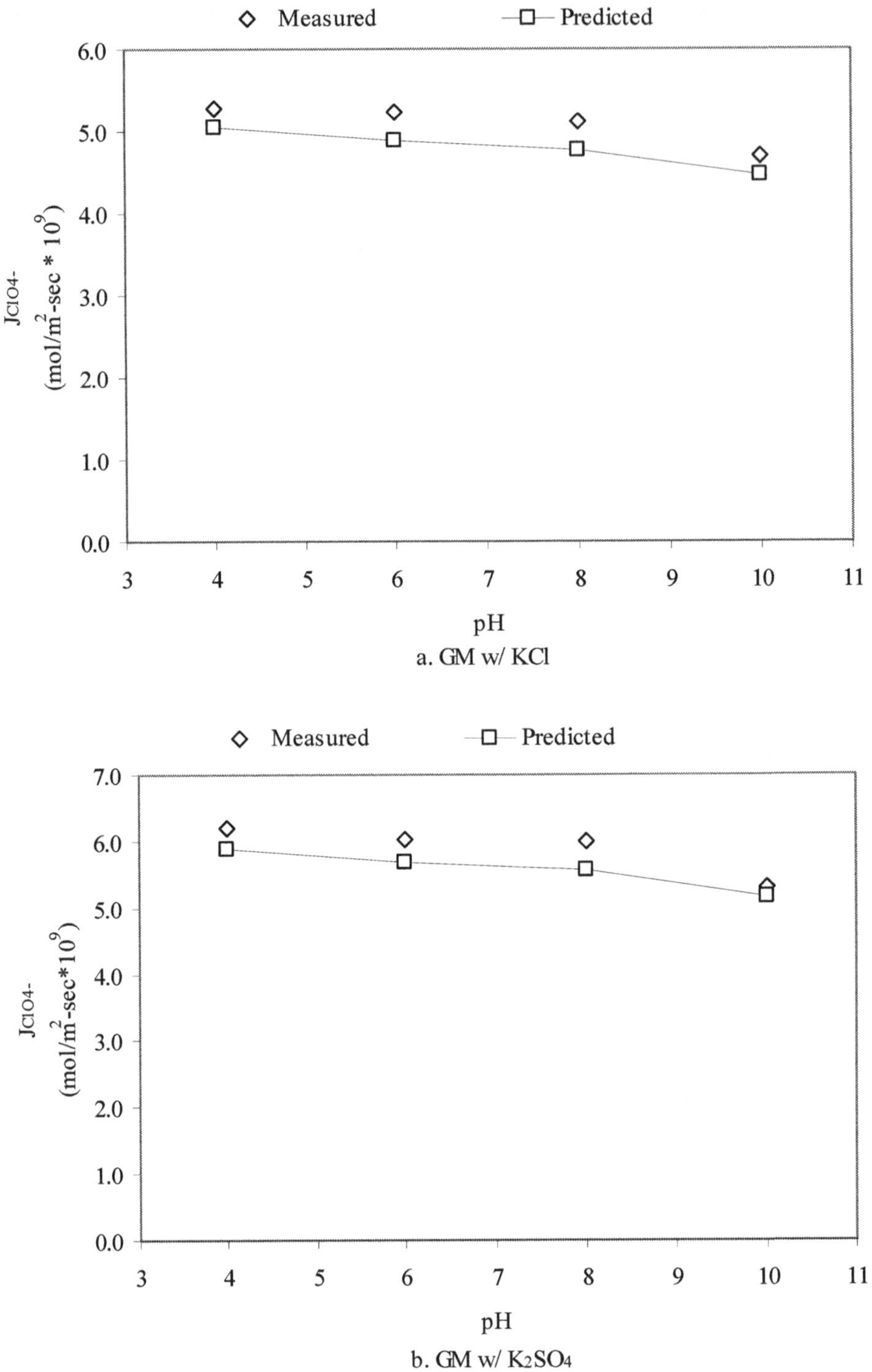

(continued)

Figure 7.12 Comparison of measured and predicted perchlorate flux as a function of pH for the MX07 and GM membranes

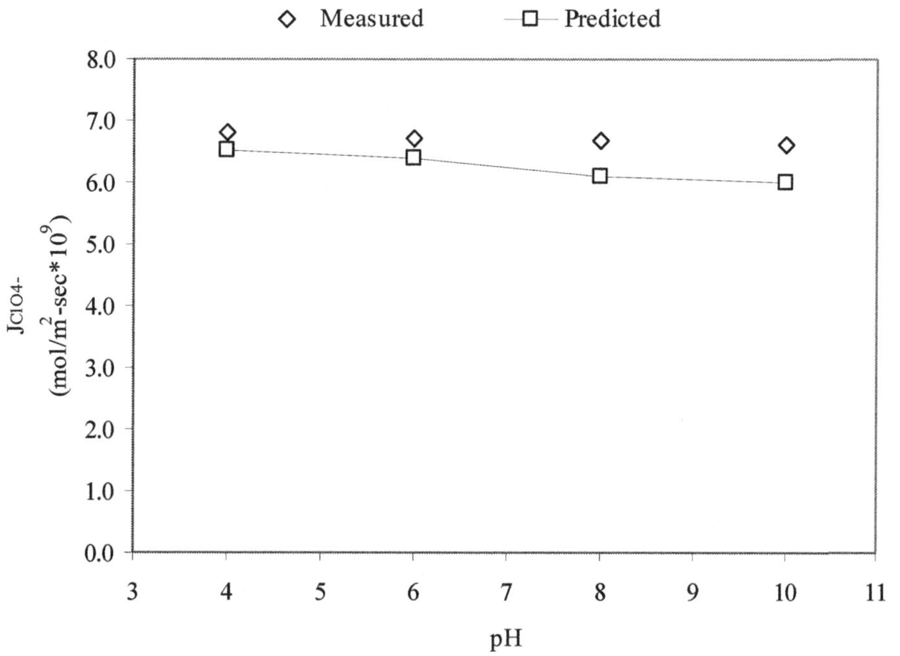

c. GM w/ $CaCl_2$

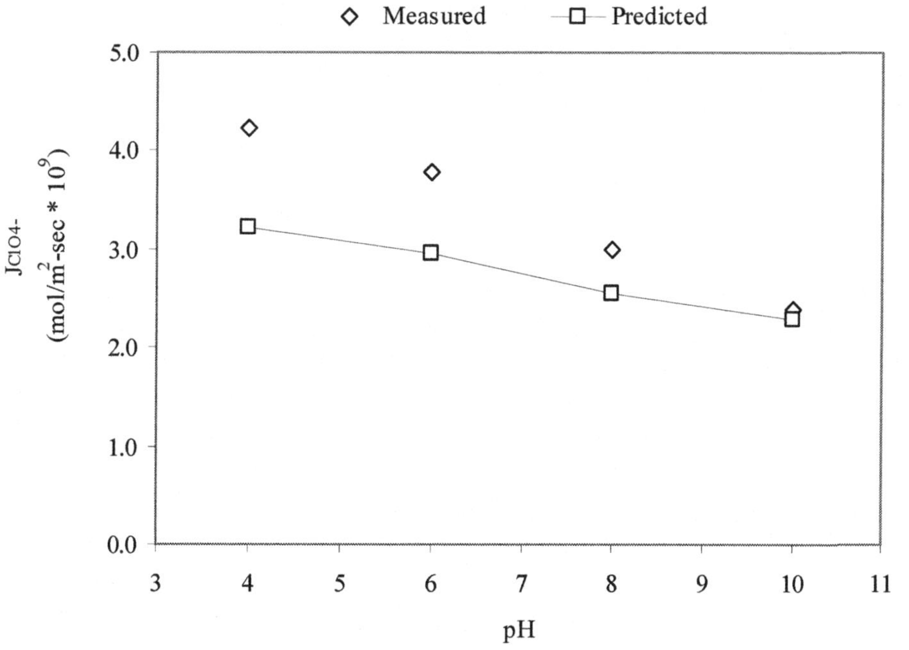

d. MX07 w/ KCl

Figure 7.12 (Continued)

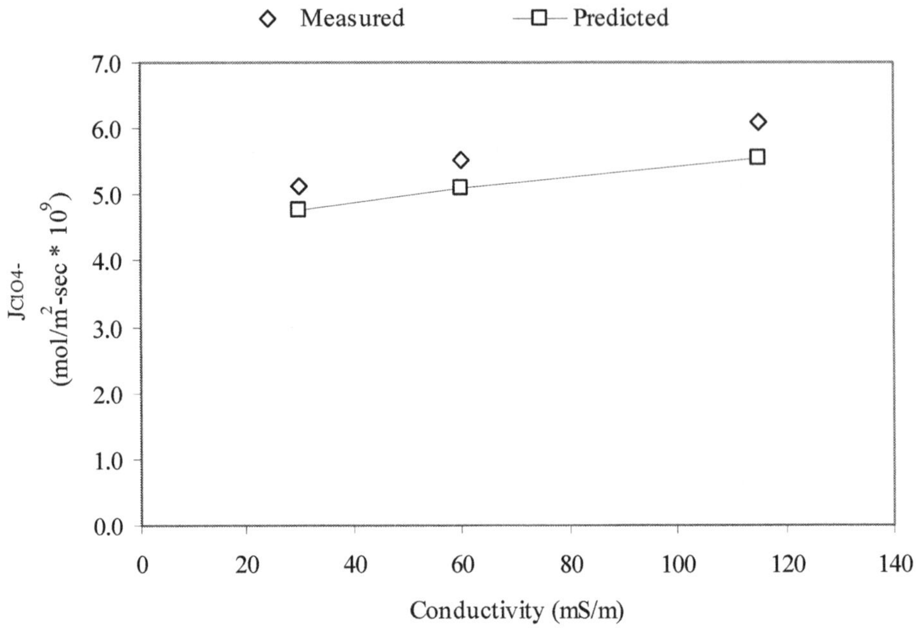

a. GM w/ KCl at pH 8

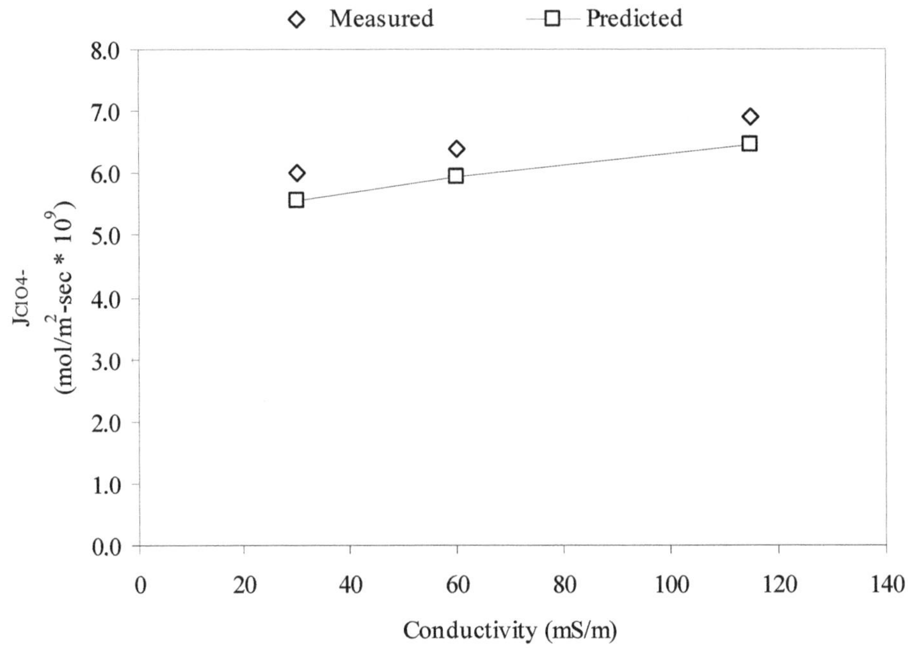

b. GM w/ K_2SO_4 at pH 8

(continued)

Figure 7.13 Comparison of measured and predicted perchlorate flux as a function of conductivity for the MX07 and GM membranes

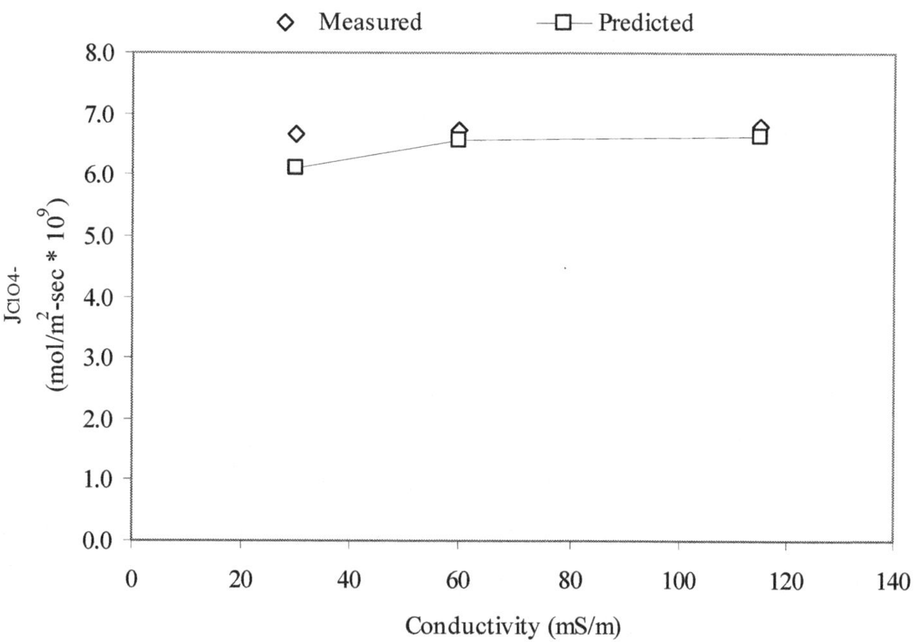

c. GM w/ $CaCl_2$ at pH 8

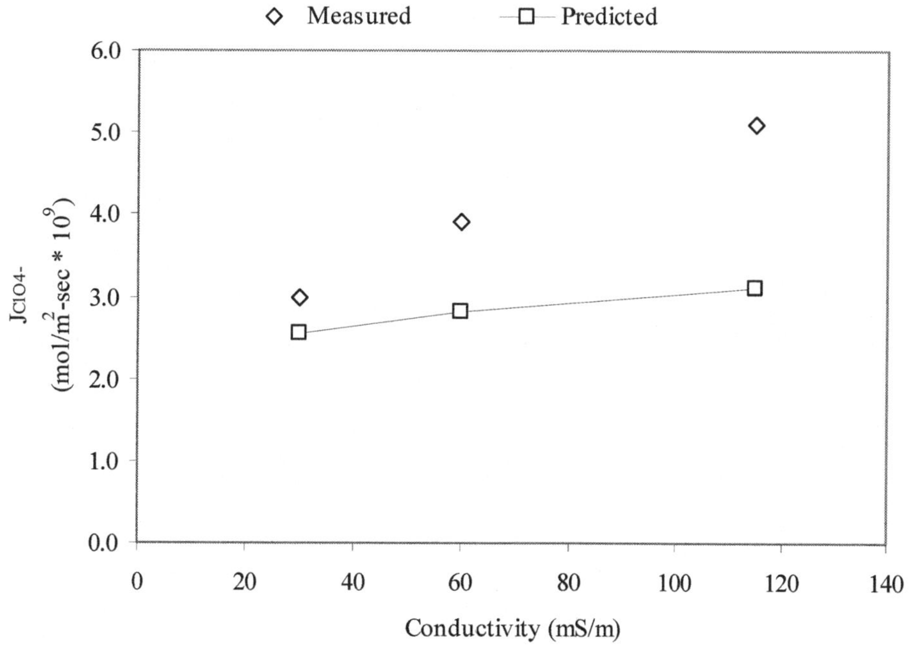

d. MX07 w/ KCl

Figure 7.13 (Continued)

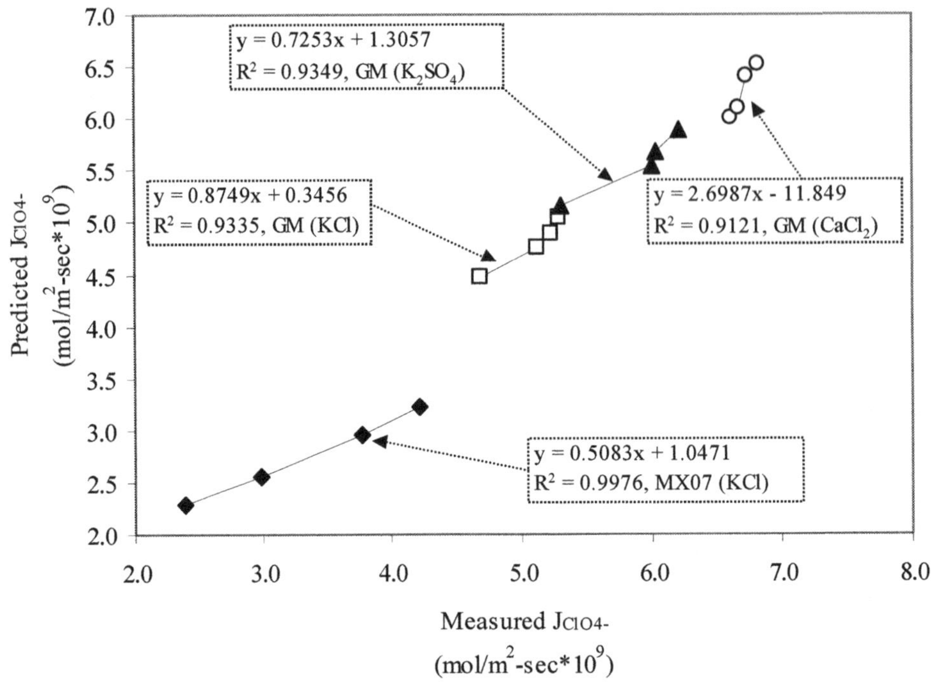

a. Obtained as a function of pH

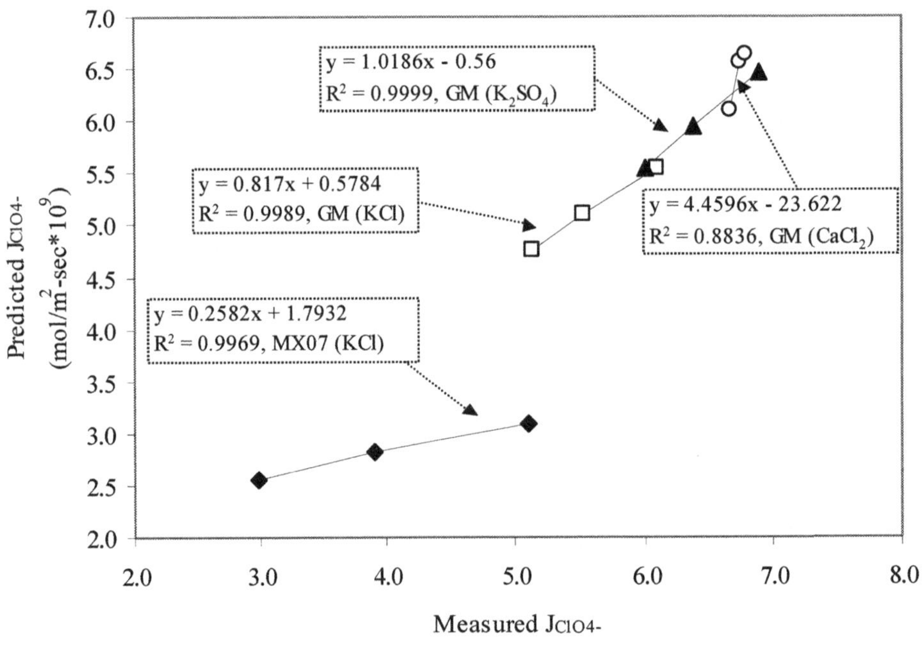

b. Obtained as a function of conductivity

Figure 7.14 Correlation between measured and predicted perchlorate flux

CHAPTER 8
EFFECTS OF J_O/k AND RECOVERY ON PERCHLORATE TRANSPORT (REJECTION) AND INORGANIC SCALE FORMATION WITH NATURAL WATERS

INTRODUCTION

Another important objective of our research was to determine the limits of membrane operation (i.e. recovery and J_o/k ratio) in treating perchlorate-contantaminated waters. While higher recovery and Jo/k ratio clearly increase the product stream and reduce the reject (concentrate) stream, perchlorate rejection decreased and inorganic scale formation increased.

In this study, we have emphasized three efforts: i) perchlorate rejection and flux-decline tests in bench-scale cross-flow flat-sheet filtration for two nanofiltration (NF) and two reverse osmosis (RO) membranes with Colorado River Water (CRW), ii) estimation of precipitative fouling with inorganic salts by equilibrium calculations, and iii) evaluation and characterization of antiscalants.

In this study, we have investigated i) rejection of water quality parameters (cations, anions, dissolved organic carbon (DOC), ultraviolet absorbance (UVA_{254}), pH, conductivity, total dissolved solids (TDS), and alkalinity) and flux decline for four different membranes (TFC-S (NF), ESNA (NF), LFC-1 (RO), and TFC-ULP (RO)) over a various range of recoveries, ii) estimation of a polarization factor, a constant (K) for the four candidate membranes corresponding to J_o/k (or recovery), which can be used to predict the concentrations of cations/anions at the membrane surface, iii) the potential scalants of inorganic precipitation by simulations with an equilibrium software program, MINEQL, iv) analysis of inorganic foulants on membrane surfaces by FTIR, XRD, SEM, and NAA, and v) inhibition tests of antiscalants, with a comparison of flux decline with/without antiscalant, and characterization of antiscalants according to functional groups.

METHODS AND MATERIALS

Membranes

Table 4.1 summarizes the physical-chemical properties of the membranes evaluated. In this chapter, we have intensively studied polyamide TFC membranes in terms of perchlorate rejection and flux-decline tests with CRW at the bench-scale. The ESNA (Hydranautics) and TFC-S (Fluid System) are designated as NF type membranes and the LFC-1 (Hydranautics) and TFC-ULP (Fluid System) are identified as RO type membranes, based on manufacturer designations and their nominal molecular mass cut-offs (MMCOs). The ESNA and TFC-S each have a reported MMCO of 200. The NF membranes, made of similar materials, show similar membrane charge based on their zeta potential (ZP) values. In Table 4.1, the TFC-S and TFC-ULP membranes show relatively high pure water permeabilities (PWP) while the ESNA and LFC-1 membranes show relatively low permeabilities among the membranes. Permeability is influenced by both pore size and pore density.

Table 8.1
Summary of source feed water (CRW) quality parameters

Parameters	Mean	Range
pH	7.93	7.77 – 8.3
DOC, mg/L	2.44	2.24 – 2.82
TDS, mg/L	359	326 – 420
Conductivity, mS/m	74.9	65.1 – 87.4
UVA_{254}, cm^{-1}	0.045	0.04 – 0.051
Alkalinity, mg/L	88.5	75 – 115
Temperature, °C	23	21 – 24
Silica (SiO_2), mg/L	3.2	2.33 – 4.19
Barium (Ba^{2+}), mg/L	0.073	0.06 – 1.0
Calcium (Ca^{2+}), mg/L	55.06	44 – 63.09
Magnesium (Mg^{2+}), mg/L	21.41	18.03 – 24.48
Sodium (Na^{+}), mg/L	47.02	22.2 – 81.1
Potassium (K^{+}), mg/L	3.4	3.06 – 3.97
Chloride (Cl^{-}), mg/L	60.5	50.62 – 70.13
Nitrate (NO_3^{-}), mg/L	0.28	0 – 0.86
Sulfate (SO_4^{2-}), mg/L	190	157 – 223
Perchlorate (ClO_4^{-}), mg/L	0.1	-

Source Water

A 0.45 μm-prefiltered source water, CRW, was tested for perchlorate rejection and flux-decline trends in bench-scale cross-flow flat-sheet filtrations. Table 8.1 summarizes the water quality of CRW, spiked up to 100 μg/L of perchlorate.

Experimental Systems and Membrane Preparation

Precipitation Inhibiting Test Unit

Figure 8.1 shows a schematic of the inhibition test system (Drela and Falewicz, 1998; NACE, 1995) to investigate the inhibition efficiency of antiscalants.

This test unit is used to evaluate the performance of different antiscalants under laboratory conditions. Since many factors such as reaction kinetics, fluid velocity and composition, and variable temperature and pressure can significantly affect actual conditions, temperature was maintained with a heating/cooling unit, the reactor was sealed with air-tight sealing material, the pH and conductivity were adjusted (Harrar and Locke, 1982), and the Ca^{2+} concentration (free calcuium) was measured with a calcium ion specific electrode.

Membrane Preparation

For each membrane, test coupons were prepared from a single element provided by the membrane manufacturer. Each membrane coupon was cut and prefiltered with pure water (8 L w/o recirculation). Each membrane was also soaked for at least 24 hrs in pure water (deionized organic-free water) prior to testing. Before testing, the membrane coupon was flushed with pure water for at least 12 hrs until the membrane was stable in terms of pure water permeability. A new membrane coupon was used for each experiment.

RESULTS AND DISCUSSION

Inhibition Tests of $CaCO_3$ Precipitation by Antiscalants

Background of Inhibition Test

In the membrane process, the formation of mineral scales such as calcium carbonate, calcium sulfate, barium sulfate, strontium sulfate, and magnesium sulfate on the membrane surface can severely reduce the performance and efficiency of the system (Amjad, 1985). In our previous studies, $CaCO_3$ scaling was dominantly detected on the membrane surface. For this reason, we performed inhibiting tests for $CaCO_3$ formation with antiscalants. According to other researchers, phosphonate groups were effective in the scale inhibition of $CaCO_3$ (Hasson and Semiat, 1998; Libutti et al., 1984). Among phosphonate-antiscalants, we have selected two different commercial antiscalants, Dequest 2000 (Aminotrimethylene phosphonic acid, ATMP : $N(CH_2PO_3H_2)_3$, Solutia Inc.) and Dequest 2010 (1-Hydroxythylidine 1,1-diphosphonic acid, HEDP : $CH_3C(PO_3H_2)_2OH$, Solutia Inc.) respectively (Amjad, 1988).

Inhibition Tests

Table 8.2 and Figures 8.2 through 8.4 show the inhibition efficiency of $CaCO_3$ formation under varying conditions. Experiments 1 through 4 are the tests corresponding to the selecting antiscalants at the given conditions.

As shown in Table 8.2, the concentrations of the two different antiscalants were 1.0 or 3.0 ppm; relatively high inhibition was observed with Dequest 2000 (3 ppm). For experiments 5 through 12, performed at an average conductivity (78 mS/m) of CRW and varying pH (4 through 10), inhibition efficiency was better at higher pH than at lower pH, as expected. Experiments 13 through 16, performed at a relatively high conductivity (116.8 ~ 130 mS/m) and high pH (8 and 10), the inhibition efficiency at relatively low conductivity was slightly higher than that at relatively high conductivity because high ionic strength can influence the $CaCO_3$ precipitation (Zuddas and Mucci, 1998; Logan and Rey, 1986).

Calculation of inhibition efficiency is based on the following equation (NACE, 1995)

$$Inhibition,\% = \frac{C_a - C_b}{C_c - C_b} \times 100 \qquad (8.1)$$

C_a: Ca^{2+} concentration in the treated sample after precipitation
C_b: Ca^{2+} concentration in the blank after precipitation
C_c: Ca^{2+} concentration in the blank before precipitation

Figure 8.5 shows experimental results to investigate the inhibition mechanism, either formation of a soluble complex between the antisclant and Ca^{2+}, or inhibition of crystal growth

Table 8.2
Summary of $CaCO_3$ inhibition results (synthetic water)

Exp. No	$[HCO_3^-]$, mg/L	pH (w. KOH or HCl)	Temp., °C	Conductivity, mS/m (w. KCl)	Additive	C_a*, mg/L	C_b†, mg/L	C_c‡, mg/L	Inhibition, %	Ref. Fig.
1	200	7.35	22.0	79.8	Dequest 2000, 50% sol. (3 ppm)	117.4	25.3	200	52.8 at t=10 hr	Fig. 8.3 (a)
2	200	7.28	21.6	78.0	Dequest 2000, 50% sol. (1 ppm)	83.7	25.3	193	34.9 at t=10 hr	Fig. 8.3 (a)
3	200	7.26	22.6	78.2	Dequest 2010, 60% sol. (3 ppm)	82.2	25.3	200	32.6 at t=10 hr	Fig. 8.3 (b)
4	200	7.13	21.9	77.6	Dequest 2010, 60% sol. (1 ppm)	81.6	25.3	194	33.2 at t=10 hr	Fig. 8.3 (b)
5	200	4.0	20.6	76.5	Dequest 2000, 50% sol. (3 ppm)	119	139	193	0 at t=24 hr	Fig. 8.4 (a)
6	200	4.0	20.4	81.6	Dequest 2010, 60% sol. (3 ppm)	126	139	189	0 at t=24 hr	Fig. 8.4 (a)
7	200	6.05	20.8	80.3	Dequest 2000, 50% sol. (3 ppm)	118	126	194	0 at t=24 hr	Fig. 8.4 (b)
8	200	6.03	20.5	76.5	Dequest 2010, 60% sol. (3 ppm)	133	126	193	0 at t=24 hr	Fig. 8.4 (b)
9	200	7.88	20.7	78.4	Dequest 2000, 50% sol. (3 ppm)	126	99.9	195	27.4 at t=24 hr	Fig. 8.4 (c)

(continued)

Table 8.2 (Continued)

Exp. No	$[HCO_3^-]$, mg/L	pH (w. KOH or HCl)	Temp., °C	Conductivity, mS/m (w. KCl)	Additive	C_a^*, mg/L	$C_b^\dagger$, mg/L	$C_c^\ddagger$, mg/L	Inhibition, %	Ref. Fig.
10	200	7.94	20.9	77.9	Dequest 2010, 60% sol. (3 ppm)	107	99.9	202	6.8 at t=24 hr	Fig. 8.4 (c)
11	200	9.98	20.6	84.4	Dequest 2000, 50% sol. (3 ppm)	77.4	38.4	203	23.6 at t=24 hr	Fig. 8.4 (d)
12	200	10.01	21.6	86.1	Dequest 2010, 60% sol. (3 ppm)	61.8	38.4	203	14.2 at t=24 hr	Fig. 8.4 (d)
13	200	8.24	21.6	130	Dequest 2000, 50% sol. (3 ppm)	84.1	88.4	207	0 at t=24 hr	Fig. 8.5 (a)
14	200	8.06	21.2	117	Dequest 2010, 60% sol. (3 ppm)	76.9	88.4	208	0 at t=24 hr	Fig. 8.5 (a)
15	200	9.94	20.2	118	Dequest 2000, 50% sol. (3 ppm)	69.6	38.3	207	18.48 at t=24 hr	Fig. 8.5 (b)
16	200	10.05	21.6	120	Dequest 2010, 60% sol. (3 ppm)	55.8	38.3	189	11.64 at t=24 hr	Fig. 8.5 (b)

C_a^* : Ca^{2+} concentration in the treated sample after precipitation
$C_b^\dagger$: Ca^{2+} concentration in the blank after precipitation
$C_c^\ddagger$: Ca^{2+} concentration in the blank before precipitation

of calcium carbonate. In these experiments, we evaluated the same concentration of Ca^{2+} at different concentrations of antiscalant (3 and 10 ppm). It can be seen that the concentration of free calcium did not change over 24 hrs; based on these results, we can assume that inhibition of crystal growth is the dominant mechanism.

Perchlorate Rejection With Various Natural Waters

The TFC-S and ESNA membranes were selected to perform perchlorate rejection tests using a cross-flow flat-sheet filtration unit because these membranes showed good performance in terms of perchlorate rejection with synthetic waters in preliminary tests. The experimental operating conditions are described in Table 4.4. Four source waters were used for this study: SG-GW, LM-SW, BHWB-GW, and RGW. The ground waters, SG-GW, BHWB-GW, and RGW, contain around 70 to 75 μg/L of ClO_4^- and LM-SW contains approximately 10 μg/L of ClO_4^-. The ground waters contain relatively low levels of DOC and aromatic NOM, based on their SUVA values. SUVA is defined as UVA_{254}/DOC ratio and represents an index of NOM aromaticity (see Table 4.2). Table 4.3 describes the anions and cations in the source waters. For all of the experiments, the source waters containing the perchlorate anion (present at a total concentration of 100 μg/L of ClO_4, spiked with $KClO_4$) were pretreated by 0.45 μm filtration.

Figures 8.6 and 8.7 show perchlorate rejection and flux-decline trends for each membrane and water. Both the TFC-S and ESNA membranes performed well, removing 82 to 88% of the perchlorate regardless of feed water type at 2.6% recovery and a J_o/k ratio of 0.5. These results were obtained by employing a relatively low recovery. The flux at the operating conditions ranged from 83 to 89% of the initial flux during a 24 hour testing period. The flux was not significantly affected by source water type for any of the membranes. Overall, there was no significant flux-decline effects on perchlorate rejection during the short testing period.

Comparison of Flux-Decline and Other Parameters

Figure 8.8 (a)-(d) presents the flux-decline results over time during the filtration of 0.45 μm-filtered CRW through two different RO (LFC-1 and TFC-ULP) and two different NF (ESNA and TFC-S) membranes in bench-scale cross-flow flat-sheet filtrations. The source water (CRW) was spiked with perchlorate up to 100 μg/L. The water quality composition of CRW was described in Table 8.1. These experiments were performed at recoveries ranging from 15% to 70%, corresponding to a J_o/k ratio range of 0.5 to 3.

Variable pressures of 80–270 psi (552-1863 kPa) for the LFC-1, 50–255 psi (345–1760 kPa) for the TFC-ULP, 38–215 psi (262–1484 kPa) for the ESNA, and 28–130 psi (193–897 kPa) for the TFC-S were employed for these flux decline tests. Under the specified operational conditions, each membrane showed different flux-decline trends at 15% recovery: 9% (LFC-1), 8% (TFC-ULP), 15% (ESNA), 6.5% (TFC-S); at 30% recovery: 18% (LFC-1), 14% (TFC-ULP), 19% (ESNA), 12% (TFC-S); at 50% recovery: 23% (LFC-1), 22% (TFC-ULP), 27% (ESNA), 19% (TFC-S); and 70% recovery: 25%(LFC-1), 26% (TFC-ULP), 34% (ESNA), 23% (TFC-S), respectively. Also, under the 70% recovery in the presence of antiscalant (Dequest 2000, 3 ppm), each membrane exhibited a smaller flux-decline than that at 70% recovery without antiscalant.

Figures 8.9 (a)-(d) and 8.10 (a)-(d) show the rejection of cations/anions including perchlorate, UVA, DOC, TDS, and conductivity at varying recovery conditions. As seen in both figures, each rejection trend generally decreased with increasing recovery according to theoretical expectations (Chellam and Jacangelo, 1997).

Precipitative Inorganic Scale Determination

X-ray Diffraction (XRD)

Four fouled membranes, with and without antiscalant, LFC-1, TFC-ULP, ESNA, and TFC-S, were characterized to determine precipitative inorganic scale by XRD. The results are shown in Figure 8.11. In Figure 8.11 (a, c, e, and g), the characteristic peak of calcite at 29.454 $\theta°$ is clearly shown in the respective spectrum of the composite calcite. Other calcite peaks such as the peak at 39.419 $\theta°$ and 47.146 $\theta°$ are distinguished for the fouled RO membranes (Klepetsanis and Koutsoukos, 1999). However, the NF membranes show relatively small peaks of calcite and for all of the four membranes with antiscalant (Dequest 2000, 3 ppm at recovery 70%) no calcite peaks are shown (b, d, f, and h). Over the range of 20 $\theta°$ to 27 $\theta°$, the two broad peaks are interferences associated with direct membrane analysis. Also, the XRD results did not show any other precipitative inorganic foulants such as $BaSO_4$, $CaSO_4$, and $MgSO_4$.

FTIR of Fouled Membranes and Antiscalant

Two fouled RO membranes, LFC-1, and TFC-ULP, were analyzed to investigate inorganic scale by FTIR. The results are shown in Figure 8.12. The IR transmission spectra of fouled membranes after the precipitation of $CaCO_3$ provides additional evidence of the presence of prepitative inorganic scale. The band corresponding to 877 cm^{-1} and 1430 cm^{-1} is characteristic for $CaCO_3$ (Nakayama, 2000). In Figure 8.12, star symbols are used to indicate the inorganic fouled membrane after 24 hrs at a recovery 70%, and triangle symbols are used to indicate the same conditions with antiscalant. In both figures, the spectra of the two RO membranes without antiscalant show the $CaCO_3$ peak around 877 cm^{-1} and 1430 cm^{-1}, however, spectra of membranes with antiscalant show the functional groups of phosphonate, not $CaCO_3$ peaks. Also, we investigated the anticalant functional groups, particularly phosphonate, of Dequest 2000 by FTIR measurements. Phosphonates are known to prevent the precipitation of calcium carbonate from water by inhibiting the crystallization of calcite.

The IR transmission spectra of Dequest 2000 are presented in Figure 8.13. All spectra are presented over a range of 1300 cm^{-1} to 850 cm^{-1}.

Comparison of the two different concentrations of Dequest 2000 shows similarity in the wavenumber and relative intensities of the bands. The stretching mode of the P=O bonds gives rise to a strong band in the 1300 cm^{-1} to 1130 cm^{-1} region. The bands around 1080 cm^{-1} and 930 cm^{-1} are attributed to the PO_3H_2 group of Dequest 2000. The 1080 cm^{-1} band is assigned to the P-O stretch of the ionic species (PO_3^{2-}) and the other to the P-OH stretch (Gonzalez and Lafont, 1994).

Scanning Electron Microscopy

For visualization of scaling phenomena, SEM was employed to determine the deposit characteristics of four different membranes. Figures 8.14 through 8.17 show SEM photographs af 5000 times magnitude. In the each figure, (a) is the clean membrane surface, (b) is a photograph of the fouled membrane surface at a recovery of 15%, and (c) and (d) are the membrane surface without and with adding antiscalant at a recovery of 70%.

Comparing each figure (b) and (c), the membrane at a higher recovery condition shows greater and denser fouling than at a lower recovery condition. Also, comparing each figure (c) and (d) derived under the same recovery condition, (d) indicates no scale on the membrane surface and the antiscalant shows very effective inhibition of scale. With these results, a small amount of antiscalant inhibited the crystallization of scale. By the previous results (XRD and FTIR), we know that $CaCO_3$ is a dominant foulant. Moreover, fouling morphology is supporting evidence of $CaCO_3$, with the crystal structure of $CaCO_3$ as a spiral type. As seen in figures 8.14 (c), and 8.15 (c), morphologies of the foulant show spiral growth, concurring with the results of other research.

Also, other research helps explain why the $CaCO_3$ scale was more dominant in the presence of other foulants (e.g. $CaSO_4$ and $MgSO_4$); that is, $CaCO_3$ make a weak or loose other foulant from the competition (Sudmalis and Sheikholeslami, 2000).

Neutron Activation Analysis (NAA)

Application of neutron activation analysis (NAA) is novel technique in the field of membrane fouling. NAA is potentially an important methodology in analysis of inorganic foulants. The most common type of nuclear reaction used for NAA is the neutron capture or (n, γ) reaction, which is described as follows.

When a neutron interacts with a target nucleus via an inelastic collision, a collective compound nucleus forms in an excited state. The compound nucleus immediately (in 10^{-12} sec) rearranges its internal structure to a more stable configuration thorough the emission of one or more characteristic prompt γ rays. These prompt γ rays are used for evaluation of the radioactive elements produced. The most relevant characteristics of NAA can be summarized in a few main points: i) NAA is one of the most sensitive techniques used to evaluate many elements at levels as low as nanograms and picograms, ii) NAA is one of the few analytical techniques that can give both qualitative and quantitative multielemental information, iii) NAA may be considered a typical nondestructive method of analysis, not requiring complicated handling and is directly applicable to untreated solid samples, and iv) C, H, O, N are not activated or do not produce any long-lived activities (Orvinim and Speziali, 1998; Bettinelli, and Spezia, 1998). Table 8.3 summarizes the approximate detection limits provided by NAA for trace element determination, dissolved in natural water.

Table 8.3
Detection limits by NAA (Orvinim and Speziali, 1998)

Detection Limit (pg)	Elements
$10\text{-}10^2$	As, V
102-103	Al, Br, Cl, Co, Cu, I, Ti, U, Zn
103-104	Ba, Cr, Hg, Ni, Sr
104-105	Ca, K, Mg, P, Si
105-106	F, Fe
107	Pb, S

Based on NAA results, we can quantitatively estimate the total concentration of each trace element on the membrane surface, not by calculation, and determine the mean precipitation rate on membrane (Drak and Glucina, 2000).

Table 8.4 summarized the results according to mean precipitation rate of elements. Each element on the membrane surface was measured under specific conditions of neutron flux density (2.4 × 10^{13} n/cm^2-sec), irradiation time (3 min), cooling time (5 min and 10 min), and counting time (300 sec and 1000 sec). As shown in Table 8.4, the concentration of each element increased as recovery increased. As mentioned earlier, since carbon, hydrogen, nitrogen, and oxygen compounds can not be analyzed by NAA, it is not possible to measure the concentration of sulfate directly, however, the concentration of sulfur (e.g. $CaSO_4$) implies the existence of sulfate on the fouled membrane.

In this table, as expected, the precipitation rate of inorganic elements on the membranes shows a higher rate for RO than NF membranes, and the Ca^{2+} concentration is the highest of the elements.

Table 8.4
Mean precipitation rate of elements on the fouled membrane surface (mg/m^2-hr)

Membrane	Recovery, %	Barium	Calcium	Magnesium	Sodium	Chloride	Sulfur
LFC-1	Clean	0	0	0	0	0	0
	15	0	6.31	0	0	0	173
	30	0.48	243.76	4.56	0.14	0	191
	50	1.78	458.86	11.31	1.11	3.52	200
	70	3.8	899.38	12.62	1.68	14.79	220
	70 w/ antiscalant	0	7.05	0	0.28	2.60	97
TFC-ULP	Clean	0	0	0	0	0	0
	15	0	0	0	1.82	0	34
	30	0	4.96	0	5.49	0	93
	50	1.12	256.13	8.83	6.38	3.30	284
	70	3.76	986.11	24.31	8.41	4.76	1093
	70 w/ antiscalant	0.4	36.48	0	5.63	2.84	0
ESNA	Clean	0	0	0	0	0	0
	15	0	6.19	0	0.45	3.27	0
	30	0	6.75	0	0.86	3.35	0
	50	0	7.14	3.33	1.46	3.72	80
	70	0	15.94	5.57	3.57	6.37	120
	70 w/ antiscalant	0	3.69	4.23	2.6	4.16	16
TFC-S	Clean	0	0	0	0	0	0
	15	0	0	0	0.78	0.6	35
	30	0	5.87	5.65	1.33	1.88	192
	50	0	15.06	6.93	2.34	2.03	299
	70	1.06	361.26	9.01	5.89	4.87	302
	70 w/ antiscalant	0.23	23.92	8.30	1.96	4.92	0

Prediction of Semiempirical Constant (K) in Concentration Polarization Model

The concentration of solute at the membrane surface depends upon the flux through the membrane, the retention of the membrane, the diffusion coefficient of the solute, D, and the approximate thickness of the concentration boundary layer, δ (i.e., the region near the membrane in which the concentration of solute varies). Therefore, a general model (Mallevialle et al., 1996) can be used to approximately estimate the concentration of solute at the membrane surface (Lee and Kim, 1999).

$$C_m = C_b \exp\left(\frac{D\delta}{V}\right) \tag{8.2}$$

where C_m is the concentration of solute at membrane surface (M/L^3), C_b is the bulk concentration of solute (M/L^3), D is the diffusion coefficient (L^2/t), δ is the thickness of boundary layer (L), and V is the transmembrane flux (L/t).

The ratio of concentration in the boundary layer to that in the bulk solution is termed the concentration-polarization factor, PF. The value of PF can be estimated as an exponential function of recovery (r) (Wiesner and Aptel, 1996):

$$PF=\exp(Kr) \tag{8.3}$$

$$C_m = C_b PF = C_b \exp(Kr) \tag{8.4}$$

where K is a semiempirical constant typically taking on values of 0.6 to 0.9 for commercially available RO membranes. However, it is difficult to apply this general concept/values to our specific broader range of membranes (RO and NF) and different ions. Thus, we attempted to estimate the constant for each of the four different membranes and each of the ions by bench-scale experiments and calculations.

Correlation Between Recovery and J_o/k Ratio

Figure 8.18 represents the relationship between recovery and J_o/k ratio on ClO_4^- rejection; the J_o/k ratio is a function of recovery, which influences the polarization factor (PF) in Equation (8.4). For this study, 100 μg/L of ClO_4^- was added to CRW. Consistent trends were observed for ClO_4^- rejection, slightly decreasing for the LFC-1 and TFC-ULP membranes, and decreasing significantly for the ESNA and TFC-S membranes with increasing recovery (3 to 70%) and J_o/k ratio (0.5 to 2.6). However, when J_o/k ratio was fixed at 0.5, ClO_4^- rejection also significantly decreased for ESNA and TFC-S membranes with increasing recovery ranging from 3 to 15%.

Prediction of Constant (K) for Four Different Membranes and Ions

If we know the bulk (feed) and permeate concentrations of each species, we can calculate the concentration of each species at the membrane surface by the concentration polarization model (see Equation 8.5).

$$\ln\frac{C_m - C_P}{C_b - C_p} = \frac{J\delta}{D} \text{ or } \ln\frac{C_m - C_P}{C_b - C_p} = \frac{J}{k} \quad (8.5)$$

$$k = \frac{D}{\delta} \quad (8.6)$$

$$k = 1.117\left(\frac{UD^2}{hL}\right)^{0.33} \quad (8.7)$$

In Equation (8.5), C_m, C_p, and C_b are species concentrations at the membrane surface, in the permeate, and in the bulk water, respectively. Also, in Equation (8.6), the mass transfer coefficient, k, is a function of the diffusion coefficient (D) for a given species and the thickness of the boundary layer (δ) as well as the cross flow velocity (U), diffusion coefficient (D), equivalent hydraulic diameter (h), and the channel length (L) of the membrane system. In a equation (8.7), k is mass transfer coefficient determined at the bench scale. Through the equations, C_m can be calculated, and the polarization factor (PF) can be obtained with Equation (8.4).

According to Lisdonk and Rietman (2001), the polarization factor (PF) is an exponential function of water flux, cross flow velocity, and diffusion coefficient, and thus is related to the J_o/k ratio.

When neglecting the permeate concentration, Equation (8.5) can also be expressed as:

$$\ln\frac{C_m - C_P}{C_b - C_p} \approx \ln\frac{C_m}{C_b} = \ln PF = K\frac{J}{UD} = Kr \quad (8.8)$$

In Equation (8.8), ln PF is a function of J/UD, and the slope K is a constant for a given ionic species. Also, in the equation J/UD, the denominator, U and D, embodies the mass transfer coefficient (k), and J/k implies the recovery (r). The evidence of a relationship between the J/k ratio and recovery is shown in Figure 8.18.

Table 8.5 (a)-(j) and Figure 8.19 (a)-(j) describe the overall procedure of calculating the concentration of each ion at the membrane surface by the concentration polarization model, and present plots of the slope (constant, K) for each ion, and each of the four different membranes.

MINEQL Simulations

We have estimated the concentration of solute at the membrane surface based on the water quality characteristics of CRW to estimate precipitative fouling by inorganic salts for four different membranes. Values of C_m were estimated at four different recoveries (J_o/k ratio), 15%, 30%, 50%, and 70% (J_o/k ratio; 0.6, 1.2, 2.0, and 2.7). Table 8.5 and Figure 8.19 represent C_b and C_m values, and values of the constant (K) for different cations/anions in CRW with the different membranes.

Table 8.5

Overall calculation of the concentration of each ion at the membrane surface by the concentration polarization model

(a) Barium

Membrane	Recovery, %	Q, mL/min	U, cm/sec	k, cm/sec	Perm. mL/min	J_o	J_o/k	δ	D, cm^2/sec	C_b, mg/L	C_p, mg/L	C_b, M	C_p, M	C_b-C_p, mg/L	C_b-C_p, M	exp(Jδ / D)	C_m, M	$\ln(C_m/C_b)$
LFC-1	15	20	0.49	0.0006191	3	0.00036	0.581	0.027364	1.69E-05	0.083	0.002	6.04E-07	1.46E-08	0.081	5.9E-07	1.7873979	1.07E-06	0.570
LFC-1	30	20	0.49	0.0006191	6	0.000719	1.162	0.027364	1.69E-05	0.079	0.001	5.75E-07	7.28E-09	0.078	5.68E-07	3.1947911	1.82E-06	1.153
LFC-1	50	20	0.49	0.0006191	10	0.001198	1.936	0.027364	1.69E-05	0.064	0.001	4.66E-07	7.28E-09	0.063	4.59E-07	6.9300672	3.19E-06	1.922
LFC-1	70	20	0.49	0.0006191	14	0.001678	2.710	0.027364	1.69E-05	0.08	0.001	5.83E-07	7.28E-09	0.079	5.75E-07	15.032542	8.66E-06	2.698
TFC-ULP	15	20	0.49	0.0006191	3	0.00036	0.581	0.027364	1.69E-05	0.1	0.007	7.28E-07	5.1E-08	0.093	6.77E-07	1.7873979	1.26E-06	0.549
TFC-ULP	30	20	0.49	0.0006191	6	0.000719	1.162	0.027364	1.69E-05	0.089	0.017	6.48E-07	1.24E-07	0.072	5.24E-07	3.1947911	1.8E-06	1.021
TFC-ULP	50	20	0.49	0.0006191	10	0.001198	1.936	0.027364	1.69E-05	0.086	0.038	6.26E-07	2.77E-07	0.048	3.5E-07	6.9300672	2.7E-06	1.461
TFC-ULP	70	20	0.49	0.0006191	14	0.001678	2.710	0.027364	1.69E-05	0.088	0.03	6.41E-07	2.18E-07	0.058	4.22E-07	15.032542	6.57E-06	2.327
ESNA	15	20	0.49	0.0006191	3	0.00036	0.581	0.027364	1.69E-05	0.075	0.022	5.46E-07	1.6E-07	0.053	3.86E-07	1.7873979	8.5E-07	0.442
ESNA	30	20	0.49	0.0006191	6	0.000719	1.162	0.027364	1.69E-05	0.062	0.018	4.51E-07	1.31E-07	0.044	3.2E-07	3.1947911	1.15E-06	0.939
ESNA	50	20	0.49	0.0006191	10	0.001198	1.936	0.027364	1.69E-05	0.065	0.017	4.73E-07	1.24E-07	0.048	3.5E-07	6.9300672	2.55E-06	1.683
ESNA	70	20	0.49	0.0006191	14	0.001678	2.710	0.027364	1.69E-05	0.058	0.02	4.22E-07	1.46E-07	0.038	2.77E-07	15.032542	4.31E-06	2.322
TFC-S	15	20	0.49	0.0006191	3	0.00036	0.581	0.027364	1.69E-05	0.056	0.018	4.08E-07	1.31E-07	0.038	2.77E-07	1.7873979	6.26E-07	0.428
TFC-S	30	20	0.49	0.0006191	6	0.000719	1.162	0.027364	1.69E-05	0.055	0.02	4.01E-07	1.46E-07	0.035	2.55E-07	3.1947911	9.6E-07	0.874
TFC-S	50	20	0.49	0.0006191	10	0.001198	1.936	0.027364	1.69E-05	0.053	0.013	3.86E-07	9.47E-08	0.04	2.91E-07	6.9300672	2.11E-06	1.700
TFC-S	70	20	0.49	0.0006191	14	0.001678	2.710	0.027364	1.69E-05	0.053	0.012	3.86E-07	8.74E-08	0.041	2.99E-07	15.032542	4.58E-06	2.473

(continued)

Table 8.5 (Continued)

(b) Calcium

Membrane	Recovery, %	Q, mL/min	U, cm/sec	k, cm/sec	Perm. mL/min	J_o	J_o/k	δ	D, cm^2/sec	C_b, mg/L	C_p, mg/L	C_b, M	C_p, M	C_b-C_p, mg/L	C_b-C_p, M	$exp(J\delta / D)$	C_m, M	$ln(C_m/C_b)$
LFC-1	15	20	0.49	0.000592	3	0.00036	0.607	0.026746	1.584E-05	62.843	1.557	0.001571	3.89E-05	61.286	0.001532	1.8350545	0.002851	0.5957356
LFC-1	30	20	0.49	0.000592	6	0.000719	1.214	0.026746	1.584E-05	63.087	1.129	0.001577	2.82E-05	61.958	0.001549	3.3674251	0.005244	1.2014871
LFC-1	50	20	0.49	0.000592	10	0.001198	2.024	0.026746	1.584E-05	53.924	1.321	0.001348	3.3E-05	52.603	0.001315	7.5653652	0.009982	2.0020921
LFC-1	70	20	0.49	0.000592	14	0.001678	2.833	0.026746	1.584E-05	62.887	2.341	0.001572	5.85E-05	60.546	0.001514	16.996592	0.025785	2.7973491
TFC-ULP	15	20	0.49	0.000592	3	0.00036	0.607	0.026746	1.584E-05	71.408	2.065	0.001785	5.16E-05	69.343	0.001734	1.8350545	0.003233	0.5938273
TFC-ULP	30	20	0.49	0.000592	6	0.000719	1.214	0.026746	1.584E-05	64.571	11.046	0.001614	0.000276	53.525	0.001338	3.3674251	0.004782	1.0860118
TFC-ULP	50	20	0.49	0.000592	10	0.001198	2.024	0.026746	1.584E-05	61.986	11.429	0.00155	0.000286	50.557	0.001264	7.5653652	0.009848	1.8492169
TFC-ULP	70	20	0.49	0.000592	14	0.001678	2.833	0.026746	1.584E-05	62.585	13.616	0.001565	0.00034	48.969	0.001224	16.996592	0.021148	2.6039017
ESNA	15	20	0.49	0.000592	3	0.00036	0.607	0.026746	1.584E-05	49.61	18.1	0.00124	0.000453	31.51	0.000788	1.8350545	0.001898	0.4255215
ESNA	30	20	0.49	0.000592	6	0.000719	1.214	0.026746	1.584E-05	47.14	16.7	0.001179	0.000418	30.44	0.000761	3.3674251	0.00298	0.9277179
ESNA	50	20	0.49	0.000592	10	0.001198	2.024	0.026746	1.584E-05	46.78	14.74	0.00117	0.000369	32.04	0.000801	7.5653652	0.006428	1.7041428
ESNA	70	20	0.49	0.000592	14	0.001678	2.833	0.026746	1.584E-05	43.92	19.95	0.001098	0.000499	23.97	0.000599	16.996592	0.010684	2.275253
TFC-S	15	20	0.49	0.000592	3	0.00036	0.607	0.026746	1.584E-05	47.442	13.729	0.001186	0.000343	33.713	0.000843	1.8350545	0.00189	0.4658716
TFC-S	30	20	0.49	0.000592	6	0.000719	1.214	0.026746	1.584E-05	46.934	20.268	0.001173	0.000507	26.666	0.000667	3.3674251	0.002752	0.8523174
TFC-S	50	20	0.49	0.000592	10	0.001198	2.024	0.026746	1.584E-05	47.35	12.865	0.001184	0.000322	34.485	0.000862	7.5653652	0.006844	1.7546726
TFC-S	70	20	0.49	0.000592	14	0.001678	2.833	0.026746	1.584E-05	47.956	11.781	0.001199	0.000295	36.175	0.000904	16.996592	0.015666	2.5700767

(continued)

Table 8.5 (Continued)

(c) Chloride

Membrane	Recovery, %	Q, mL/min	U, cm/sec	k, cm/sec	Perm. mL/min	J_o	J_o/k	δ	D, cm^2/sec	C_b, mg/L	C_p, mg/L	C_b, M	C_p, M	C_b-C_p, mg/L	C_b-C_p, M	$\exp(J\delta / D)$	C_m, M	$\ln(C_m/C_b)$
LFC-1	15	20	0.49	0.000698	3	0.00036	0.515	0.02911	2.032E-05	65.7	4.46	0.001853	0.000126	61.24	0.001727	1.6737245	0.003017	0.487346
LFC-1	30	20	0.49	0.000698	6	0.000719	1.030	0.02911	2.032E-05	63.43	4.33	0.001789	0.000122	59.1	0.001667	2.8013536	0.004792	0.985214
LFC-1	50	20	0.49	0.000698	10	0.001198	1.717	0.02911	2.032E-05	59.01	5.28	0.001664	0.000149	53.73	0.001516	5.5668974	0.008586	1.640601
LFC-1	70	20	0.49	0.000698	14	0.001678	2.404	0.02911	2.032E-05	65.43	7.28	0.001846	0.000205	58.15	0.00164	11.062633	0.01835	2.296871
TFC-ULP	15	20	0.49	0.000698	3	0.00036	0.515	0.02911	2.032E-05	70.13	6.01	0.001978	0.00017	64.12	0.001809	1.6737245	0.003197	0.479946
TFC-ULP	30	20	0.49	0.000698	6	0.000719	1.030	0.02911	2.032E-05	64	14.73	0.001805	0.000415	49.27	0.00139	2.8013536	0.004309	0.869937
TFC-ULP	50	20	0.49	0.000698	10	0.001198	1.717	0.02911	2.032E-05	66.46	16.42	0.001875	0.000463	50.04	0.001411	5.5668974	0.008321	1.490333
TFC-ULP	70	20	0.49	0.000698	14	0.001678	2.404	0.02911	2.032E-05	66.39	19.58	0.001873	0.000552	46.81	0.00132	11.062633	0.015159	2.091237
ESNA	15	20	0.49	0.000698	3	0.00036	0.515	0.02911	2.032E-05	56.6	29.55	0.001596	0.000834	27.05	0.000763	1.6737245	0.002111	0.279133
ESNA	30	20	0.49	0.000698	6	0.000719	1.030	0.02911	2.032E-05	51.22	30.55	0.001445	0.000862	20.67	0.000583	2.8013536	0.002495	0.546352
ESNA	50	20	0.49	0.000698	10	0.001198	1.717	0.02911	2.032E-05	54.2	27.26	0.001529	0.000769	26.94	0.00076	5.5668974	0.004999	1.18478
ESNA	70	20	0.49	0.000698	14	0.001678	2.404	0.02911	2.032E-05	50.62	32.13	0.001428	0.000906	18.49	0.000522	11.062633	0.006676	1.542354
TFC-S	15	20	0.49	0.000698	3	0.00036	0.515	0.02911	2.032E-05	60.13	31.54	0.001696	0.00089	28.59	0.000806	1.6737245	0.002239	0.277886
TFC-S	30	20	0.49	0.000698	6	0.000719	1.030	0.02911	2.032E-05	58.85	35.71	0.00166	0.001007	23.14	0.000653	2.8013536	0.002836	0.535497
TFC-S	50	20	0.49	0.000698	10	0.001198	1.717	0.02911	2.032E-05	57.69	27.06	0.001627	0.000763	30.63	0.000864	5.5668974	0.005573	1.23103
TFC-S	70	20	0.49	0.000698	14	0.001678	2.404	0.02911	2.032E-05	58.06	25.58	0.001638	0.000722	32.48	0.000916	11.062633	0.010857	1.891492

(continued)

Table 8.5 (Continued)

(d) Carbonate

Membrane	Recovery, %	Q, mL/min	U, cm/sec	k, cm/sec	Perm. mL/min	J_o	J_o/k	δ	D, cm^2/sec	C_b, mg/L	C_p, mg/L	C_b, M	C_p, M	C_b-C_p, mg/L	C_b-C_p, M	$\exp(J\delta / D)$	C_m, M	$\ln(C_m/C_b)$
LFC-1	15	20	0.49	0.000655	3	0.00036	0.549	0.0281749	1.846E-05	0.694	0.003	1.16E-05	5E-08	0.691	1.15E-05	1.7310719	2E-05	0.5469135
LFC-1	30	20	0.49	0.000655	6	0.000719	1.097	0.0281749	1.846E-05	0.657	0.005	1.1E-05	8.33E-08	0.652	1.09E-05	2.9966099	3.26E-05	1.092398
LFC-1	50	20	0.49	0.000655	10	0.001198	1.829	0.0281749	1.846E-05	0.44	0.003	7.33E-06	5E-08	0.437	7.28E-06	6.2285032	4.54E-05	1.8233961
LFC-1	70	20	0.49	0.000655	14	0.001678	2.561	0.0281749	1.846E-05	0.501	0.003	8.35E-06	5E-08	0.498	8.3E-06	12.946047	0.000108	2.5552497
TFC-ULP	15	20	0.49	0.000655	3	0.00036	0.549	0.0281749	1.846E-05	1.393	0.038	2.32E-05	6.33E-07	1.355	2.26E-05	1.7310719	3.97E-05	0.5371533
TFC-ULP	30	20	0.49	0.000655	6	0.000719	1.097	0.0281749	1.846E-05	1.067	0.027	1.78E-05	4.5E-07	1.04	1.73E-05	2.9966099	5.24E-05	1.0804777
TFC-ULP	50	20	0.49	0.000655	10	0.001198	1.829	0.0281749	1.846E-05	0.934	0.014	1.56E-05	2.33E-07	0.92	1.53E-05	6.2285032	9.57E-05	1.8164735
TFC-ULP	70	20	0.49	0.000655	14	0.001678	2.561	0.0281749	1.846E-05	2.115	0.039	3.53E-05	6.5E-07	2.076	3.46E-05	12.946047	0.000449	2.5436287
ESNA	15	20	0.49	0.000655	3	0.00036	0.549	0.0281749	1.846E-05	0.857	0.179	1.43E-05	2.98E-06	0.678	1.13E-05	1.7310719	2.25E-05	0.4563954
ESNA	30	20	0.49	0.000655	6	0.000719	1.097	0.0281749	1.846E-05	0.623	0.184	1.04E-05	3.07E-06	0.439	7.32E-06	2.9966099	2.5E-05	0.8783483
ESNA	50	20	0.49	0.000655	10	0.001198	1.829	0.0281749	1.846E-05	0.398	0.191	6.63E-06	3.18E-06	0.207	3.45E-06	6.2285032	2.47E-05	1.3135482
ESNA	70	20	0.49	0.000655	14	0.001678	2.561	0.0281749	1.846E-05	0.664	0.31	1.11E-05	5.17E-06	0.354	5.9E-06	12.946047	8.15E-05	1.9972584
TFC-S	15	20	0.49	0.000655	3	0.00036	0.549	0.0281749	1.846E-05	0.392	0.035	6.53E-06	5.83E-07	0.357	5.95E-06	1.7310719	1.09E-05	0.5103041
TFC-S	30	20	0.49	0.000655	6	0.000719	1.097	0.0281749	1.846E-05	0.531	0.038	8.85E-06	6.33E-07	0.493	8.22E-06	2.9966099	2.53E-05	1.0486256
TFC-S	50	20	0.49	0.000655	10	0.001198	1.829	0.0281749	1.846E-05	0.463	0.04	7.72E-06	6.67E-07	0.423	7.05E-06	6.2285032	4.46E-05	1.7538493
TFC-S	70	20	0.49	0.000655	14	0.001678	2.561	0.0281749	1.846E-05	0.394	0.045	6.57E-06	7.5E-07	0.349	5.82E-06	12.946047	7.61E-05	2.449422

(continued)

Table 8.5 (Continued)

(e) Potassium

Membrane	Recovery, %	Q, mL/min	U, cm/sec	k, cm/sec	Perm. mL/min	J_o	J_o/k	δ	D, cm^2/sec	C_b, mg/L	C_p, mg/L	C_b, M	C_p, M	C_b-C_p, mg/L	C_b-C_p, M	$exp(J\delta / D)$	C_m, M	$ln(C_m/C_b)$
LFC-1	15	20	0.49	0.000681	3	0.00036	0.528	0.02874	1.957E-05	3.971	0.315	0.000102	8.06E-06	3.656	9.35E-05	1.69553	0.000167	0.494914
LFC-1	30	20	0.49	0.000681	6	0.000719	1.056	0.02874	1.957E-05	3.594	0.193	9.19E-05	4.94E-06	3.401	8.7E-05	2.874823	0.000255	1.0203421
LFC-1	50	20	0.49	0.000681	10	0.001198	1.760	0.02874	1.957E-05	3.693	0.286	9.45E-05	7.31E-06	3.407	8.71E-05	5.812351	0.000514	1.6937173
LFC-1	70	20	0.49	0.000681	14	0.001678	2.464	0.02874	1.957E-05	3.694	0.366	9.45E-05	9.36E-06	3.328	8.51E-05	11.75148	0.00101	2.3689557
TFC-ULP	15	20	0.49	0.000681	3	0.00036	0.528	0.02874	1.957E-05	3.64	0.459	9.31E-05	1.17E-05	3.181	8.14E-05	1.69553	0.00015	0.4748821
TFC-ULP	30	20	0.49	0.000681	6	0.000719	1.056	0.02874	1.957E-05	3.634	0.882	9.29E-05	2.26E-05	2.752	7.04E-05	2.874823	0.000225	0.8836803
TFC-ULP	50	20	0.49	0.000681	10	0.001198	1.760	0.02874	1.957E-05	3.27	1.035	8.36E-05	2.65E-05	2.235	5.72E-05	5.812351	0.000359	1.4560945
TFC-ULP	70	20	0.49	0.000681	14	0.001678	2.464	0.02874	1.957E-05	3.399	1.093	8.69E-05	2.8E-05	2.306	5.9E-05	11.75148	0.000721	2.1155538
ESNA	15	20	0.49	0.000681	3	0.00036	0.528	0.02874	1.957E-05	3.429	1.546	8.77E-05	3.95E-05	1.883	4.82E-05	1.69553	0.000121	0.3234907
ESNA	30	20	0.49	0.000681	6	0.000719	1.056	0.02874	1.957E-05	3.162	1.532	8.09E-05	3.92E-05	1.63	4.17E-05	2.874823	0.000159	0.6762373
ESNA	50	20	0.49	0.000681	10	0.001198	1.760	0.02874	1.957E-05	3.267	1.395	8.36E-05	3.57E-05	1.872	4.79E-05	5.812351	0.000314	1.3237512
ESNA	70	20	0.49	0.000681	14	0.001678	2.464	0.02874	1.957E-05	3.087	1.748	7.9E-05	4.47E-05	1.339	3.42E-05	11.75148	0.000447	1.7340424
TFC-S	15	20	0.49	0.000681	3	0.00036	0.528	0.02874	1.957E-05	3.066	1.571	7.84E-05	4.02E-05	1.495	3.82E-05	1.69553	0.000105	0.2920311
TFC-S	30	20	0.49	0.000681	6	0.000719	1.056	0.02874	1.957E-05	3.136	1.699	8.02E-05	4.35E-05	1.437	3.68E-05	2.874823	0.000149	0.6200895
TFC-S	50	20	0.49	0.000681	10	0.001198	1.760	0.02874	1.957E-05	3.064	1.394	7.84E-05	3.57E-05	1.67	4.27E-05	5.812351	0.000284	1.2872802
TFC-S	70	20	0.49	0.000681	14	0.001678	2.464	0.02874	1.957E-05	3.202	1.367	8.19E-05	3.5E-05	1.835	4.69E-05	11.75148	0.000586	1.9687125

(continued)

Table 8.5 (Continued)

(f) Magnesium

Membrane	Recovery, %	Q, mL/min	U, cm/sec	k, cm/sec	Perm. mL/min	J_o	J_o/k	δ	D, cm^2/sec	C_b, mg/L	C_p, mg/L	C_b, M	C_p, M	C_b-C_p, mg/L	C_b-C_p, M	$exp(J\delta / D)$	C_m, M	$ln(C_m/C_b)$
LFC-1	15	20	0.49	0.000549	3	0.00036	0.655	0.025721	1.412E-05	24.182	0.553	0.000995	2.28E-05	23.629	0.000972	1.924992	0.001894	0.643872
LFC-1	30	20	0.49	0.000549	6	0.000719	1.310	0.025721	1.412E-05	23.853	0.282	0.000981	1.16E-05	23.571	0.00097	3.705592	0.003605	1.301174
LFC-1	50	20	0.49	0.000549	10	0.001198	2.183	0.025721	1.412E-05	21.672	0.362	0.000892	1.49E-05	21.31	0.000877	8.873523	0.007795	2.16814
LFC-1	70	20	0.49	0.000549	14	0.001678	3.056	0.025721	1.412E-05	23.953	0.462	0.000986	1.9E-05	23.491	0.000967	21.2488	0.020556	3.03775
TFC-ULP	15	20	0.49	0.000549	3	0.00036	0.655	0.025721	1.412E-05	24.475	0.763	0.001007	3.14E-05	23.712	0.000976	1.924992	0.001909	0.639828
TFC-ULP	30	20	0.49	0.000549	6	0.000719	1.310	0.025721	1.412E-05	22.122	4.32	0.00091	0.000178	17.802	0.000732	3.705592	0.002892	1.156014
TFC-ULP	50	20	0.49	0.000549	10	0.001198	2.183	0.025721	1.412E-05	23.158	4.803	0.000953	0.000198	18.355	0.000755	8.873523	0.006899	1.979696
TFC-ULP	70	20	0.49	0.000549	14	0.001678	3.056	0.025721	1.412E-05	23.271	5.744	0.000957	0.000236	17.527	0.000721	21.2488	0.015559	2.788141
ESNA	15	20	0.49	0.000549	3	0.00036	0.655	0.025721	1.412E-05	20.29	7.75	0.000835	0.000319	12.54	0.000516	1.924992	0.001312	0.452145
ESNA	30	20	0.49	0.000549	6	0.000719	1.310	0.025721	1.412E-05	18.82	7.14	0.000774	0.000294	11.68	0.000481	3.705592	0.002075	0.985494
ESNA	50	20	0.49	0.000549	10	0.001198	2.183	0.025721	1.412E-05	18.92	6.74	0.000778	0.000277	12.18	0.000501	8.873523	0.004724	1.803142
ESNA	70	20	0.49	0.000549	14	0.001678	3.056	0.025721	1.412E-05	17.73	8.97	0.000729	0.000369	8.76	0.00036	21.2488	0.008028	2.398303
TFC-S	15	20	0.49	0.000549	3	0.00036	0.655	0.025721	1.412E-05	20.375	5.814	0.000838	0.000239	14.561	0.000599	1.924992	0.001392	0.507447
TFC-S	30	20	0.49	0.000549	6	0.000719	1.310	0.025721	1.412E-05	20.044	8.376	0.000825	0.000345	11.668	0.00048	3.705592	0.002124	0.945841
TFC-S	50	20	0.49	0.000549	10	0.001198	2.183	0.025721	1.412E-05	19.49	6.162	0.000802	0.000254	13.328	0.000548	8.873523	0.005119	1.853828
TFC-S	70	20	0.49	0.000549	14	0.001678	3.056	0.025721	1.412E-05	19.943	5.36	0.000821	0.000221	14.583	0.0006	21.2488	0.01297	2.760428

(continued)

Table 8.5 (Continued)

(g) Sodium

Membrane	Recovery, %	Q, mL/min	U, cm/sec	k, cm/sec	Perm. mL/min	J_o	J_o/k	δ	D, cm^2/sec	C_b, mg/L	C_p, mg/L	C_b, M	C_p, M	C_b-C_p, mg/L	C_b-C_p, M	$\exp(J\delta / D)$	C_m, M	$\ln(C_m/C_b)$
LFC-1	15	20	0.49	0.000529	3	0.00036	0.680	0.025229	1.33E-05	27.482	2.12	0.001195	9.22E-05	25.362	0.001103	1.9737801	0.00227	0.6411491
LFC-1	30	20	0.49	0.000529	6	0.000719	1.360	0.025229	1.33E-05	26.769	1.847	0.001164	8.03E-05	24.922	0.001084	3.8958081	0.004304	1.3072522
LFC-1	50	20	0.49	0.000529	10	0.001198	2.267	0.025229	1.33E-05	25.345	2.075	0.001102	9.03E-05	23.27	0.001012	9.6456001	0.009853	2.1902876
LFC-1	70	20	0.49	0.000529	14	0.001678	3.173	0.025229	1.33E-05	25.469	2.295	0.001108	9.98E-05	23.174	0.001008	23.881465	0.024172	3.0828098
TFC-ULP	15	20	0.49	0.000529	3	0.00036	0.680	0.025229	1.33E-05	81.09	8.32	0.003527	0.000362	72.77	0.003165	1.9737801	0.006609	0.6280049
TFC-ULP	30	20	0.49	0.000529	6	0.000719	1.360	0.025229	1.33E-05	75.59	19.22	0.003288	0.000836	56.37	0.002452	3.8958081	0.010388	1.1504142
TFC-ULP	50	20	0.49	0.000529	10	0.001198	2.267	0.025229	1.33E-05	74.73	21.42	0.003251	0.000932	53.31	0.002319	9.6456001	0.023298	1.9695563
TFC-ULP	70	20	0.49	0.000529	14	0.001678	3.173	0.025229	1.33E-05	74.25	24.12	0.00323	0.001049	50.13	0.002181	23.881465	0.053123	2.8002316
ESNA	15	20	0.49	0.000529	3	0.00036	0.680	0.025229	1.33E-05	66.43	33.75	0.00289	0.001468	32.68	0.001421	1.9737801	0.004274	0.3913984
ESNA	30	20	0.49	0.000529	6	0.000719	1.360	0.025229	1.33E-05	61.01	32.66	0.002654	0.001421	28.35	0.001233	3.8958081	0.006225	0.8525489
ESNA	50	20	0.49	0.000529	10	0.001198	2.267	0.025229	1.33E-05	62.54	30.69	0.00272	0.001335	31.85	0.001385	9.6456001	0.014698	1.6869506
ESNA	70	20	0.49	0.000529	14	0.001678	3.173	0.025229	1.33E-05	58.3	37.15	0.002536	0.001616	21.15	0.00092	23.881465	0.023586	2.2301121
TFC-S	15	20	0.49	0.000529	3	0.00036	0.680	0.025229	1.33E-05	23.92	12.084	0.00104	0.000526	11.836	0.000515	1.9737801	0.001542	0.3932859
TFC-S	30	20	0.49	0.000529	6	0.000719	1.360	0.025229	1.33E-05	23.866	14.391	0.001038	0.000626	9.475	0.000412	3.8958081	0.002232	0.7653096
TFC-S	50	20	0.49	0.000529	10	0.001198	2.267	0.025229	1.33E-05	22.17	12.443	0.000964	0.000541	9.727	0.000423	9.6456001	0.004622	1.567203
TFC-S	70	20	0.49	0.000529	14	0.001678	3.173	0.025229	1.33E-05	23.425	11.976	0.001019	0.000521	11.449	0.000498	23.881465	0.012414	2.50007

(continued)

Table 8.5 (Continued)

(h) Nitrate

Membrane	Recovery, %	Q, mL/min	U, cm/sec	k, cm/sec	Perm. mL/min	J_o	J_o/k	δ	D, cm^2/sec	C_b, mg/L	C_p, mg/L	C_b, M	C_p, M	C_b-C_p, mg/L	C_b-C_p, M	$exp(J\delta / D)$	C_m, M	$ln(C_m/C_b)$
LFC-1	15	20	0.49	0.000668	3	0.00036	0.538	0.028463	0.00001902	0.065	0.016	9.03E-07	2.22E-07	0.049	6.81E-07	1.712619	1.39E-06	0.429966
LFC-1	30	20	0.49	0.000668	6	0.000719	1.076	0.028463	0.00001902	0.036	0	5E-07	0	0.036	5E-07	2.933062	1.47E-06	1.076047
LFC-1	50	20	0.49	0.000668	10	0.001198	1.793	0.028463	0.00001902	0.01	0	1.39E-07	0	0.01	1.39E-07	6.009921	8.35E-07	1.793412
LFC-1	70	20	0.49	0.000668	14	0.001678	2.511	0.028463	0.00001902	0.026	0	3.61E-07	0	0.026	3.61E-07	12.31449	4.45E-06	2.510776
TFC-ULP	15	20	0.49	0.000668	3	0.00036	0.538	0.028463	0.00001902	0.866	0.209	1.17E-05	2.82E-06	0.657	8.88E-06	1.712619	1.8E-05	0.432195
TFC-ULP	30	20	0.49	0.000668	6	0.000719	1.076	0.028463	0.00001902	0.778	0.174	1.05E-05	2.35E-06	0.604	8.16E-06	2.933062	2.63E-05	0.916584
TFC-ULP	50	20	0.49	0.000668	10	0.001198	1.793	0.028463	0.00001902	0.288	0.159	3.89E-06	2.15E-06	0.129	1.74E-06	6.009921	1.26E-05	1.176816
TFC-ULP	70	20	0.49	0.000668	14	0.001678	2.511	0.028463	0.00001902	0.257	0.103	3.47E-06	1.39E-06	0.154	2.08E-06	12.31449	2.7E-05	2.051542
ESNA	15	20	0.49	0.000668	3	0.00036	0.538	0.028463	0.00001902	0.34	0.18	4.59E-06	2.43E-06	0.16	2.16E-06	1.712619	6.14E-06	0.289193
ESNA	30	20	0.49	0.000668	6	0.000719	1.076	0.028463	0.00001902	0.27	0.13	3.65E-06	1.76E-06	0.14	1.89E-06	2.933062	7.31E-06	0.694311
ESNA	50	20	0.49	0.000668	10	0.001198	1.793	0.028463	0.00001902	0.36	0.2	4.86E-06	2.7E-06	0.16	2.16E-06	6.009921	1.57E-05	1.171439
ESNA	70	20	0.49	0.000668	14	0.001678	2.511	0.028463	0.00001902	0.31	0.17	4.19E-06	2.3E-06	0.14	1.89E-06	12.31449	2.56E-05	1.809889
TFC-S	15	20	0.49	0.000668	3	0.00036	0.538	0.028463	0.00001902	0.209	0.095	2.82E-06	1.28E-06	0.114	1.54E-06	1.712619	3.92E-06	0.328369
TFC-S	30	20	0.49	0.000668	6	0.000719	1.076	0.028463	0.00001902	0	0	0	0	0	0	2.933062	0	
TFC-S	50	20	0.49	0.000668	10	0.001198	1.793	0.028463	0.00001902	0.032	0	4.32E-07	0	0.032	4.32E-07	6.009921	2.6E-06	1.793412
TFC-S	70	20	0.49	0.000668	14	0.001678	2.511	0.028463	0.00001902	0.203	0.088	2.74E-06	1.19E-06	0.115	1.55E-06	12.31449	2.03E-05	2.002788

(continued)

Table 8.5 (Continued)

(i) Sulfate

Membrane	Recovery, %	Q, mL/min	U, cm/sec	k, cm/sec	Perm. mL/min	J_o	J_o/k	δ	D, cm^2/sec	C_b, mg/L	C_p, mg/L	C_b, M	C_p, M	C_b-C_p, mg/L	C_b-C_p, M	$\exp(J\delta / D)$	C_m, M	$\ln(C_m/C_b)$
LFC-1	15	20	0.49	0.00072	3	0.00036	0.499	0.02958	0.0000213	209.5	5.7	0.002182	5.94E-05	203.8	0.002123	1.647545	0.003557	0.4885351
LFC-1	30	20	0.49	0.00072	6	0.000719	0.999	0.02958	0.0000213	204	2.6	0.002125	2.71E-05	201.4	0.002098	2.714404	0.005722	0.9904903
LFC-1	50	20	0.49	0.00072	10	0.001198	1.664	0.02958	0.0000213	188.8	3.3	0.001967	3.44E-05	185.5	0.001932	5.281909	0.010241	1.6500167
LFC-1	70	20	0.49	0.00072	14	0.001678	2.330	0.02958	0.0000213	201	8.3	0.002094	8.65E-05	192.7	0.002007	10.27797	0.020717	2.2920143
TFC-ULP	15	20	0.49	0.00072	3	0.00036	0.499	0.02958	0.0000213	223.3	7.8	0.002326	8.13E-05	215.5	0.002245	1.647545	0.00378	0.4854622
TFC-ULP	30	20	0.49	0.00072	6	0.000719	0.999	0.02958	0.0000213	207	40.6	0.002156	0.000423	166.4	0.001733	2.714404	0.005128	0.8663226
TFC-ULP	50	20	0.49	0.00072	10	0.001198	1.664	0.02958	0.0000213	213.1	44.8	0.00222	0.000467	168.3	0.001753	5.281909	0.009727	1.4774422
TFC-ULP	70	20	0.49	0.00072	14	0.001678	2.330	0.02958	0.0000213	212.3	54	0.002211	0.000563	158.3	0.001649	10.27797	0.01751	2.0691453
ESNA	15	20	0.49	0.00072	3	0.00036	0.499	0.02958	0.0000213	169.6	63.6	0.001767	0.000663	106	0.001104	1.647545	0.002482	0.3398349
ESNA	30	20	0.49	0.00072	6	0.000719	0.999	0.02958	0.0000213	158.5	57.6	0.001651	0.0006	100.9	0.001051	2.714404	0.003453	0.7378232
ESNA	50	20	0.49	0.00072	10	0.001198	1.664	0.02958	0.0000213	157.1	53.9	0.001636	0.000561	103.2	0.001075	5.281909	0.00624	1.3383674
ESNA	70	20	0.49	0.00072	14	0.001678	2.330	0.02958	0.0000213	148.2	74.8	0.001544	0.000779	73.4	0.000765	10.27797	0.008638	1.7219022
TFC-S	15	20	0.49	0.00072	3	0.00036	0.499	0.02958	0.0000213	184.05	51.51	0.001917	0.000537	132.54	0.001381	1.647545	0.002811	0.3827537
TFC-S	30	20	0.49	0.00072	6	0.000719	0.999	0.02958	0.0000213	181.03	72.66	0.001886	0.000757	108.37	0.001129	2.714404	0.003821	0.7062085
TFC-S	50	20	0.49	0.00072	10	0.001198	1.664	0.02958	0.0000213	178.58	59.4	0.00186	0.000619	119.18	0.001241	5.281909	0.007176	1.3500565
TFC-S	70	20	0.49	0.00072	14	0.001678	2.330	0.02958	0.0000213	177.14	53.35	0.001845	0.000556	123.79	0.001289	10.27797	0.013809	2.0127252

(continued)

Table 8.5 (Continued)

(j) Perchlorate

Membrane	Recovery, %	Q, mL/min	U, cm/sec	k, cm/sec	Perm. mL/min	J_o	J_o/k	δ	D, cm^2/sec	C_b, mg/L	C_p, mg/L	C_b, M	C_p, M	C_b-C_p, mg/L	C_b-C_p, M	$\exp(J\delta / D)$	C_m, M	$\ln(C_m/C_b)$
LFC-1	15	20	0.49	0.000642	3	0.00036	0.560	0.027892	1.79E-05	0.1	0.00649	1.01E-06	6.52E-08	0.09351	9.4E-07	1.749971	1.71E-06	0.595736
LFC-1	30	20	0.49	0.000642	6	0.000719	1.119	0.027892	1.79E-05	0.1	0.00651	1.01E-06	6.54E-08	0.09349	9.4E-07	3.062399	2.94E-06	1.201487
LFC-1	50	20	0.49	0.000642	10	0.001198	1.865	0.027892	1.79E-05	0.1	0.0083	1.01E-06	8.34E-08	0.0917	9.22E-07	6.458071	6.04E-06	2.002092
LFC-1	70	20	0.49	0.000642	14	0.001678	2.611	0.027892	1.79E-05	0.1	0.01439	1.01E-06	1.45E-07	0.08561	8.6E-07	13.61896	1.19E-05	2.797349
TFC-ULP	15	20	0.49	0.000642	3	0.00036	0.560	0.027892	1.79E-05	0.1	0.00841	1.01E-06	8.45E-08	0.09159	9.21E-07	1.749971	1.7E-06	0.593827
TFC-ULP	30	20	0.49	0.000642	6	0.000719	1.119	0.027892	1.79E-05	0.1	0.01093	1.01E-06	1.1E-07	0.08907	8.95E-07	3.062399	2.85E-06	1.086012
TFC-ULP	50	20	0.49	0.000642	10	0.001198	1.865	0.027892	1.79E-05	0.1	0.02337	1.01E-06	2.35E-07	0.07663	7.7E-07	6.458071	5.21E-06	1.849217
TFC-ULP	70	20	0.49	0.000642	14	0.001678	2.611	0.027892	1.79E-05	0.1	0.02912	1.01E-06	2.93E-07	0.07088	7.12E-07	13.61896	9.99E-06	2.603902
ESNA	15	20	0.49	0.000642	3	0.00036	0.560	0.027892	1.79E-05	0.1	0.06129	1.01E-06	6.16E-07	0.03871	3.89E-07	1.749971	1.3E-06	0.425522
ESNA	30	20	0.49	0.000642	6	0.000719	1.119	0.027892	1.79E-05	0.1	0.08364	1.01E-06	8.41E-07	0.01636	1.64E-07	3.062399	1.34E-06	0.927718
ESNA	50	20	0.49	0.000642	10	0.001198	1.865	0.027892	1.79E-05	0.1	0.06768	1.01E-06	6.8E-07	0.03232	3.25E-07	6.458071	2.78E-06	1.704143
ESNA	70	20	0.49	0.000642	14	0.001678	2.611	0.027892	1.79E-05	0.1	0.08693	1.01E-06	8.74E-07	0.01307	1.31E-07	13.61896	2.66E-06	2.275253
TFC-S	15	20	0.49	0.000642	3	0.00036	0.560	0.027892	1.79E-05	0.1	0.05189	1.01E-06	5.22E-07	0.04811	4.84E-07	1.749971	1.37E-06	0.465872
TFC-S	30	20	0.49	0.000642	6	0.000719	1.119	0.027892	1.79E-05	0.1	0.0598	1.01E-06	6.01E-07	0.0402	4.04E-07	3.062399	1.84E-06	0.852317
TFC-S	50	20	0.49	0.000642	10	0.001198	1.865	0.027892	1.79E-05	0.1	0.08023	1.01E-06	8.06E-07	0.01977	1.99E-07	6.458071	2.09E-06	1.754673
TFC-S	70	20	0.49	0.000642	14	0.001678	2.611	0.027892	1.79E-05	0.1	0.09022	1.01E-06	9.07E-07	0.00978	9.83E-08	13.61896	2.25E-06	2.570077

With these values of C_m determined from the concentration polarization model, MINEQL$^+$ simulations were run for the different membranes. Figures 8.20 to 8.23 show that Ba^{2+}-CO_3^{2-}, Ba^{2+}-SO_4^{2-}, Ca^{2+}-CO_3^{2-}, Ca^{2+}-SO_4^{2-}, Mg^{2+}-CO_3^{2-}, and Mg^{2+}-Ca^{2+}-CO_3^{2-} are major possible scalants in the system. Figures 8.20 (a)-(d) and 8.23 (a)-(d) show that the amount of $CaCO_3$ (argonite, calcite), $MgCO_3$ (magnesite), and $CaMg(CO_3)_2$ (dolomite) increase, and $CaSO_4$ (anhydrite) remains fixed with increasing pH, reflecting the precipitation potentials of these compounds for the two RO (LFC-1, TFC-ULP) and two NF (ESNA, TFC-S) membranes. All precipitation potentials are shown over the pH range of 5 to 12.

The degree of saturation can be determined by using the solid saturation index (SI) (Her et al., 2000).

$$SI = \log\left(\frac{Q}{K_{SP}}\right) \tag{8.9}$$

where Q is the ion activity product (e.g. $\{Ca^{2+}\}\{CO_3^{2-}\}$) and K_{SP} is the solubility product of the inorganic foulants. A SI value of zero indicates equilibrium. Values of SI are negative when inorganic foulants are undersaturated and positive when over-saturated at the membrane surface. Figures 8.20 to 8.23 show the formation of various inorganic foulants at different pH conditions for CRW. Considering that the pH range of CRW is 7.8 to 8.0, the following inorganic foulants may be formed: i) calcite ($CaCO_3$) formed at pH above 8, and at pH 8 for a J_o/k of 2.71 (recovery 70%); however, for NF membranes, the risks of calcite saturation are less than those for RO, ii) at pH 5 to 12, the SI of anhydrite ($CaSO_4$) is near zero, and iii) argonite ($CaCO_3$), magnesite ($MgCO_3$), and dolomite ($CaMg(CO_3)_2$) are formed at higher pH (above 9.5).

The results from MINEQL$^+$ simulations can be used as a guideline in the study of the formation of various scalant that may cause fouling of membranes.

SUMMARY

- Antiscalants, composed of phosphonate groups, were effective in the inhibition of $CaCO_3$ formation. Especially, the inhibition mechanism was proven, i.e. the antiscalant inhibited the scale formation.
- As results of bench scale tests with antiscalant, it was shown to effectively inhibit scale formation and decrease flux decline at the same recovery condition with a relatively low concentration of 3 ppm. However, if a high concentration of antiscalant is used, biofouling can potentially occur because biomass in natural waters may use antiscalants as their nutrient source.
- Four candidate membranes were evaluated at various recoveries; two RO membranes (LFC-1, TFC-ULP) showed higher flux decline and rejection than two NF membranes (ESNA, TFC-S). Also, rejection efficiencies decreased as recovery increased.
- Calcite ($CaCO_3$) scale was dominant at the membrane surface and detected by various measurements (FTIR, XRD, SEM, NAA). Also, the effectiveness of the antiscalant was confirmed by SEM.

- The concentration of solute at the membrane surface can be predicted by the constant (K), a semiempirical constant. Considering results for each ion and four membranes, most K values of ions with RO membranes approached 1, while values for NF membranes were smaller than for RO membranes.

By MINEQL$^+$ simulations, an understanding of possible scale fouling can contribute into an improvement in membrane fouling control.

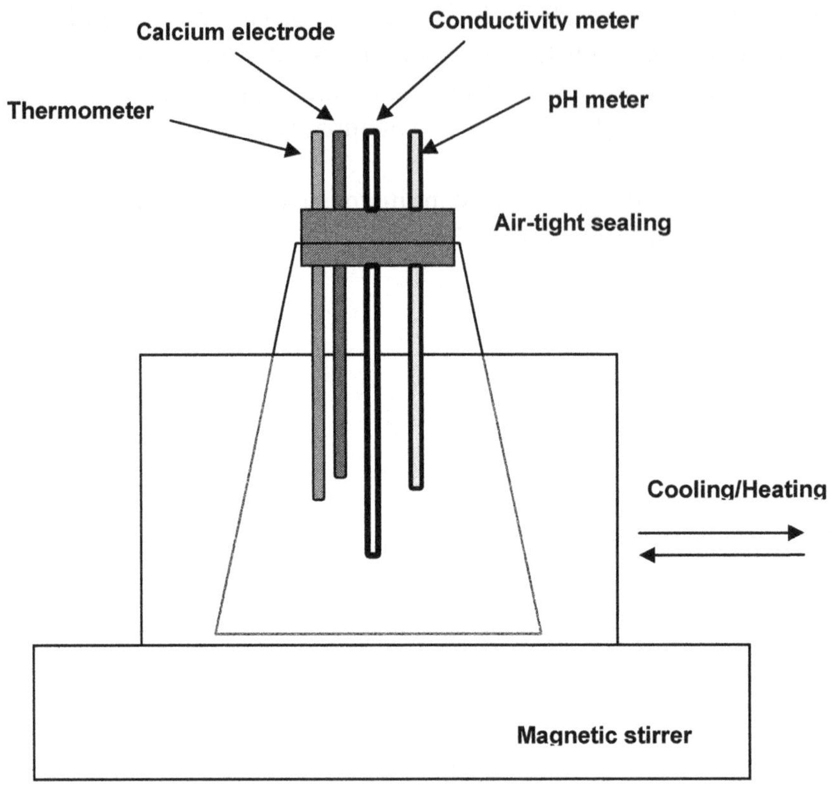

Figure 8.1 Schematic of precipitation inhibiting test unit

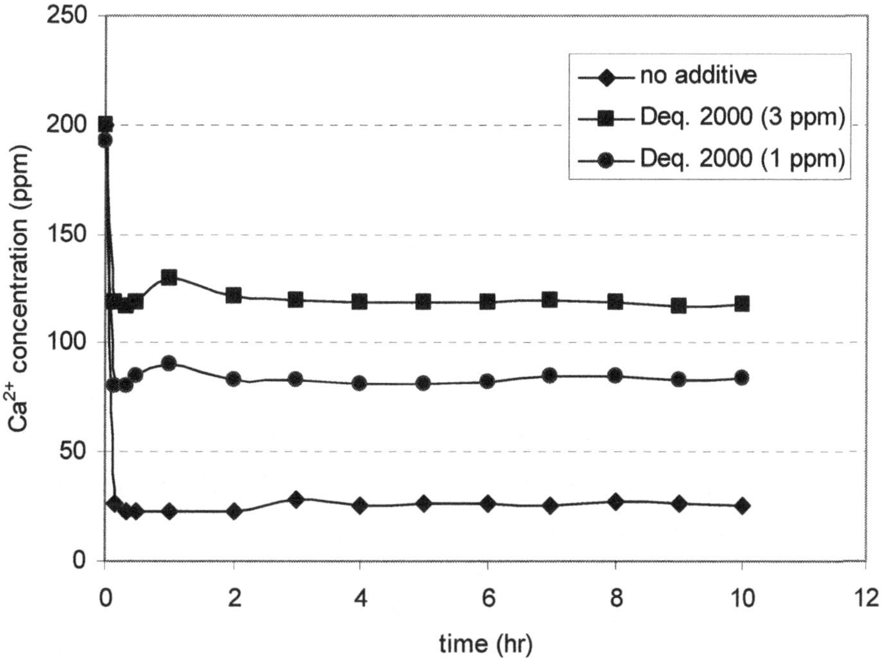

(a) Dequest 2000

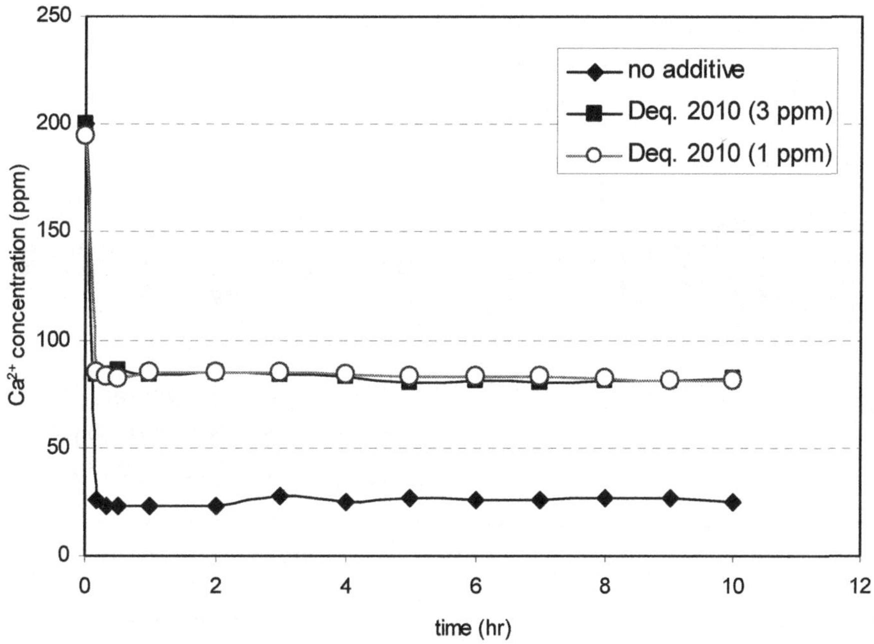

(b) Dequest 2010

Figure 8.2 Inhibition of $CaCO_3$ precipitation

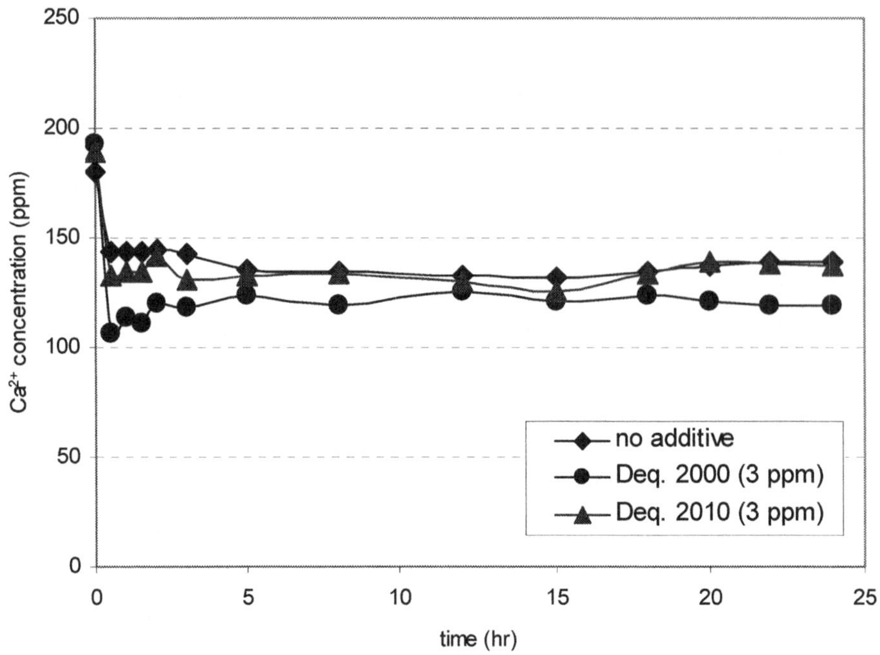

(a) pH 4

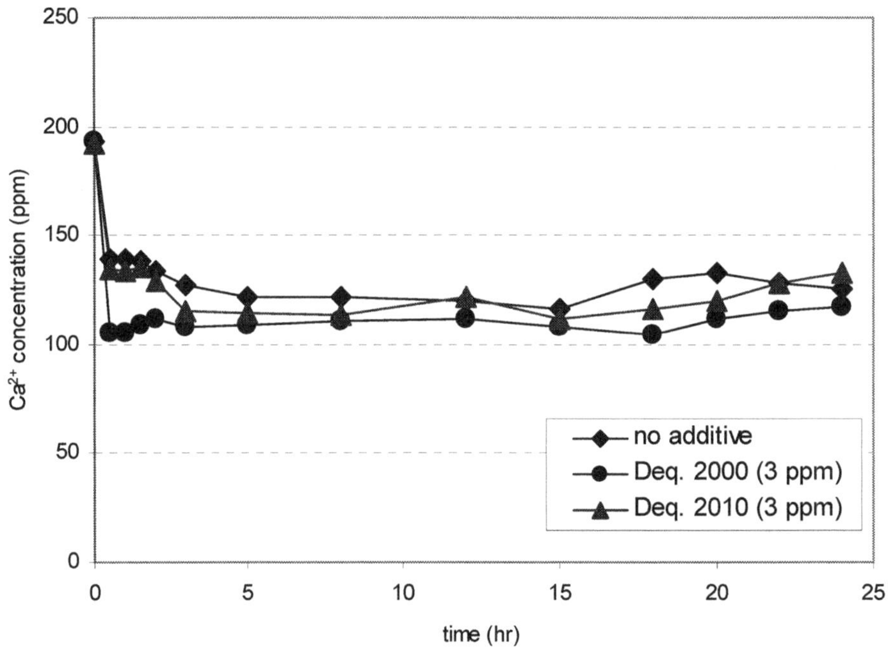

(b) pH 6

(continued)

Figure 8.3 Inhibition effect of pH at conductivity ~78 mS/m

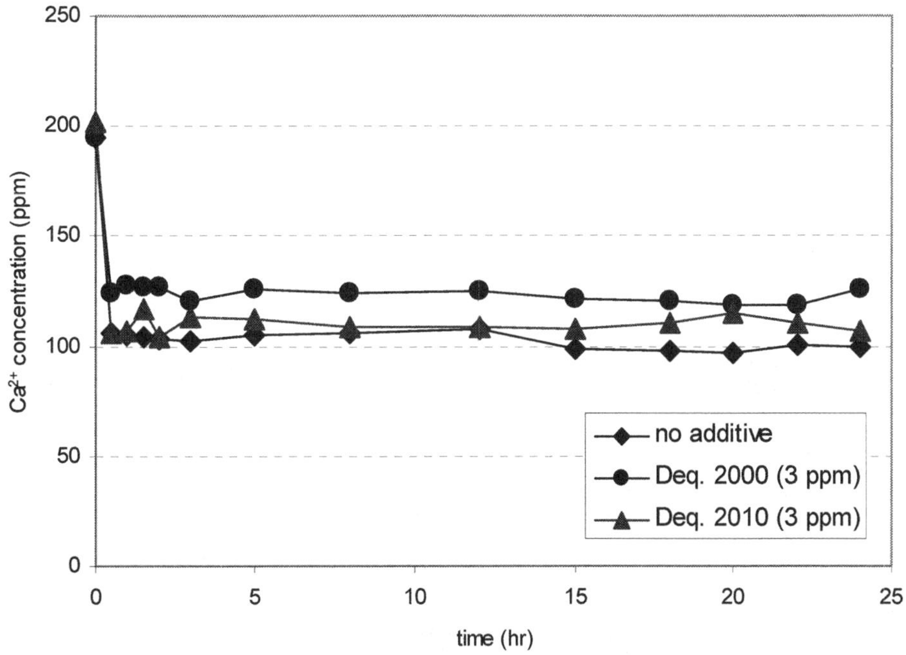

(c) pH 8

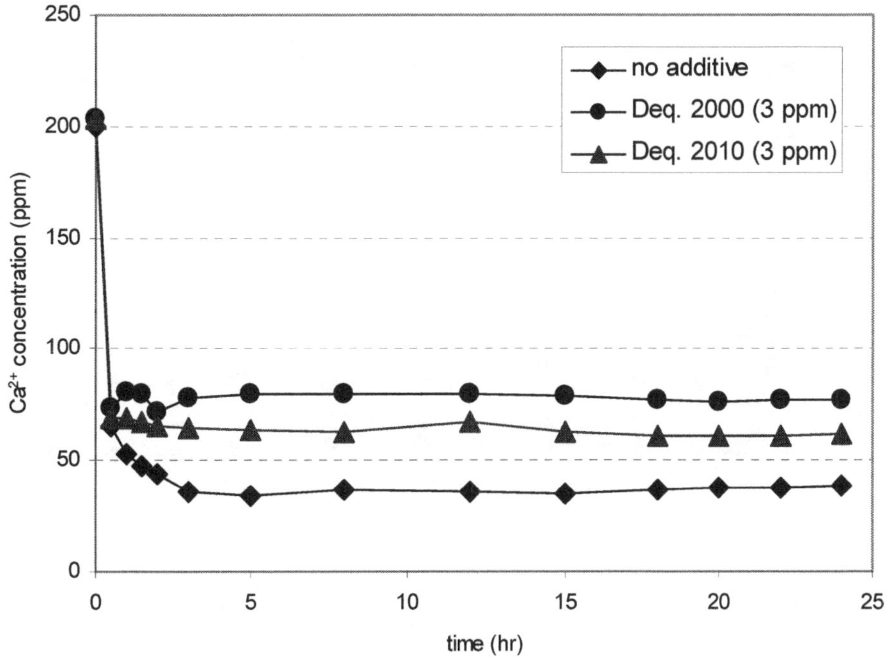

(d) pH 10

Figure 8.3 (Continued)

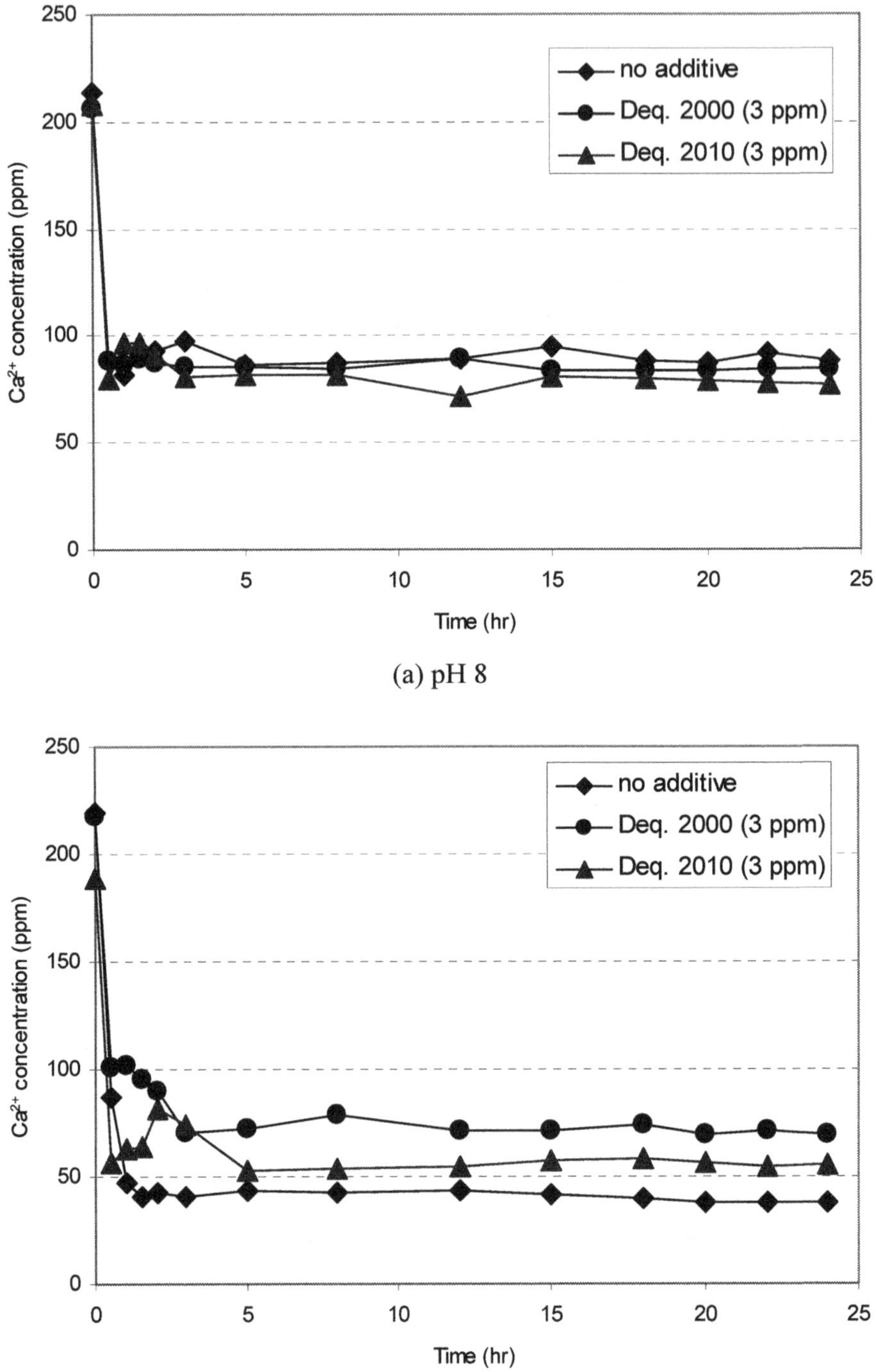

(a) pH 8

(b) pH 10

Figure 8.4 Inhibition effect of pH at conductivity ~115 mS/m

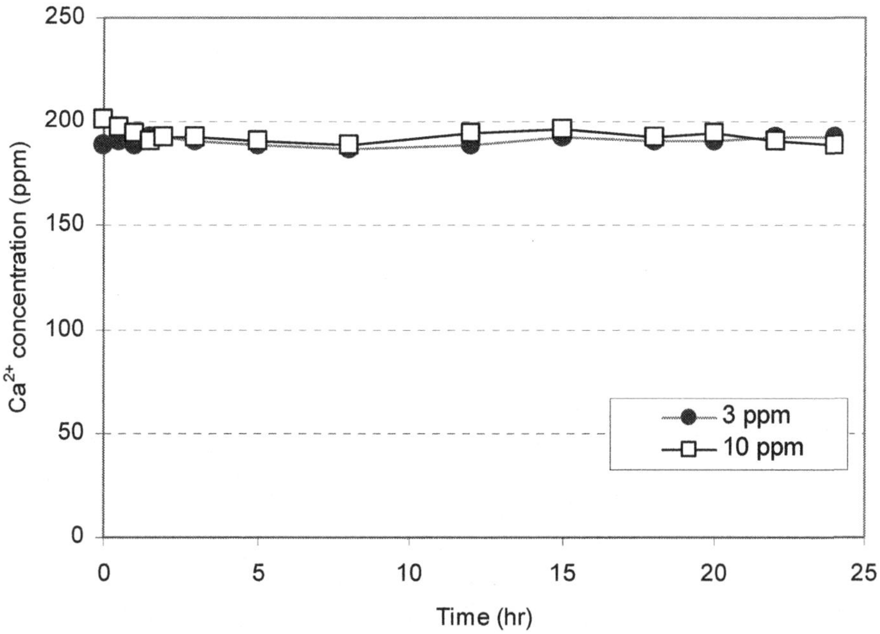

Figure 8.5 Expermemtal investigation of Ca^{2+} and antiscalant ([Ca^{2+}]=200 ppm, Dequest 2000, pH 8, and conductivity 78 mS/m)

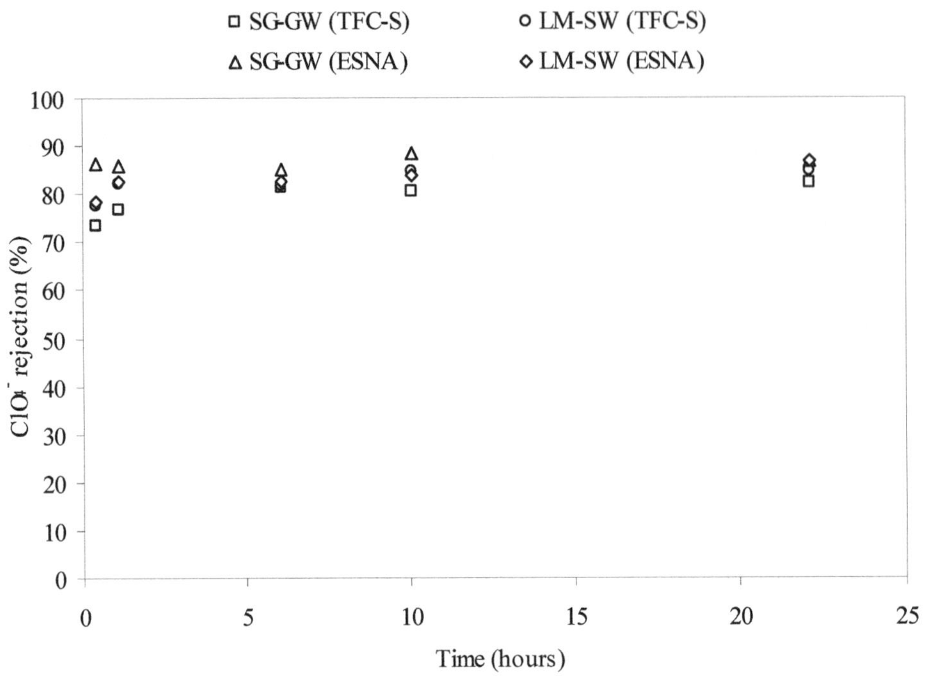

(a) SG-GW and ML-SW

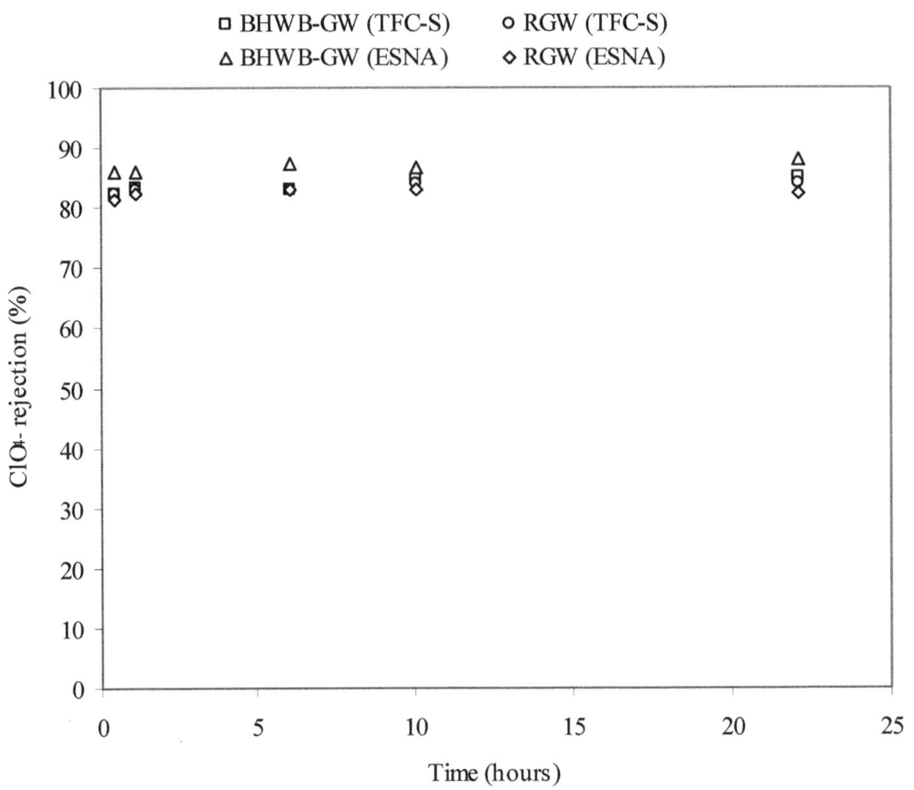

(b) BHWB-GW and RGW

Figure 8.6 Comparison of perchlorate rejection with various natural waters for the TFC-S and ESNA membranes (recovery 2.6% and J_o/k ratio 0.5)

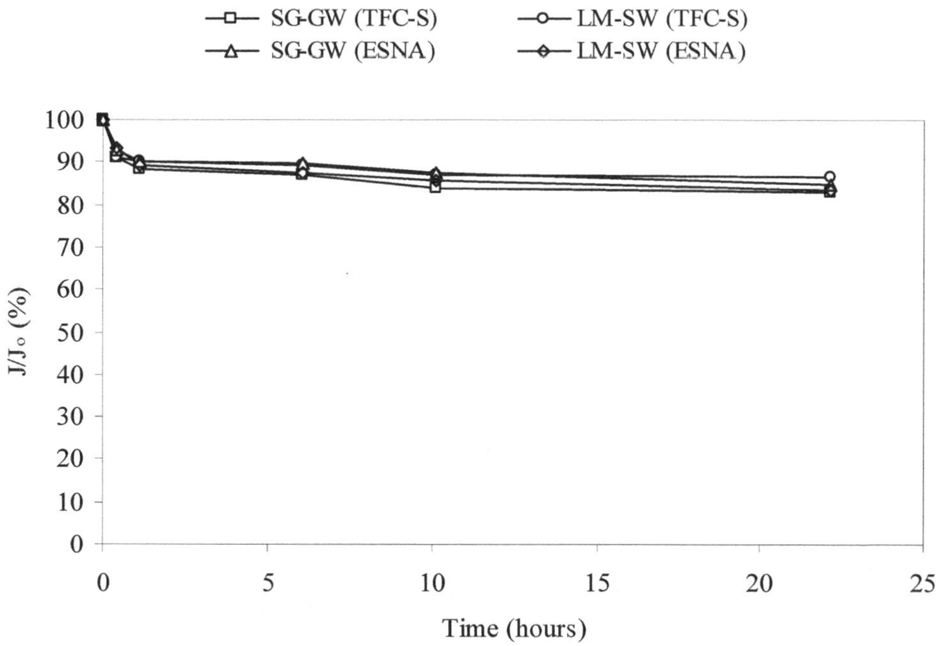

(a) SG-GW and ML-SW

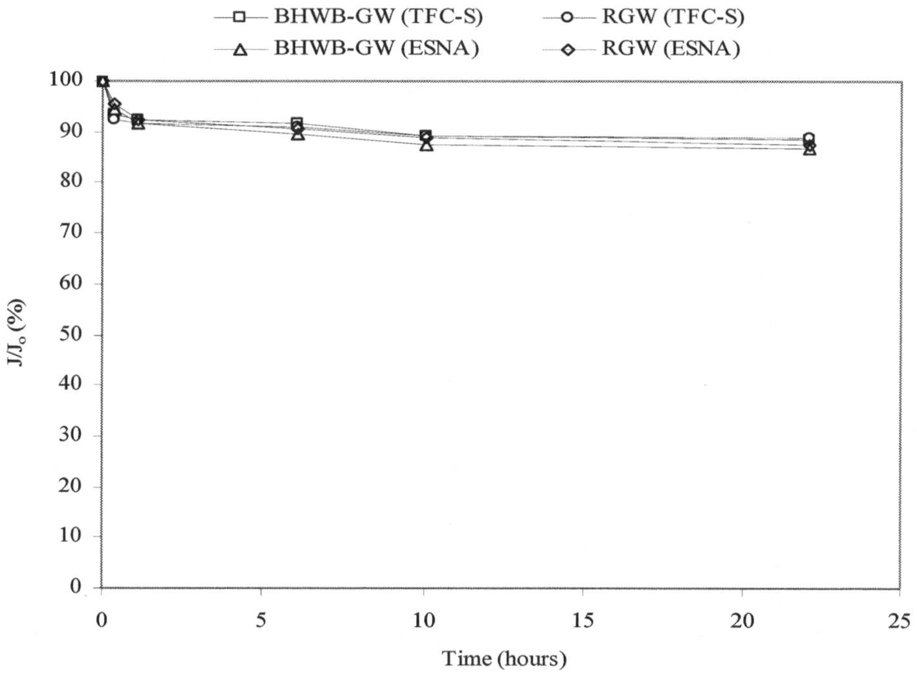

(b) BHWB-GW and RGW

Figure 8.7 Comparison of flux-decline with various natural waters for the TFC-S and ESNA membranes (recovery 2.6% and J_o/k ratio 0.5)

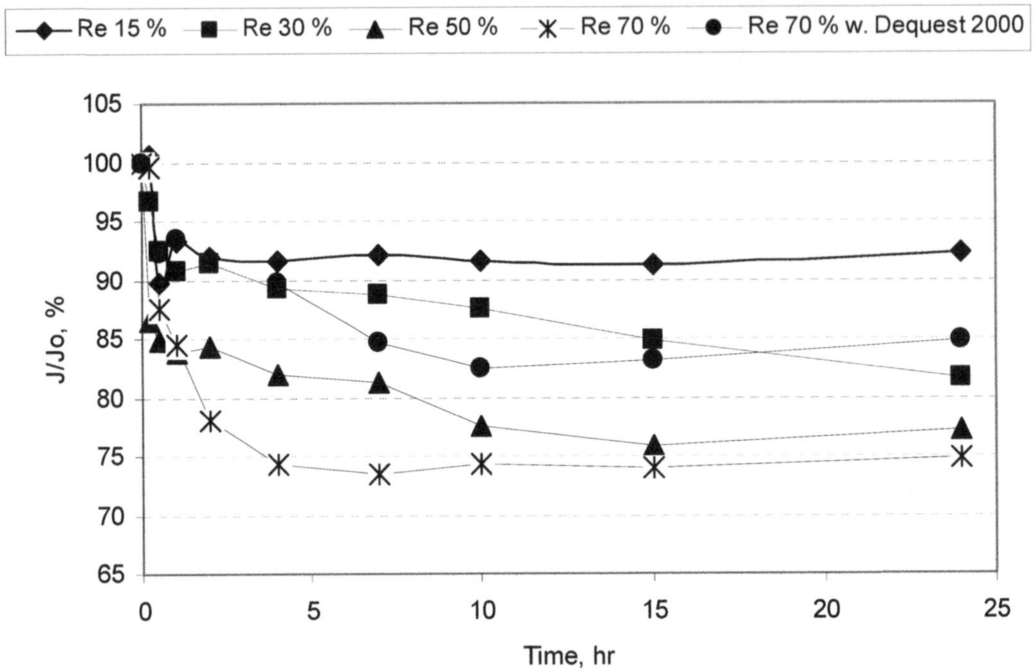

(a) LFC-1

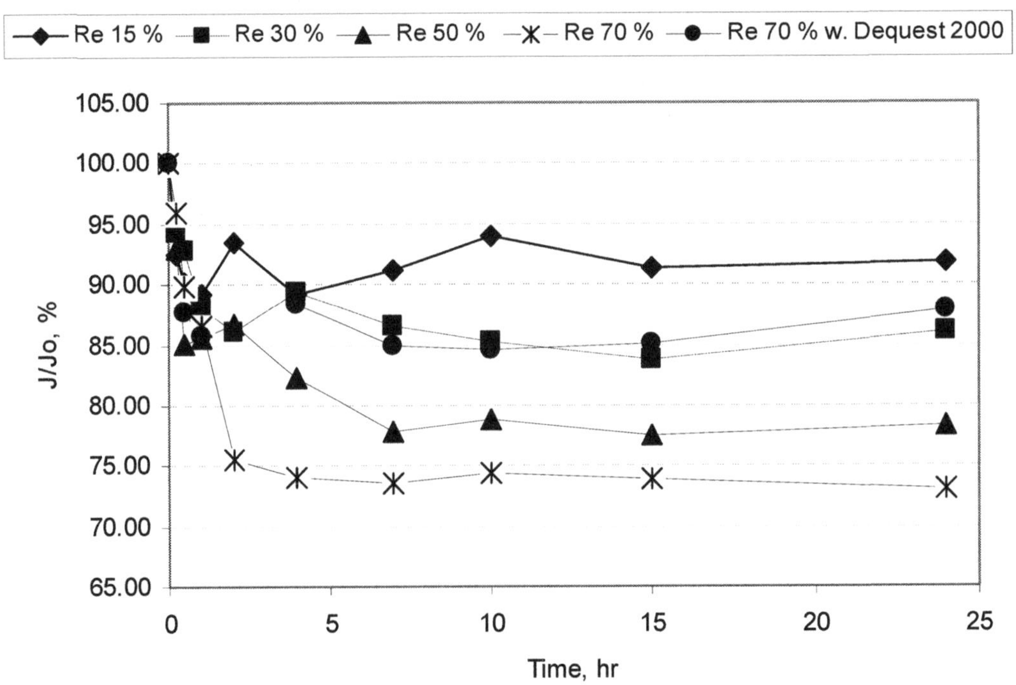

(b) TFC-ULP

(continued)

Figure 8.8 Flux-decline of four different membrane for CRW

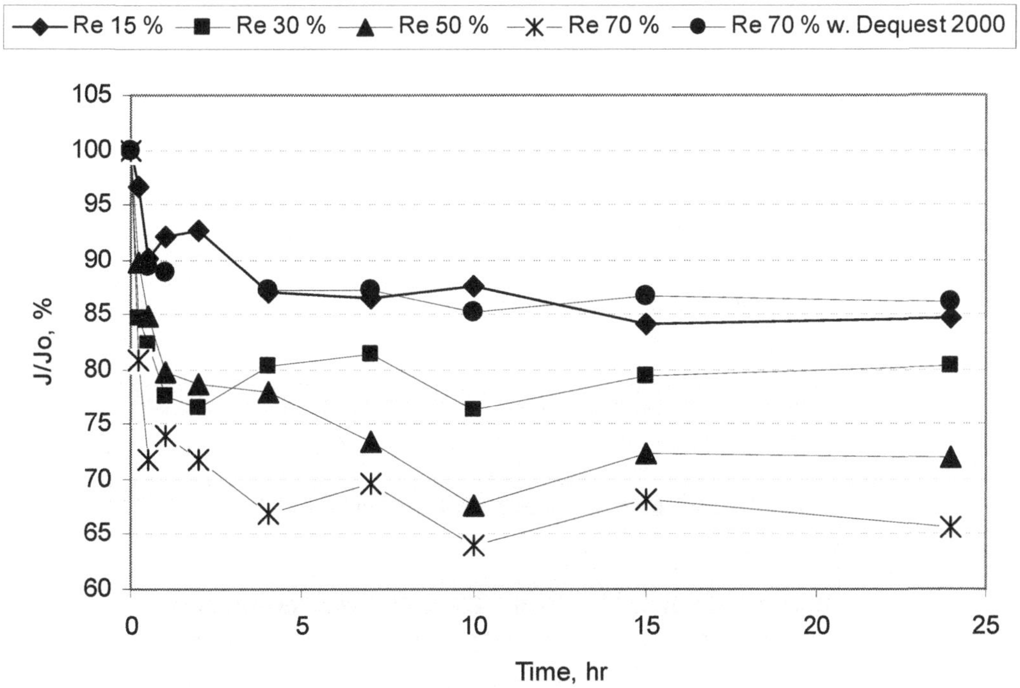

(c) ESNA

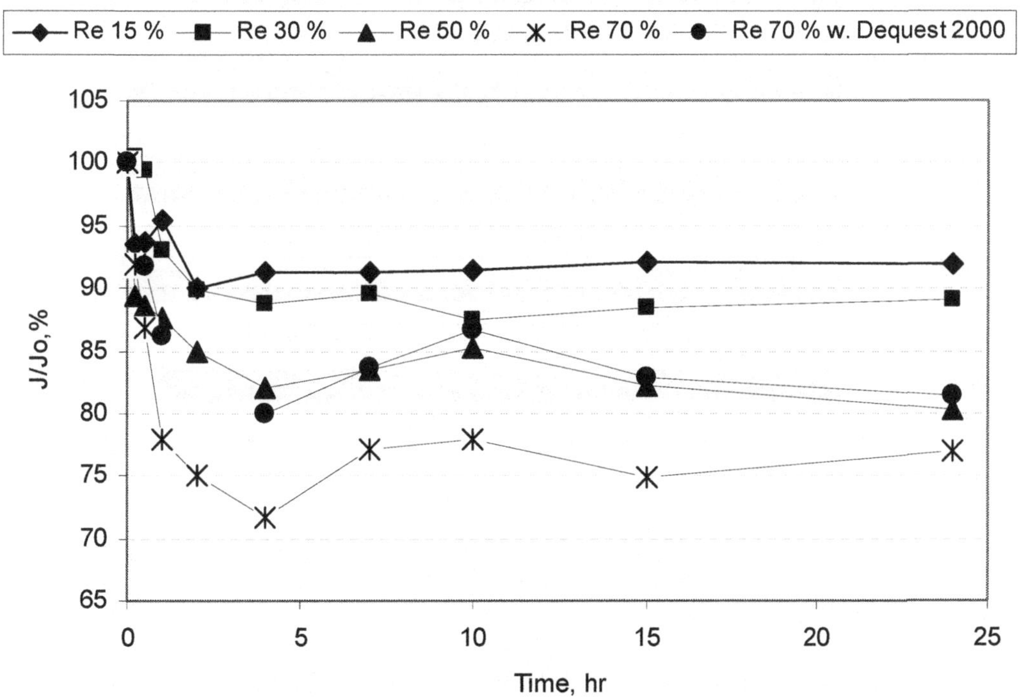

(d) TFC-S

Figure 8.8 (Continued)

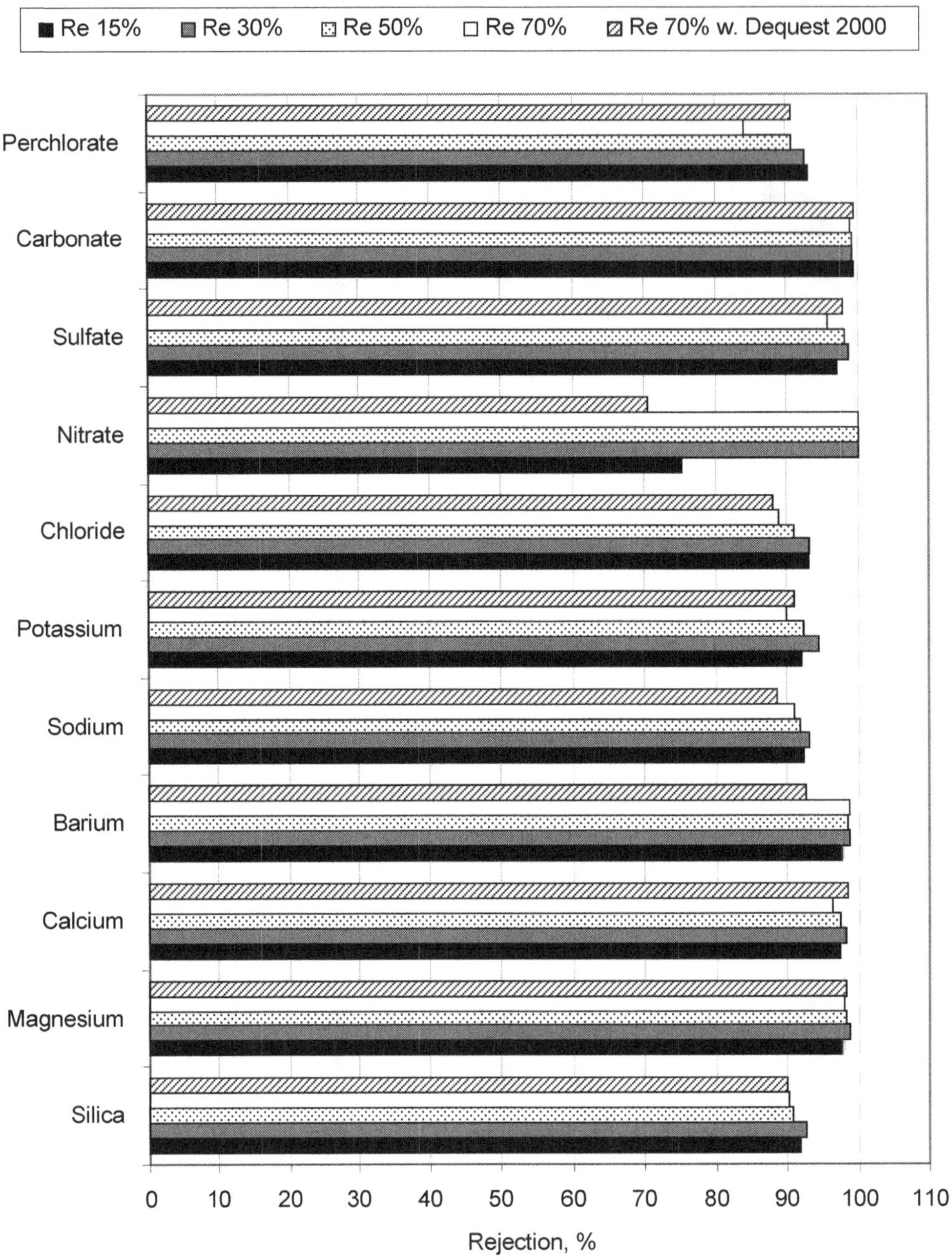

(a) LFC-1

(continued)

Figure 8.9 Cations/Anions rejection of CRW at recovery 15% to 70%

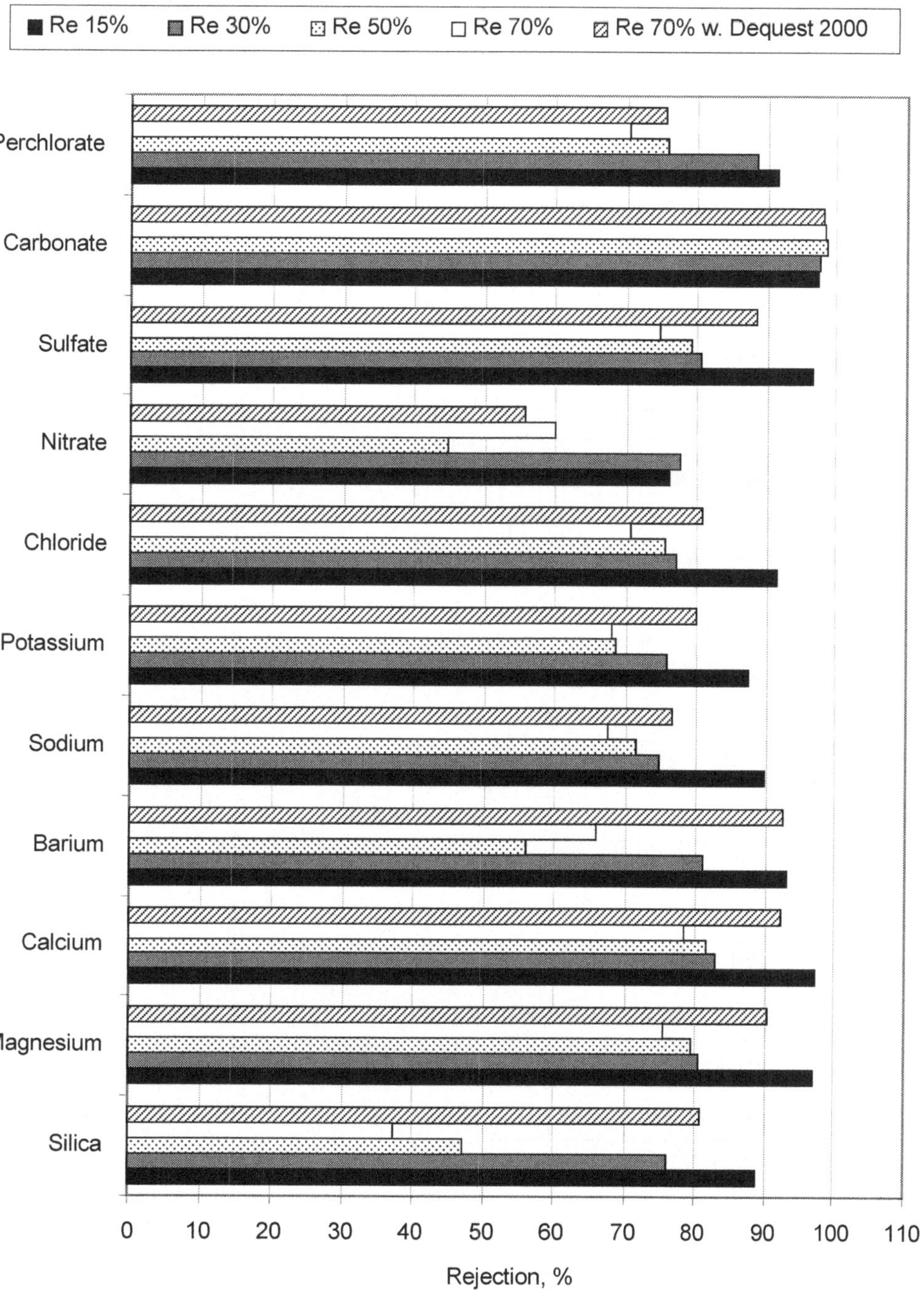

(b) TFC-ULP

(continued)

Figure 8.9 (Continued)

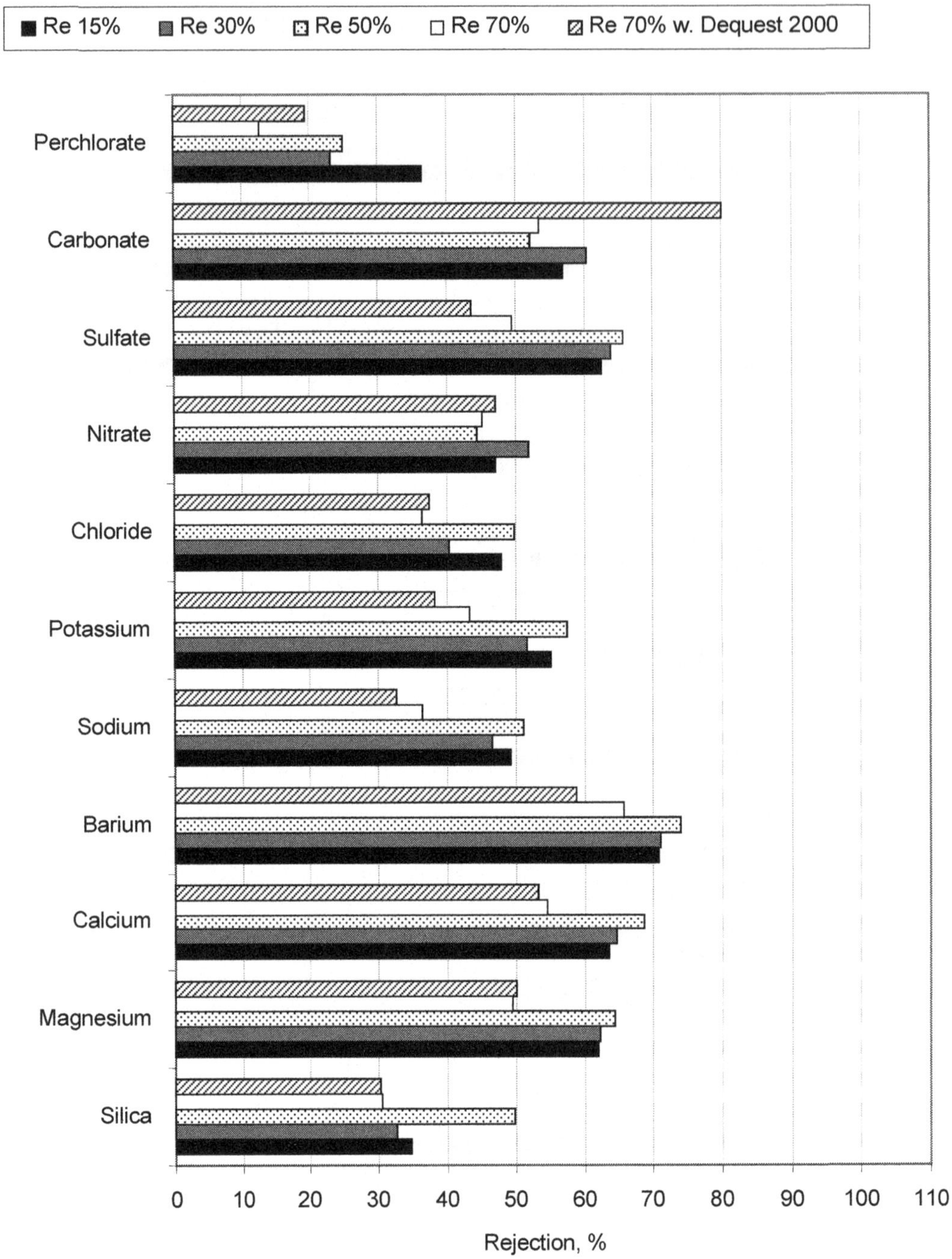

(c) ESNA

(continued)

Figure 8.9 (Continued)

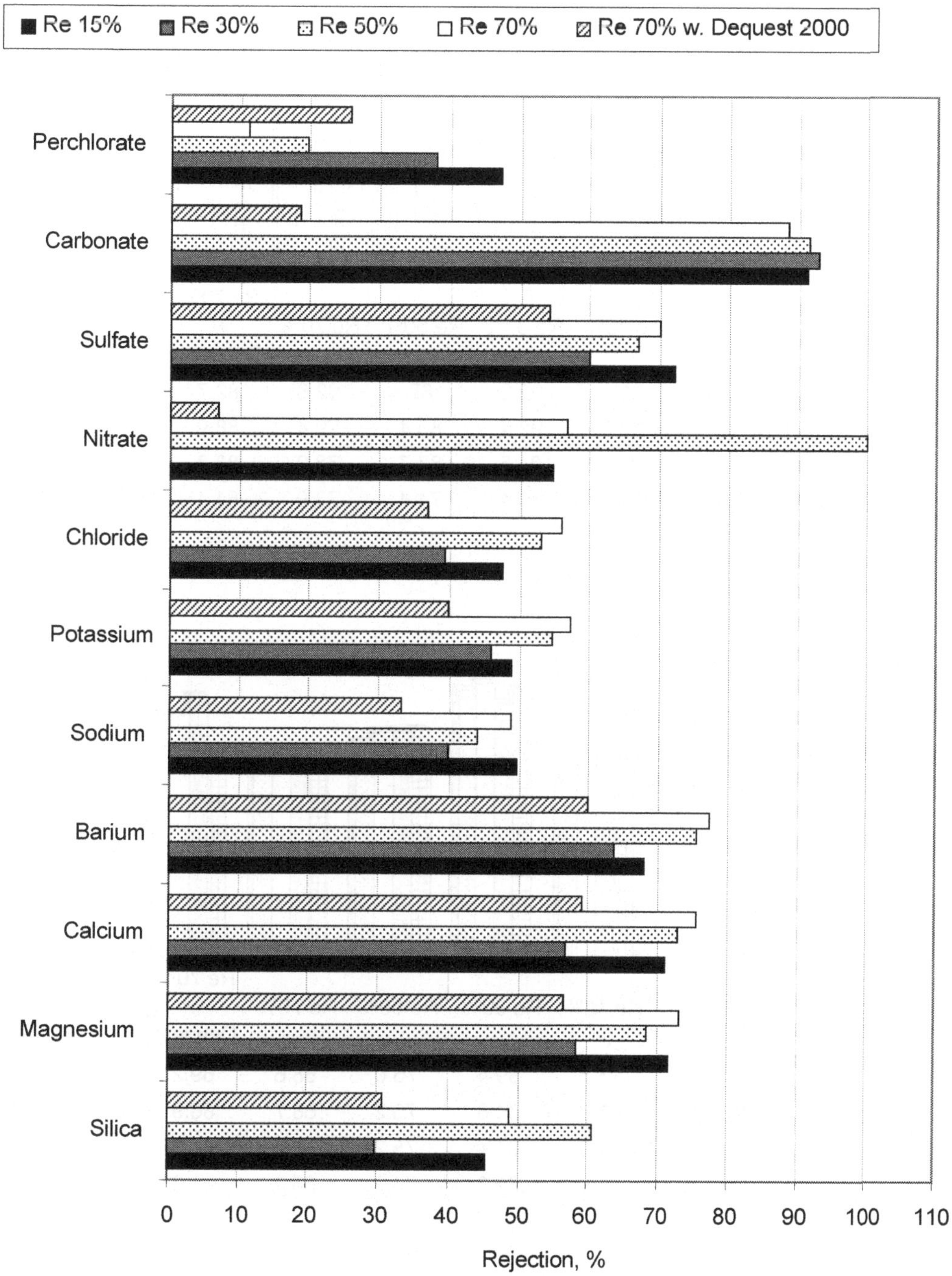

(d) TFC-S

Figure 8.9 (Continued)

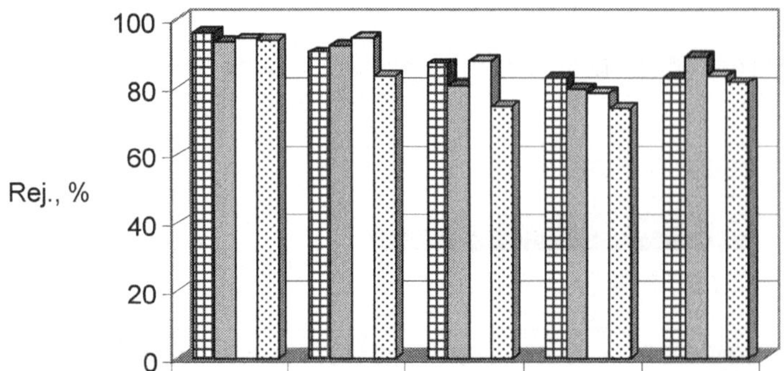

	Re 15%	Re 30%	Re 50%	Re 70%	Re 70% w. Dequest
UV A rej.	96.2	90.2	87	82.6	82.7
DOC rej.	93.3	92.4	80.4	79.4	88.9
TDS rej.	94.2	94.5	87.7	78.2	83.3
Cond., rej.	94	83.4	74.4	73.9	81.4

(a) LFC-1

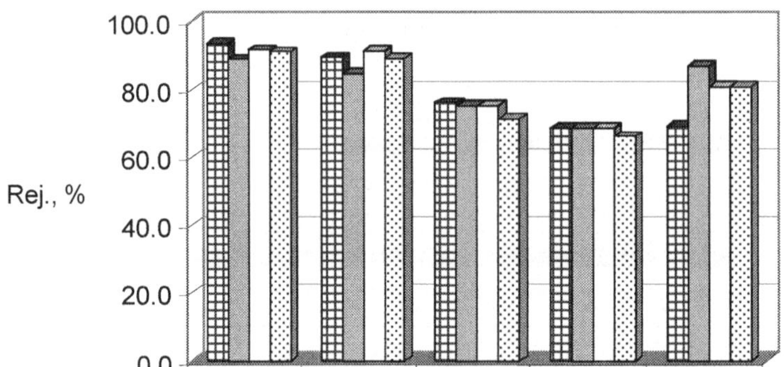

	Re 15%	Re 30%	Re 50%	Re 70%	Re 70% w. Dequest
UV A rej.	93.5	89.4	76.0	68.6	69.2
DOC rej.	88.8	84.8	75.2	68.7	86.9
TDS rej.	91.6	91.4	75.2	68.5	80.7
Cond., rej.	91.0	89.0	71.2	66.1	80.8

(b) TFC-ULP

(continued)

Figure 8.10 Rejection of UV A_{254}, DOC, TDS, and conductivity of CRW at recovery 15 to 70%

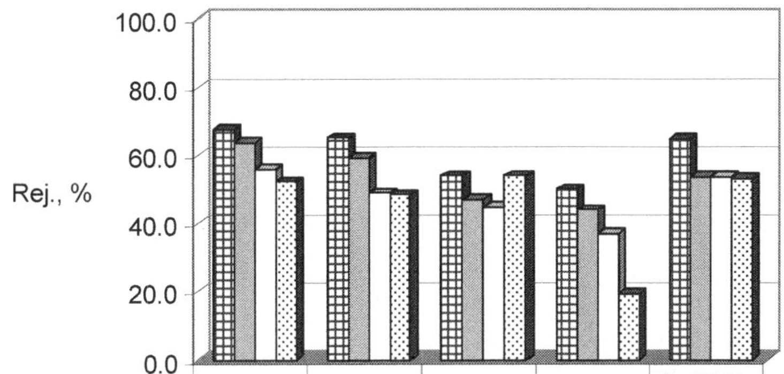

	Re 15%	Re 30%	Re 50%	Re 70%	Re 70% w. Dequest
UV A rej.	67.5	65.0	53.9	50.0	64.6
DOC rej.	63.6	59.2	47.1	43.9	53.7
TDS rej.	55.7	48.8	44.6	37.1	53.5
Cond., rej.	52.1	48.4	53.9	19.4	53.1

(c) ESNA

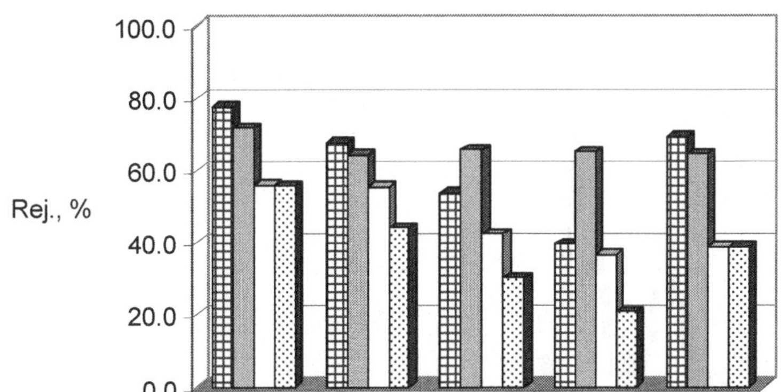

	Re 15%	Re 30%	Re 50%	Re 70%	Re 70% w. Dequest
UV A rej.	77.5	67.4	53.7	39.5	69.2
DOC rej.	71.7	64.0	65.5	65.1	64.3
TDS rej.	55.7	55.2	42.2	36.3	38.7
Cond., rej.	55.5	43.9	30.1	20.6	38.5

(d) TFC-S

Figure 8.10 (Continued)

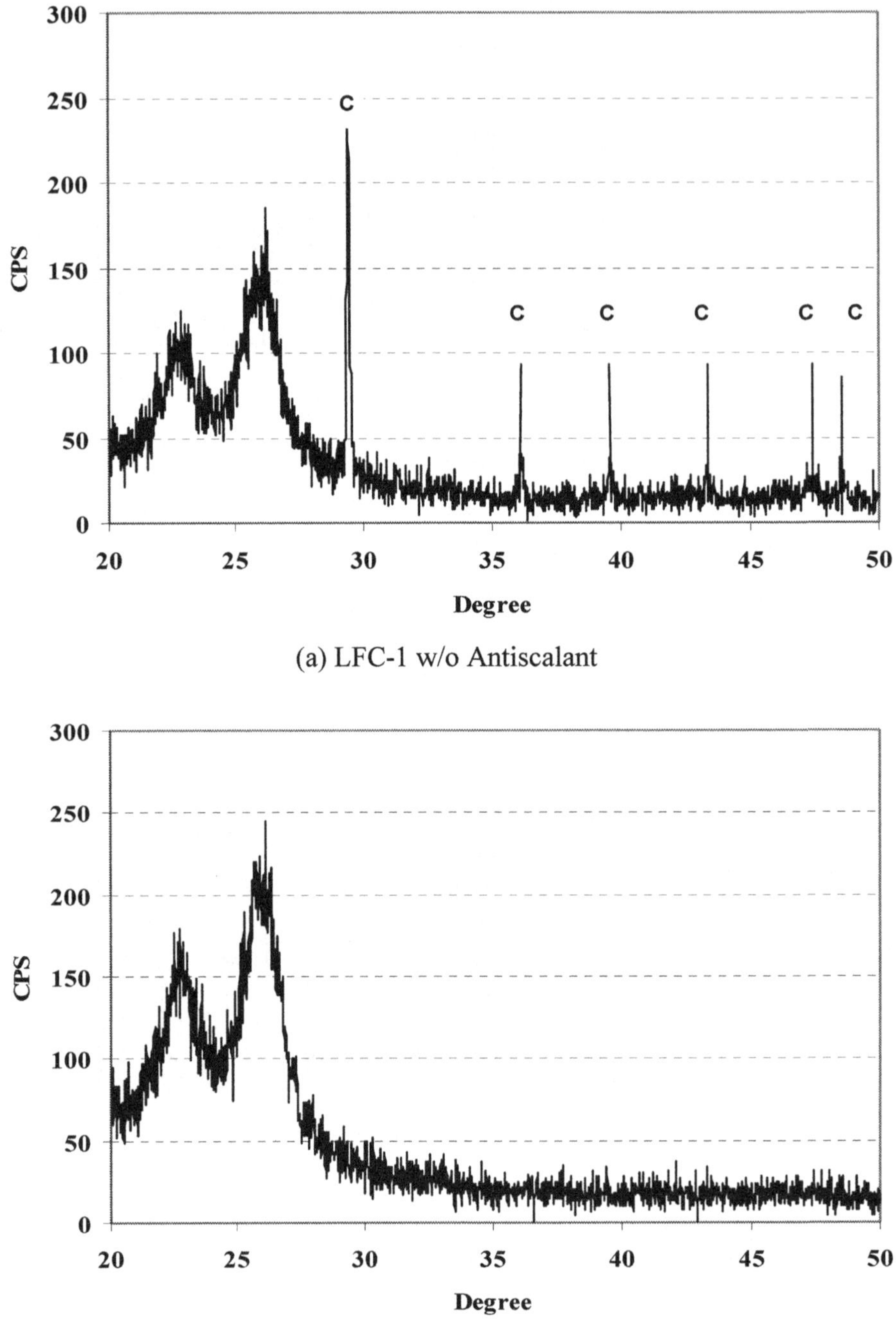

(a) LFC-1 w/o Antiscalant

(b) LFC-1 w/ Antiscalant

(continued)

Figure 8.11 Determination of calcite ($CaCO_3$) based on X-ray diffraction (XRD) analysis at recovery 70%

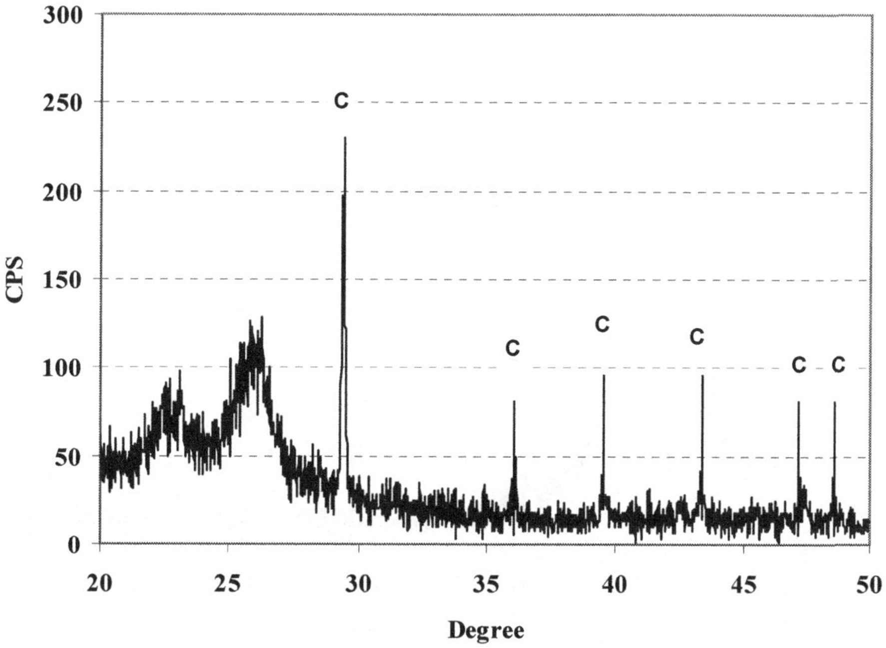

(c) TFC-ULP w/o Antiscalant

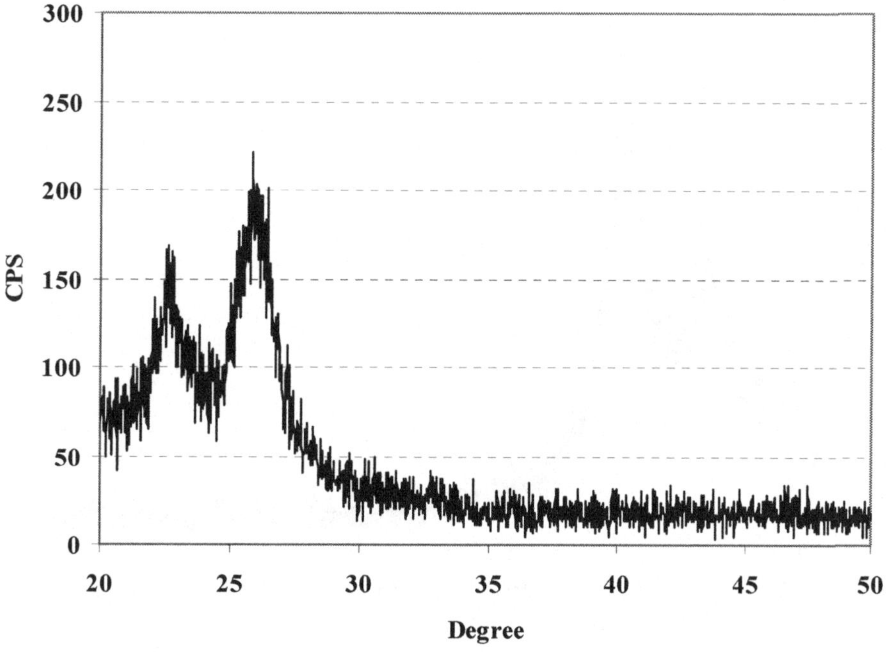

(d) TFC-ULP w/ Antiscalant

(continued)

Figure 8.11 (Continued)

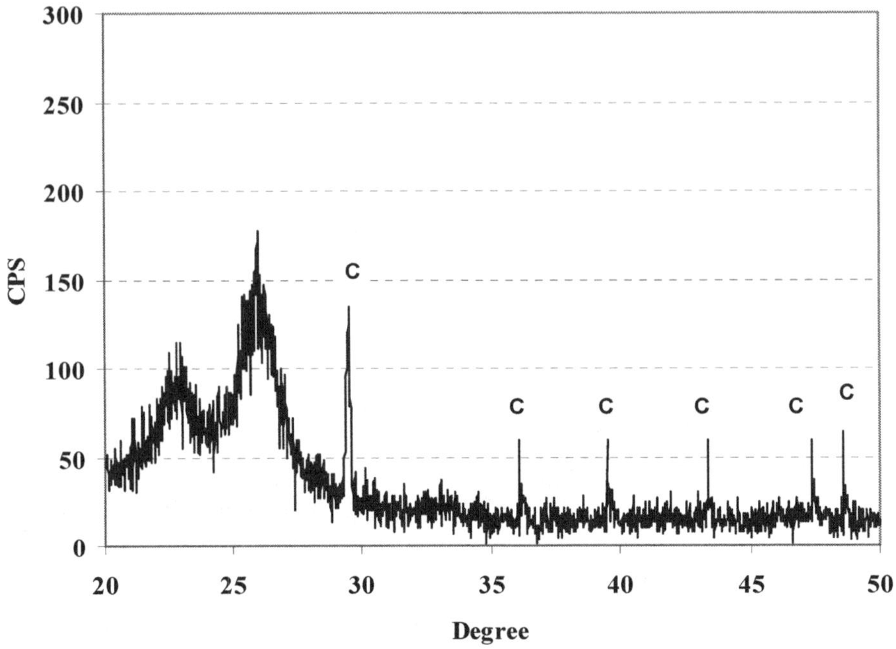

(e) ESNA w/o Antiscalant

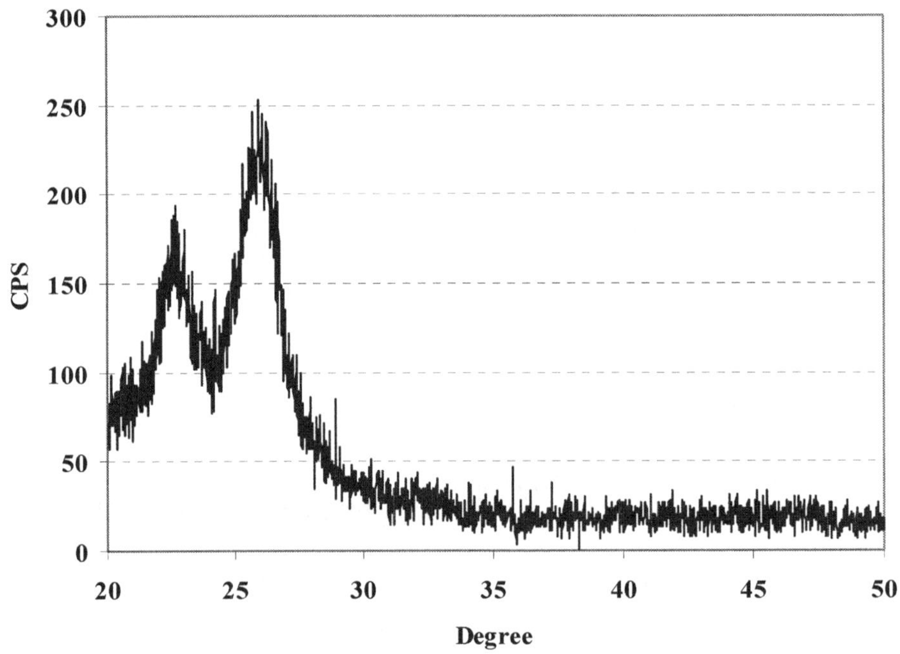

(f) ESNA w/ Antiscalant

(continued)

Figure 8.11 (Continued)

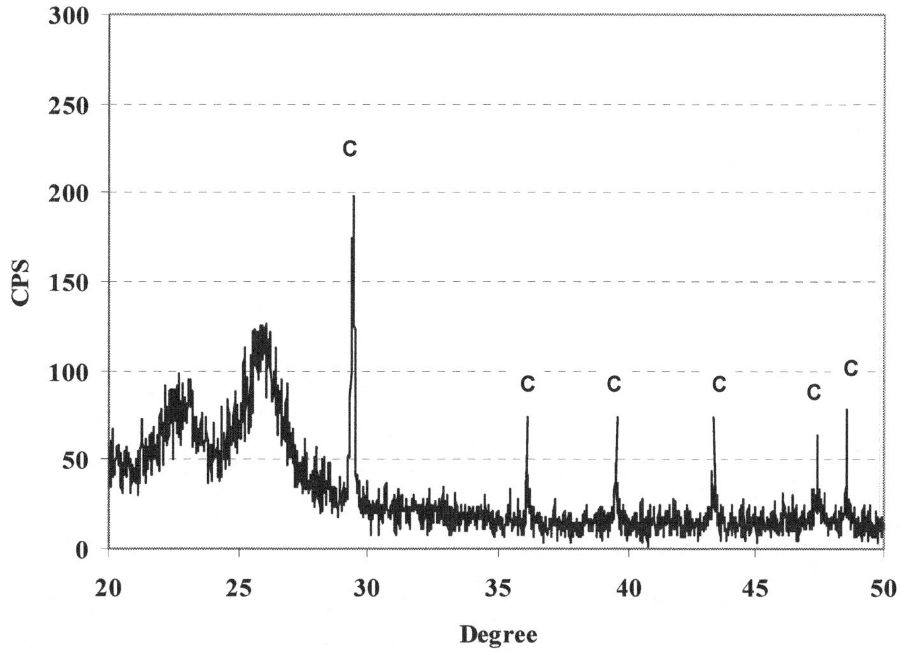

(g) TFC-S w/o Antiscalant

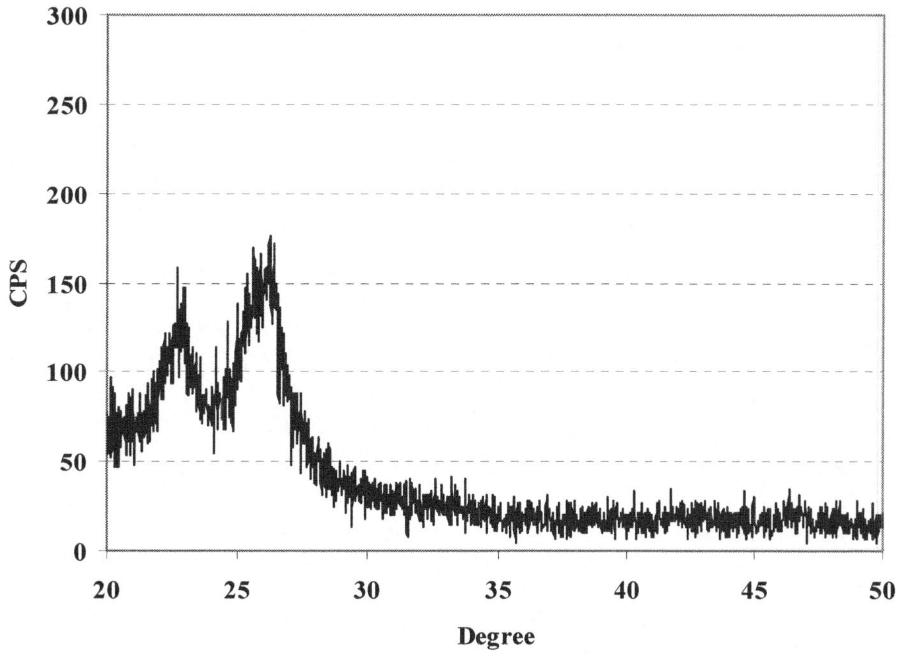

(h) TFC-S w/ Antiscalant

Figure 8.11 (Continued)

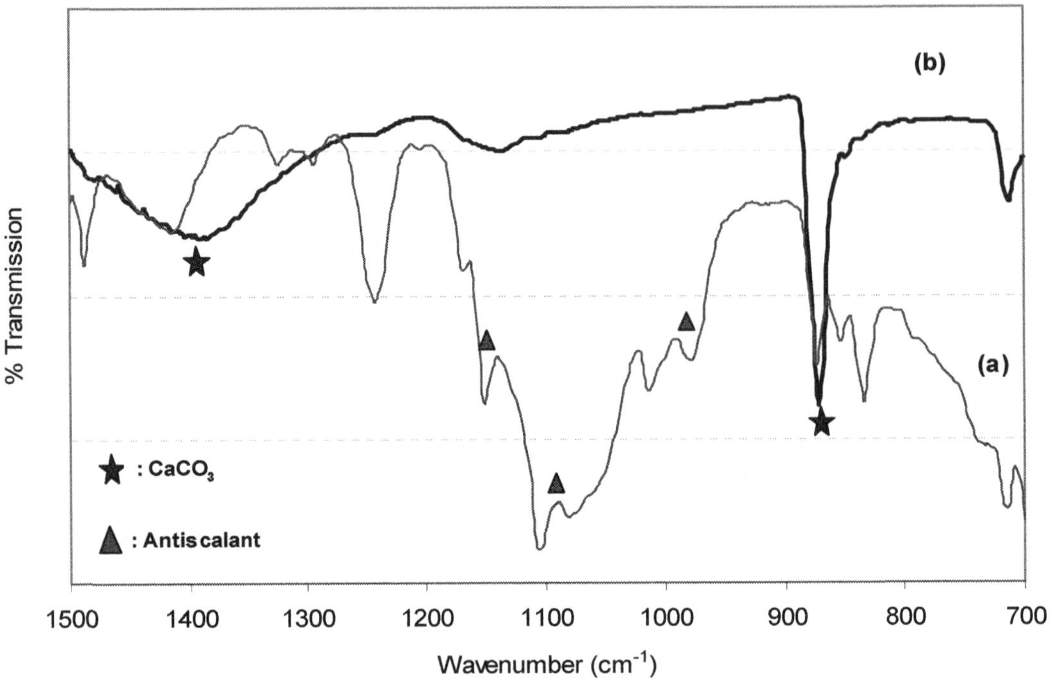

(a) LFC-1

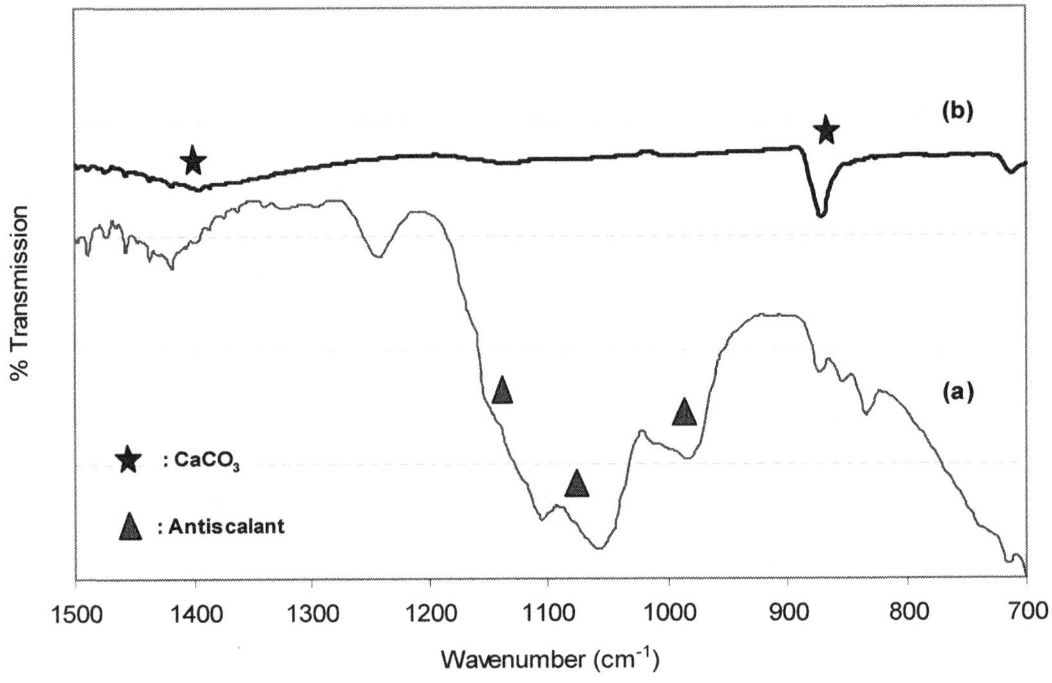

(b) TFC-ULP

Figure 8.12 Infrared transmission spectra of fouled membrane surface at recovery 70% with/without antiscalant

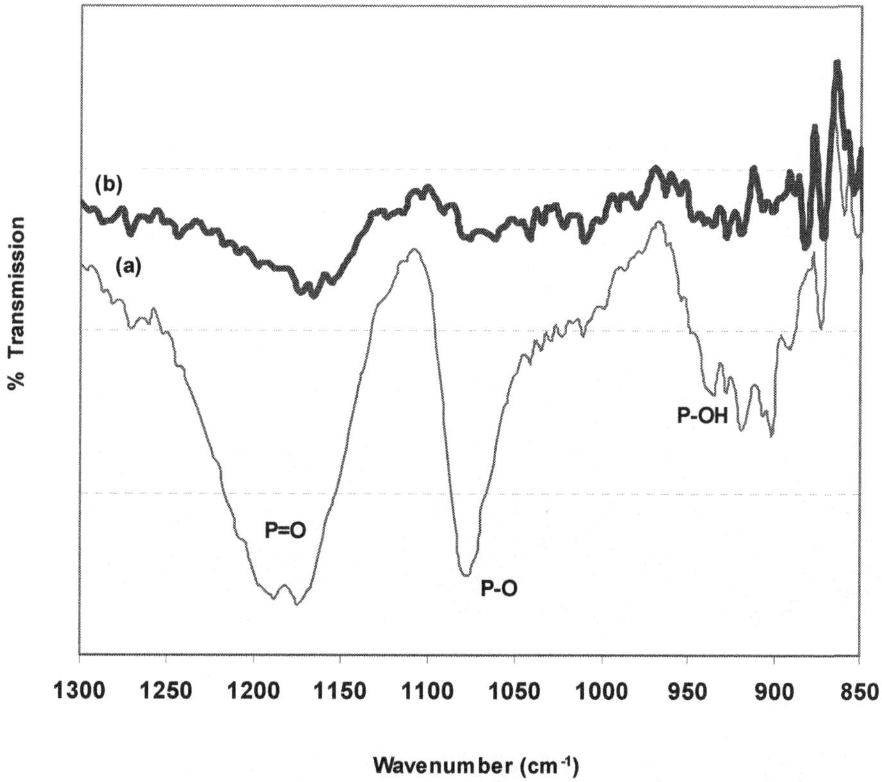

Figure 8.13 Infrared transmission spectra of antiscalant (Dequest 2000): (a) 1000 ppm (aqueous solution) and (b) 20 ppm (aqueous solution)

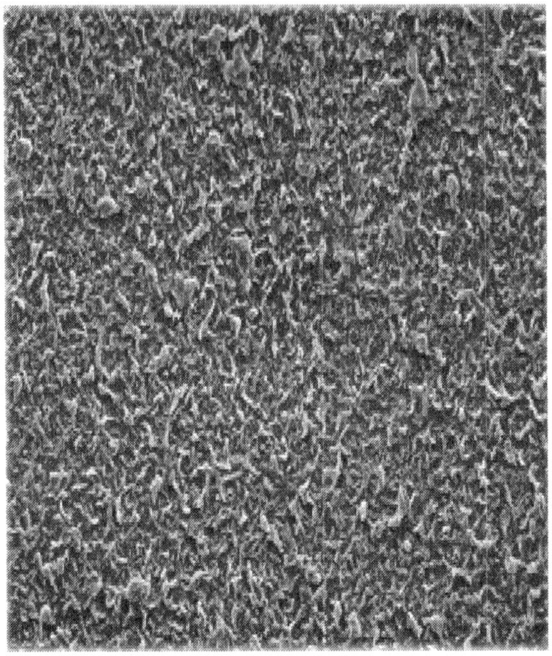

(a) Clean

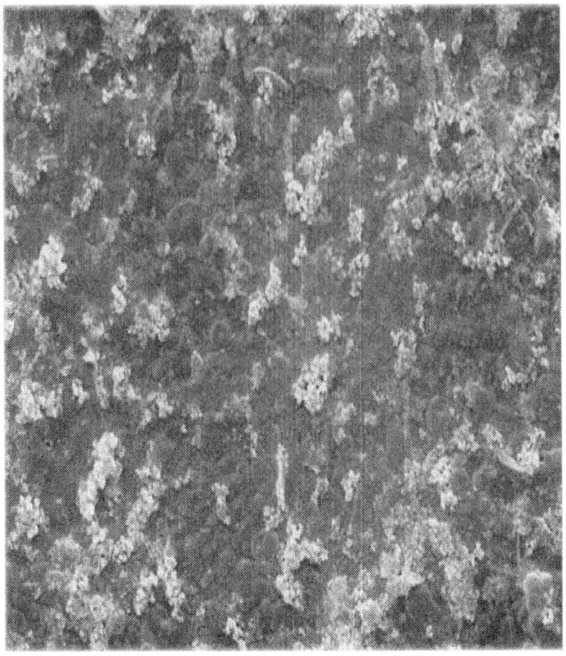

(b) at Recovery 15% (24 hr)

(c) at Recovery 70% (24 hr)

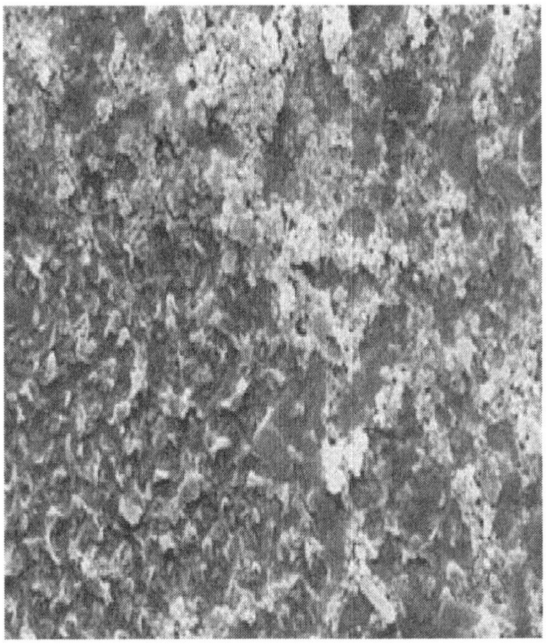

(d) at Recovery 70% with Antiscalant (24 hr)

Figure 8.14 Scanning electron micrographs of $CaCO_3$ grown on the LFC-1 membrane surface

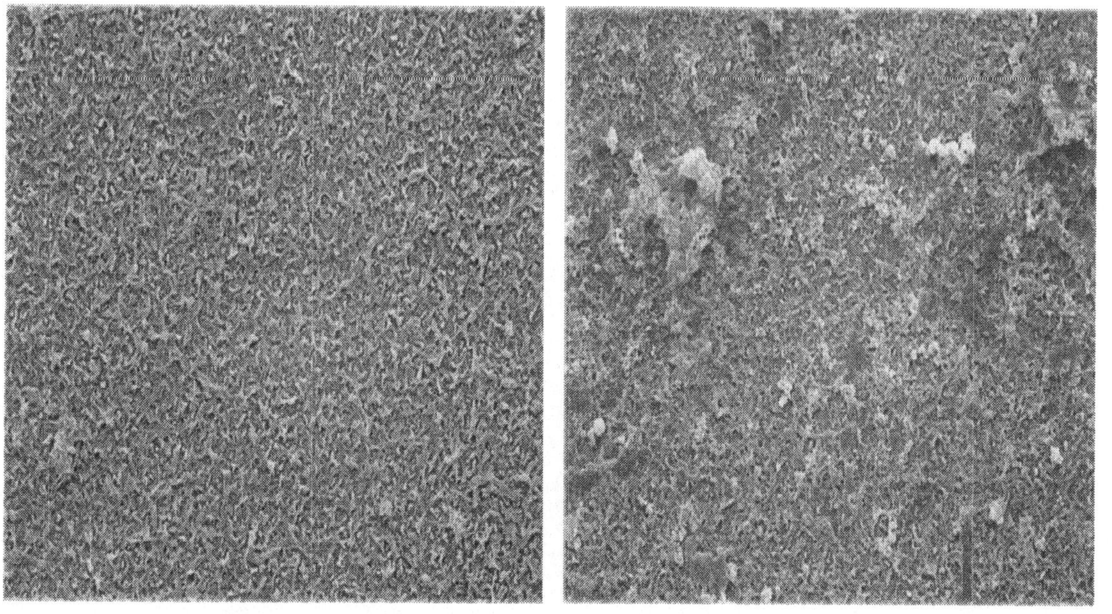

(a) Clean (b) at Recovery 15% (24 hr)

(c) at Recovery 70% (24 hr) (d) at Recovery 70% with Antiscalant (24 hr)

Figure 8.15 Scanning electron micrographs of $CaCO_3$ grown on the TFC-ULP membrane surface

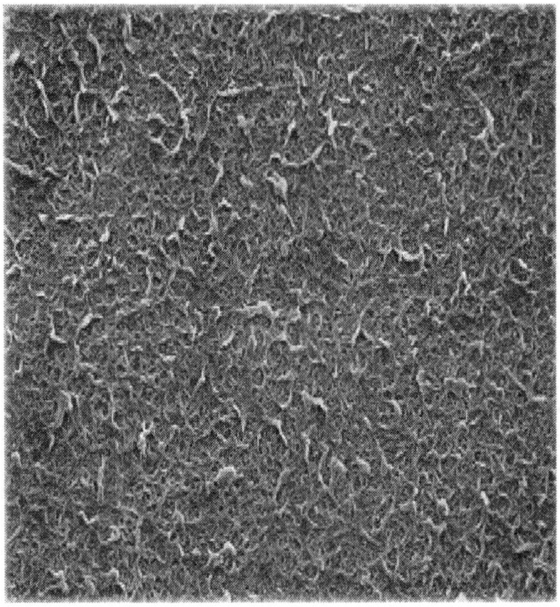

(a) Clean

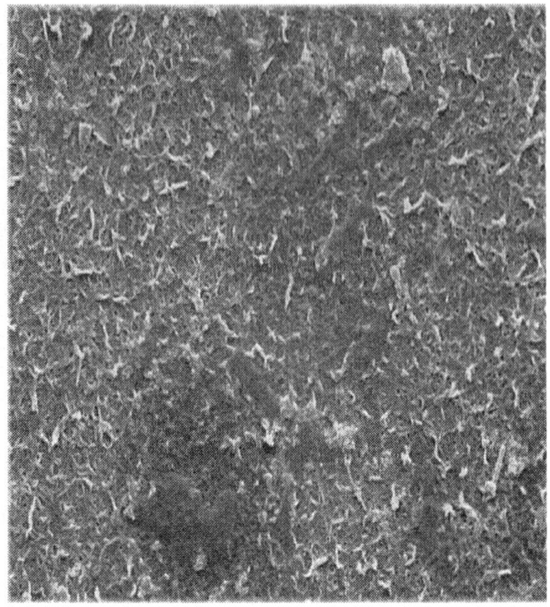

(b) at Recovery 15% (24 hr)

(c) at Recovery 70% (24 hr)

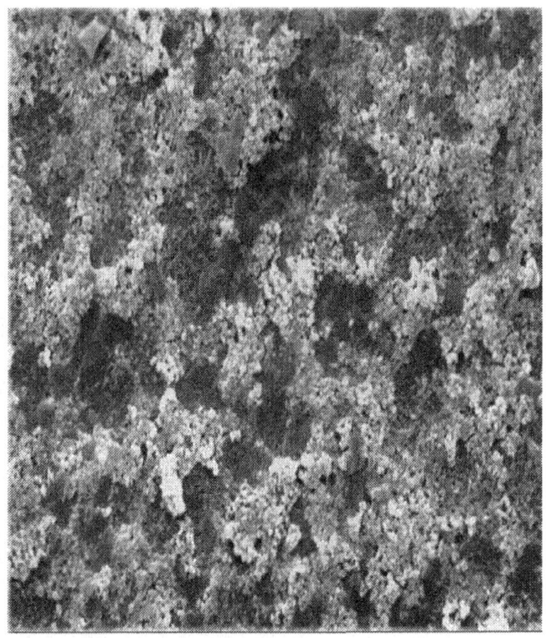

(d) at Recovery 70% with Antiscalant (24 hr)

Figure 8.16 Scanning electron micrographs of $CaCO_3$ grown on the ESNA membrane surface

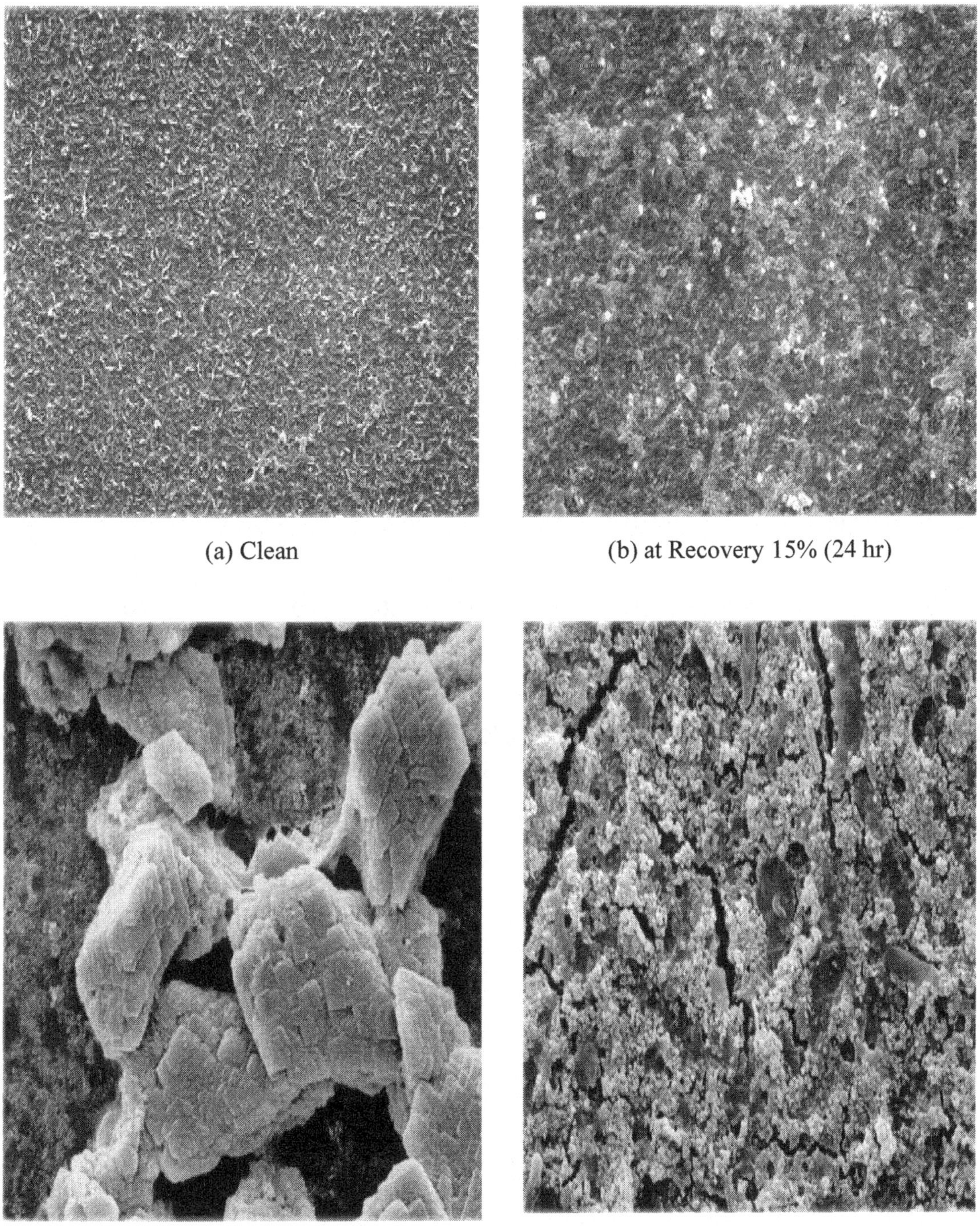

(a) Clean

(b) at Recovery 15% (24 hr)

(c) at Recovery 70% (24 hr)

(d) at Recovery 70% with Antiscalant (24 hr)

Figure 8.17 Scanning electron micrographs of $CaCO_3$ grown on the TFC-S membrane surface

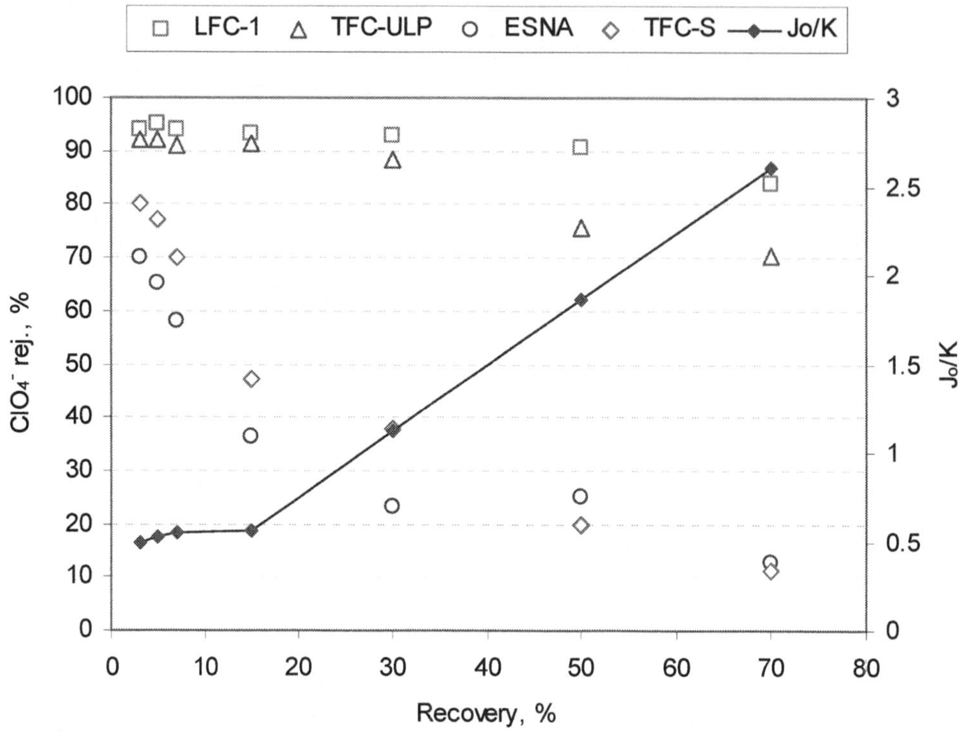

Figure 8.18 Effect of operating conditions; recovery and J_o/k ratio on ClO_4^- rejection with CRW for two RO (LFC-1 and TFC-ULP) and two NF (ESNA, TFC-S) membranes (100 μg/L ClO_4^- spiked)

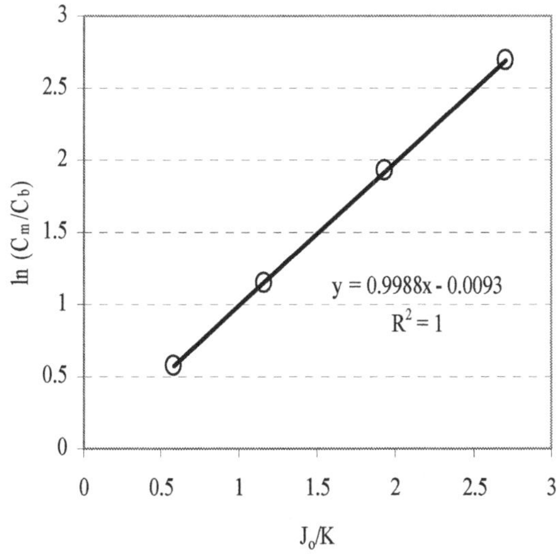

(i) LFC-1

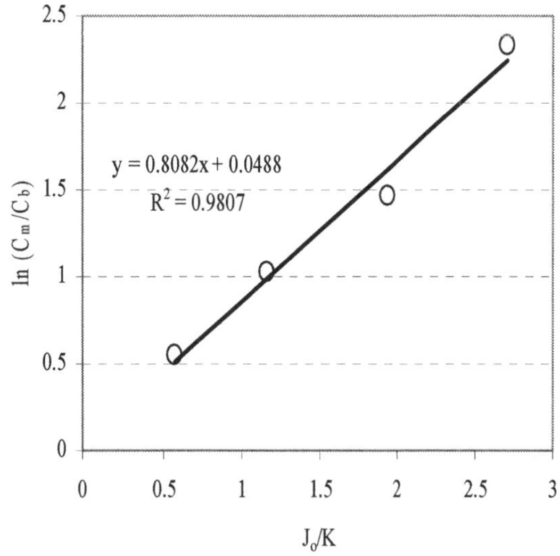

(ii) TFC-ULP

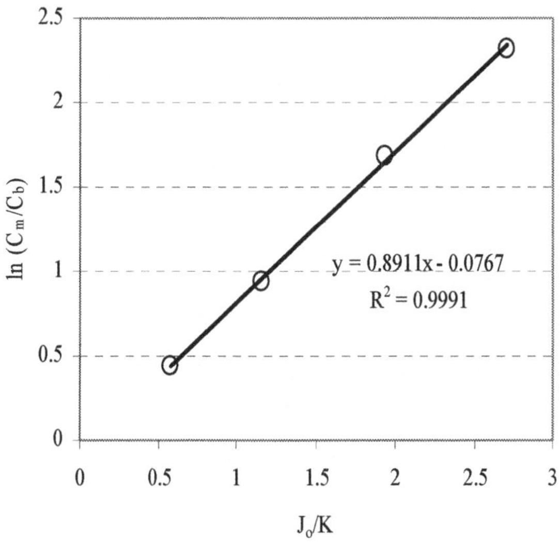

(iii) ESNA

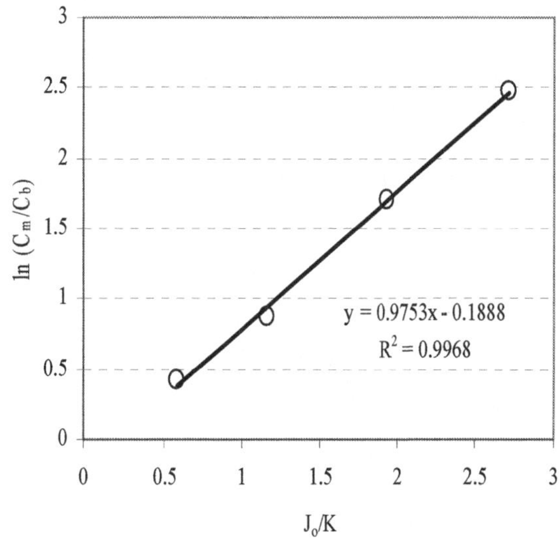

(iv) TFC-S

(a) Constant (K) of membranes for Ba^{2+}

(continued)

Figure 8.19 Constant (K) for four membranes and various ions

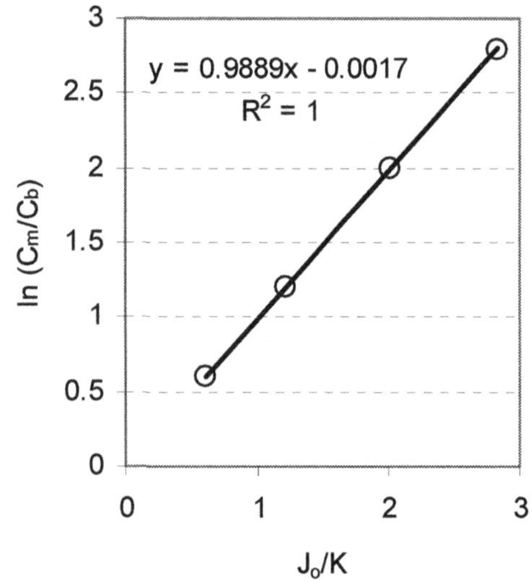

(i) LFC-1

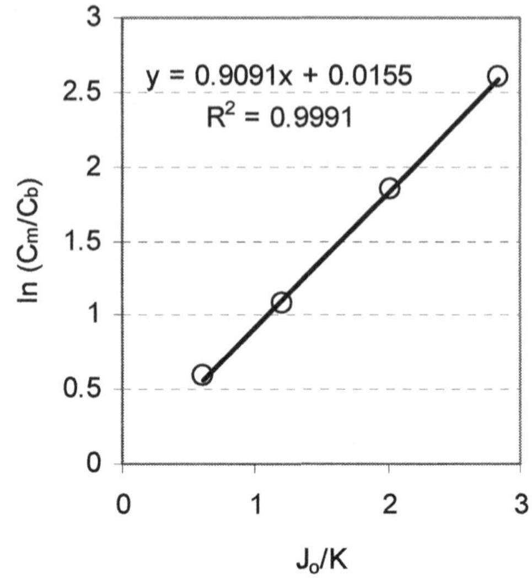

(ii) TFC-ULP

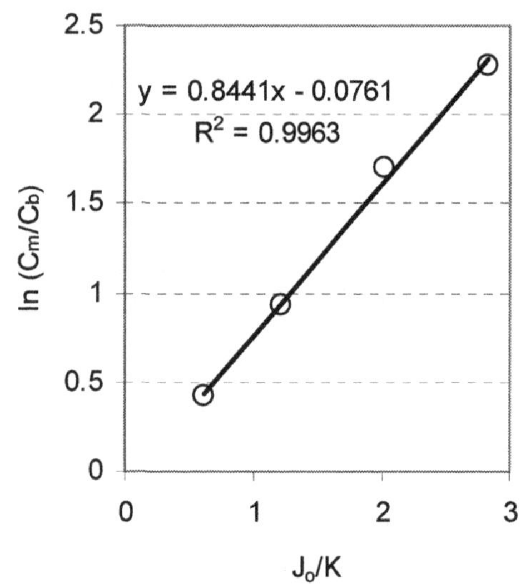

(iii) ESNA

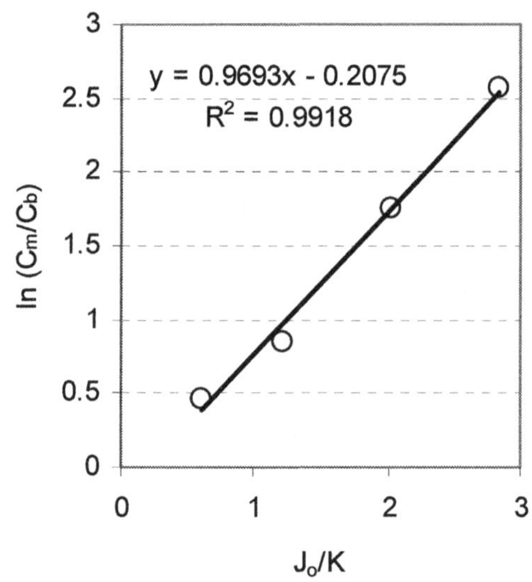

(iv) TFC-S

(b) Constant (K) of membranes for Ca^{2+}

(continued)

Figure 8.19 (Continued)

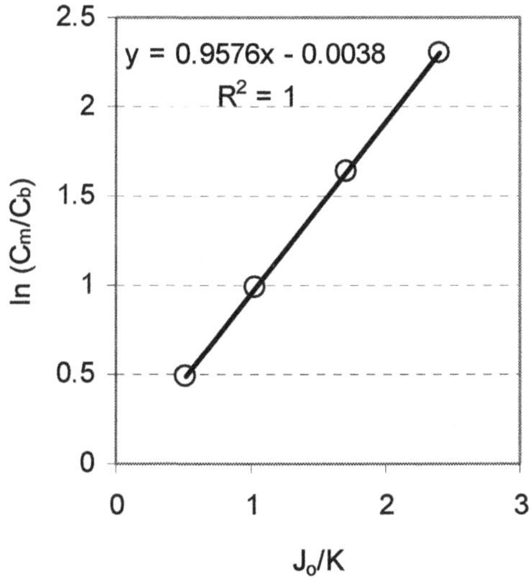

(i) LFC-1

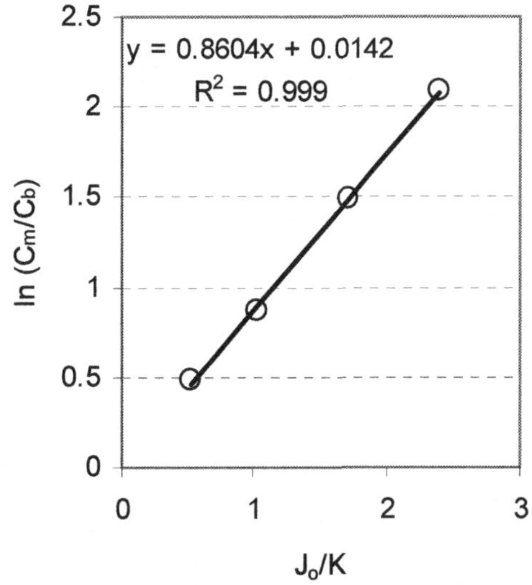

(ii) TFC-ULP

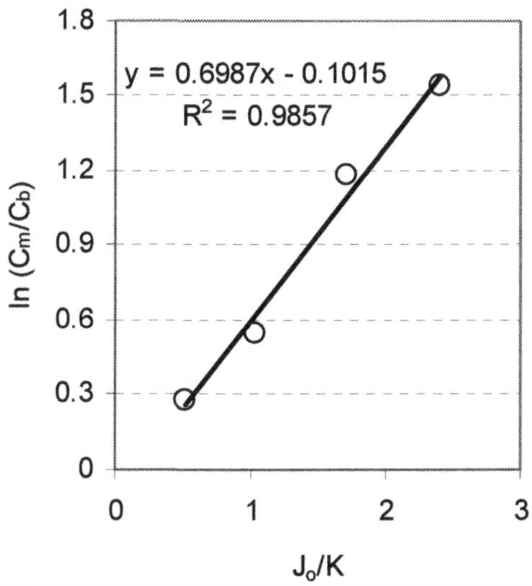

(iii) ESNA

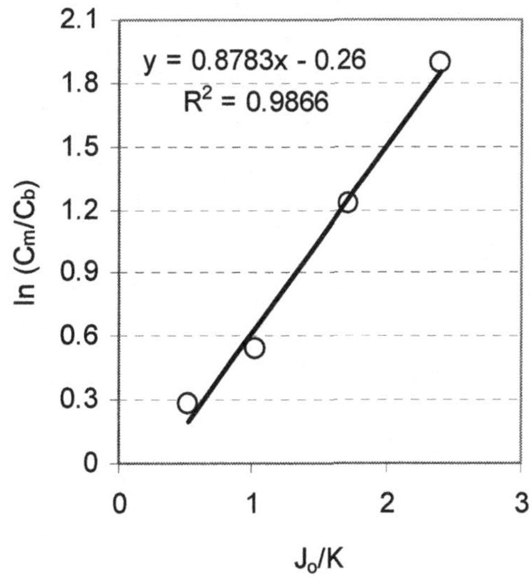

(iv) TFC-S

(c) Constant (K) of membranes for Cl^-

(continued)

Figure 8.19 (Continued)

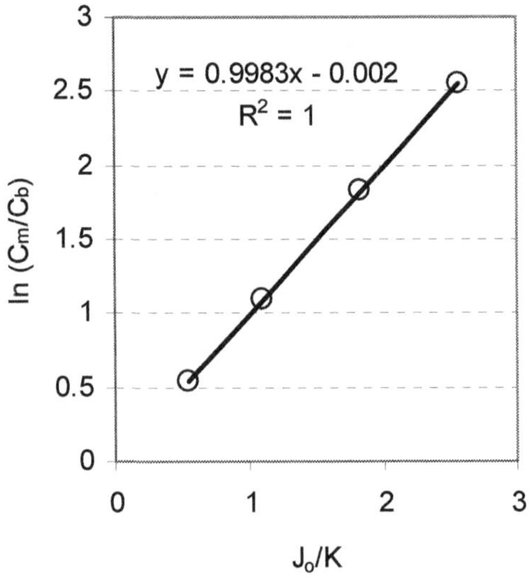

(i) LFC-1

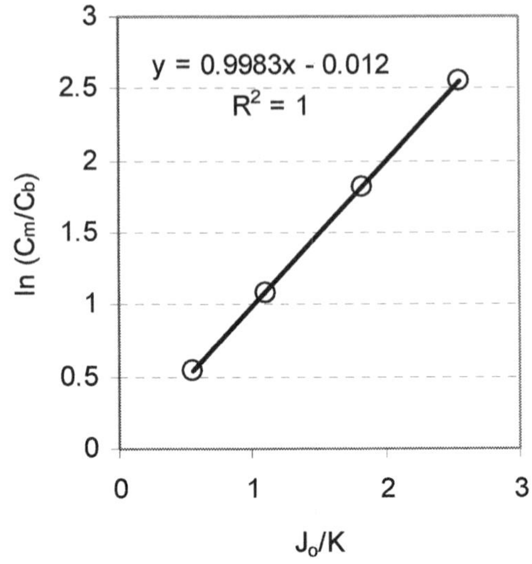

(ii) TFC-ULP

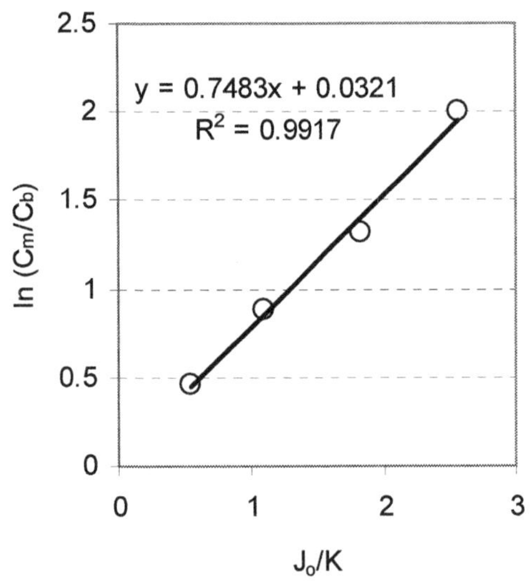

(iii) ESNA

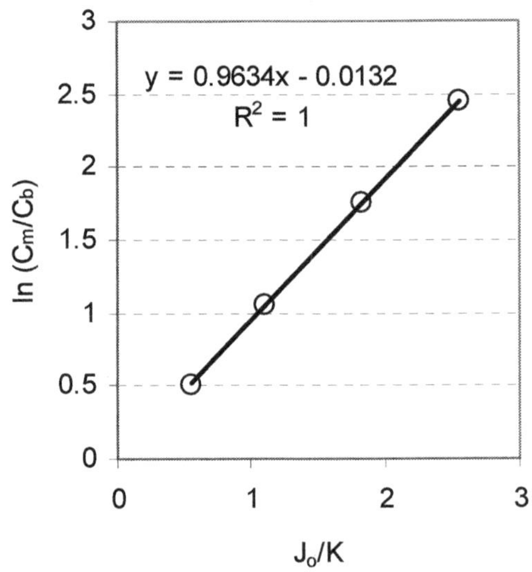

(iv) TFC-S

(d) Constant (K) of membranes for CO_3^{2-}

(continued)

Figure 8.19 (Continued)

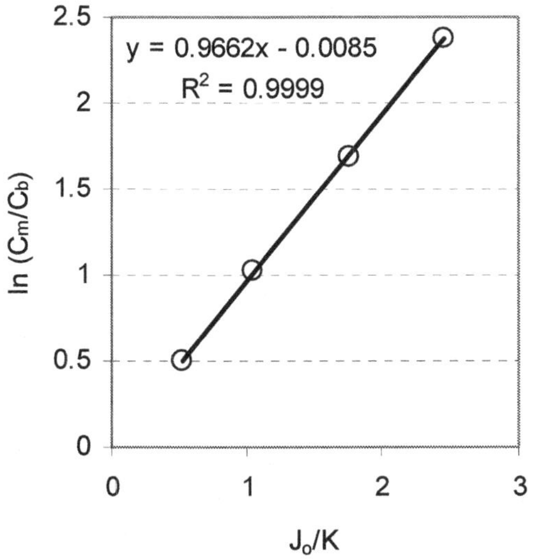

(i) LFC-1

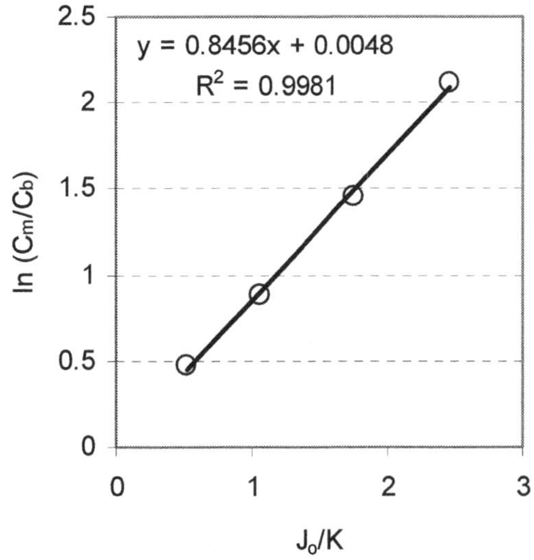

(ii) TFC-ULP

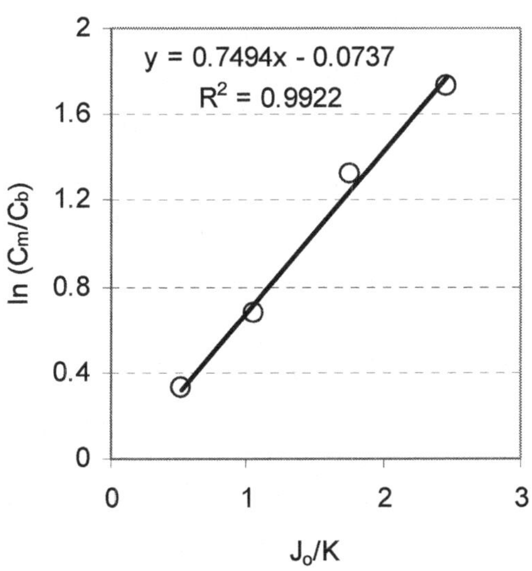

(iii) ESNA

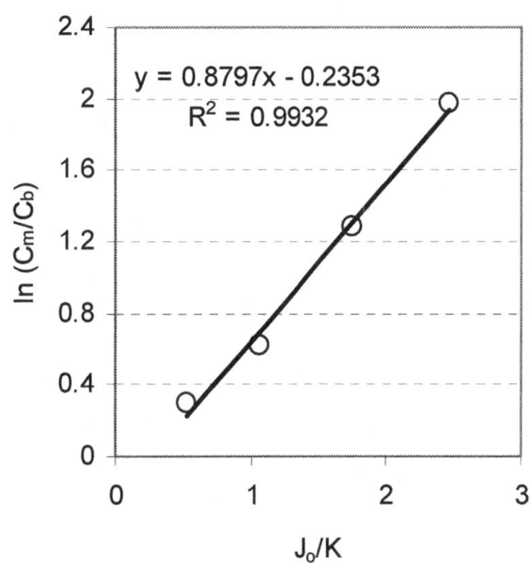

(iv) TFC-S

(e) Constant (K) of membranes for K^+

(continued)

Figure 8.19 (Continued)

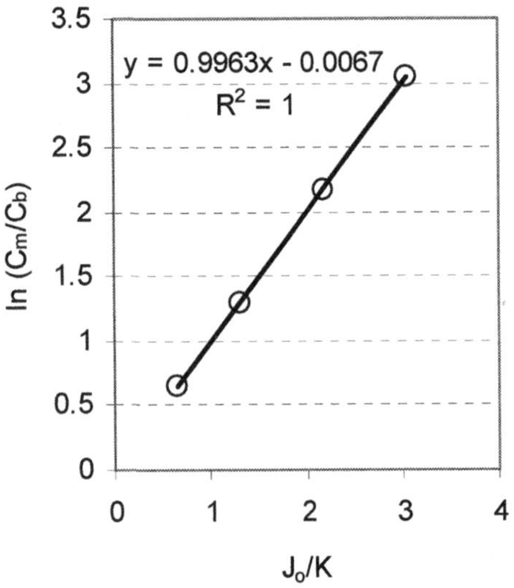

(i) LFC-1

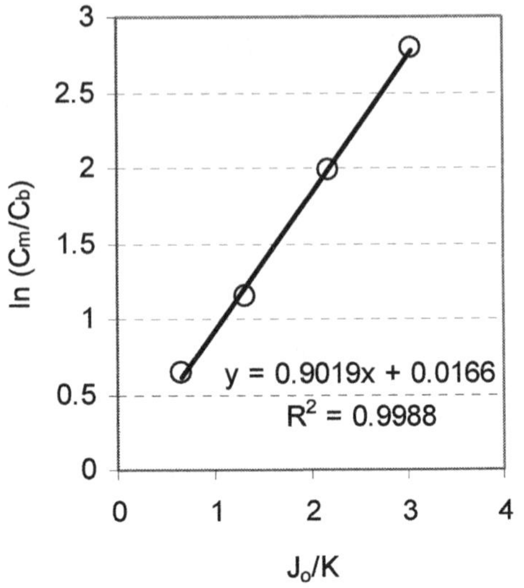

(ii) TFC-ULP

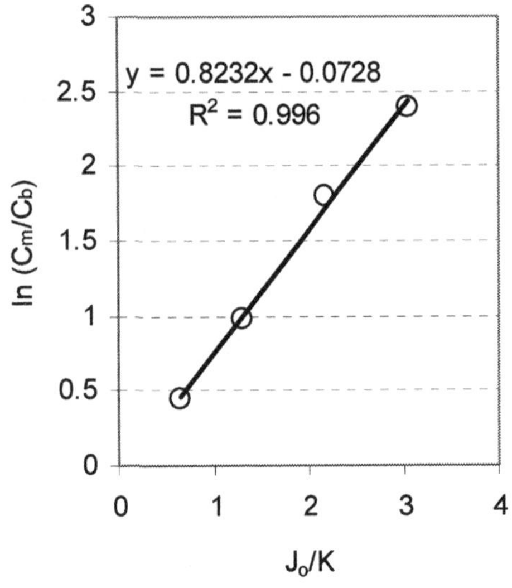

(iii) ESNA

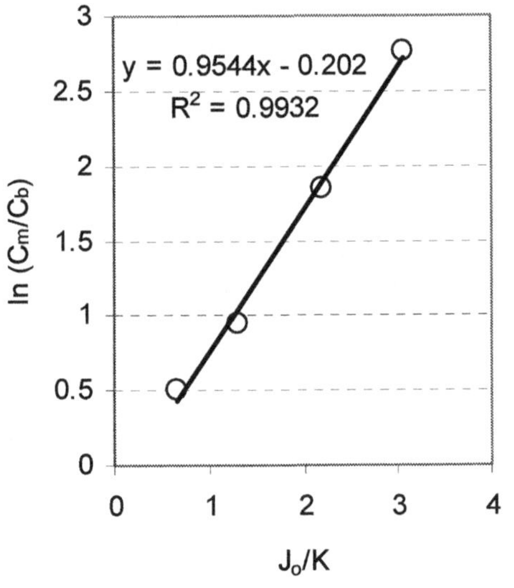

(iv) TFC-S

(f) Constant (K) of membranes for Mg^{2+}

(continued)

Figure 8.19 (Continued)

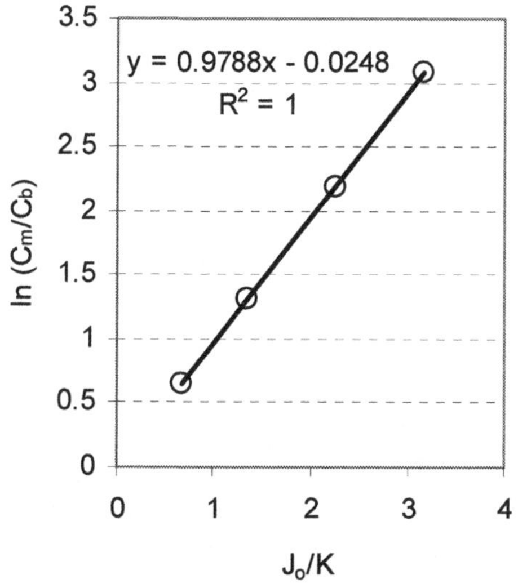

(i) LFC-1

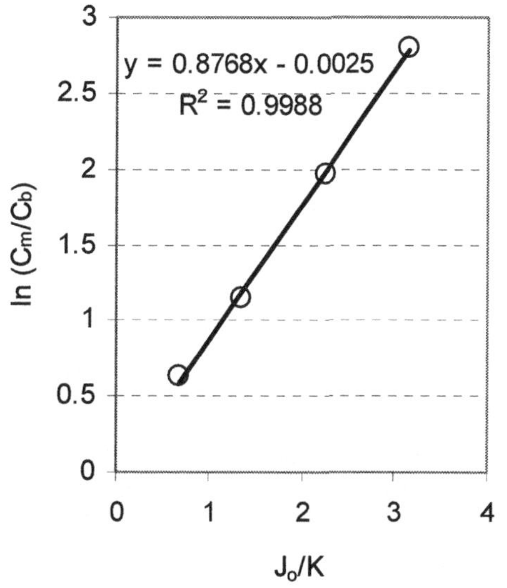

(ii) TFC-ULP

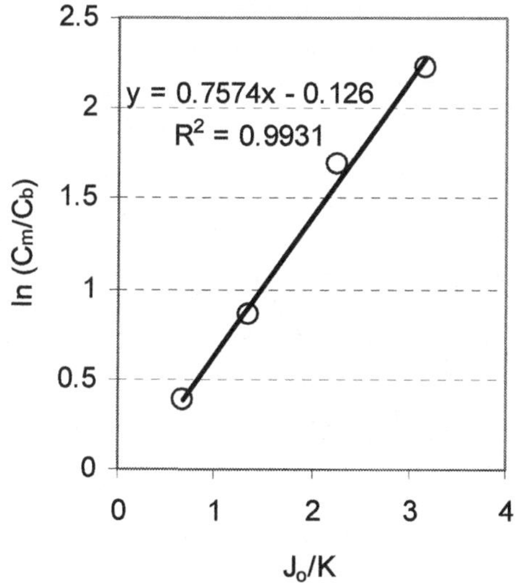

(iii) ESNA

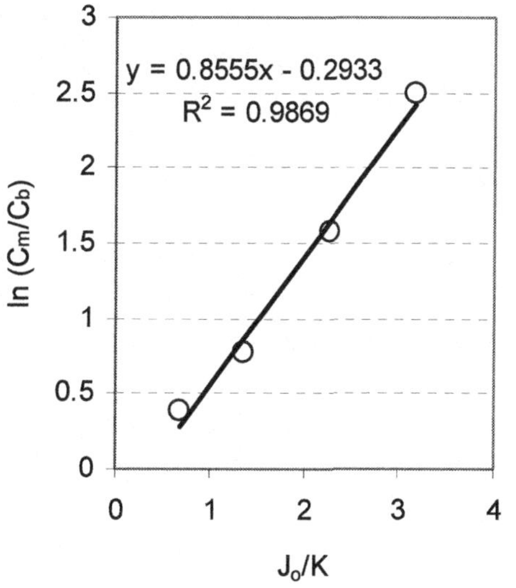

(iv) TFC-S

(g) Constant (K) of membranes for Na^+

(continued)

Figure 8.19 (Continued)

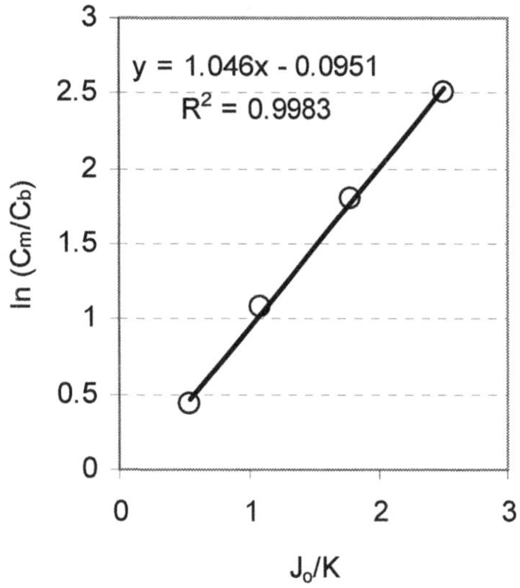

(i) LFC-1

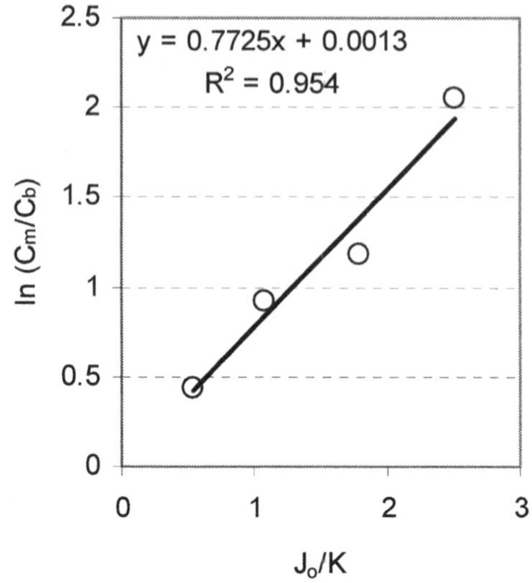

(ii) TFC-ULP

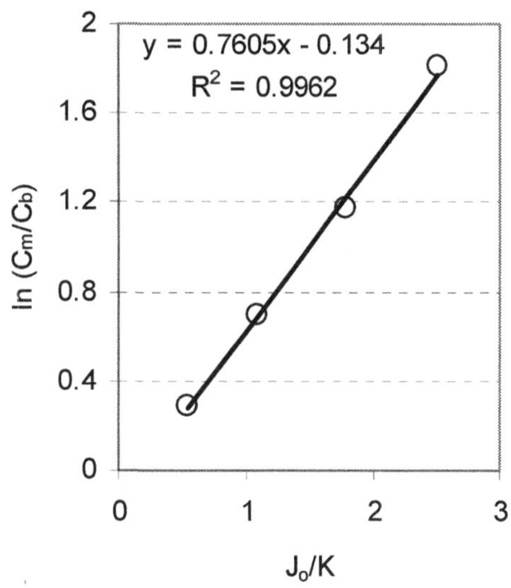

(iii) ESNA

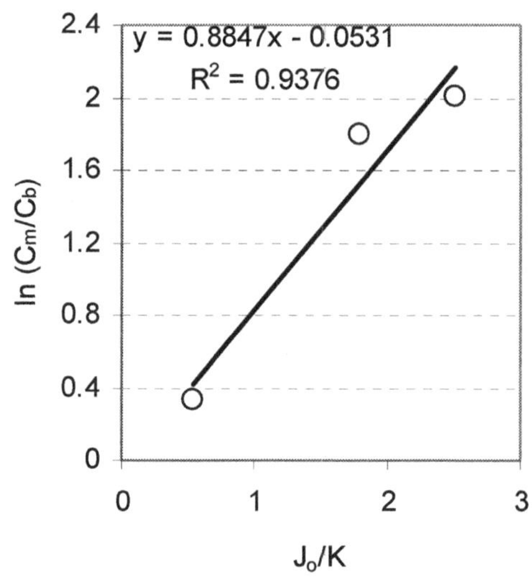

(iv) TFC-S

(h) Constant (K) of membranes for NO_3^-

(continued)

Figure 8.19 (Continued)

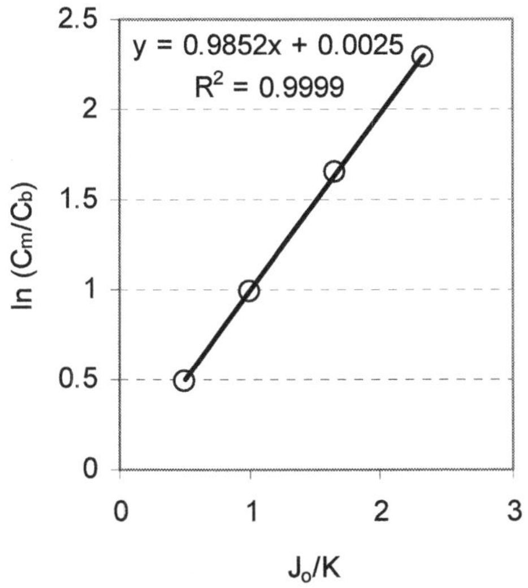

(i) LFC-1

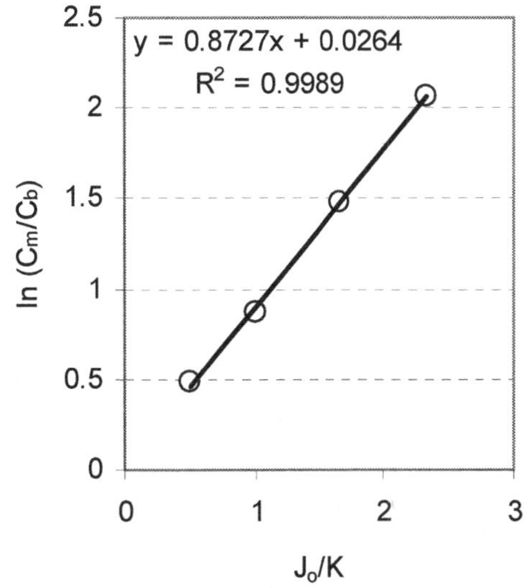

(ii) TFC-ULP

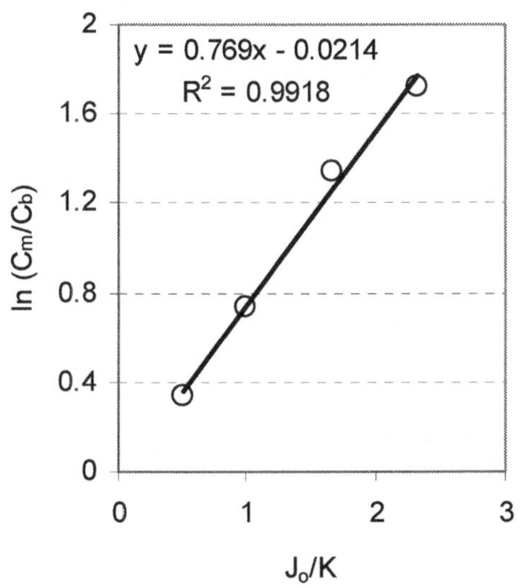

(iii) ESNA

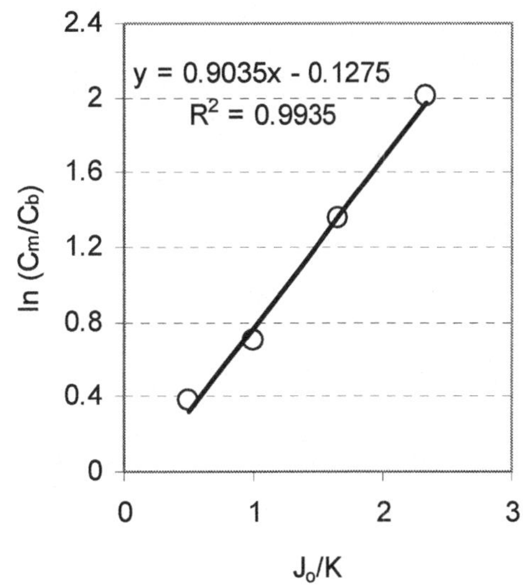

(iv) TFC-S

(i) Constant (K) of membranes for SO_4^{2-}

(continued)

Figure 8.19 (Continued)

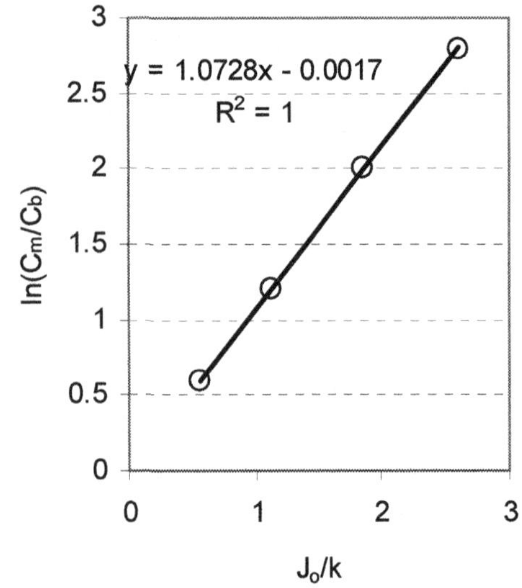

(i) LFC-1

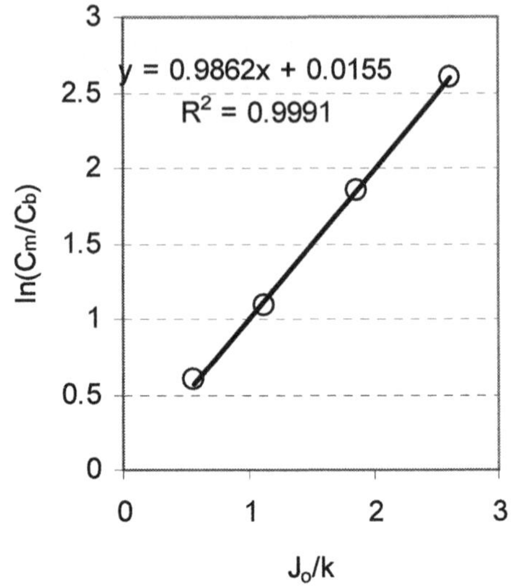

(ii) TFC-ULP

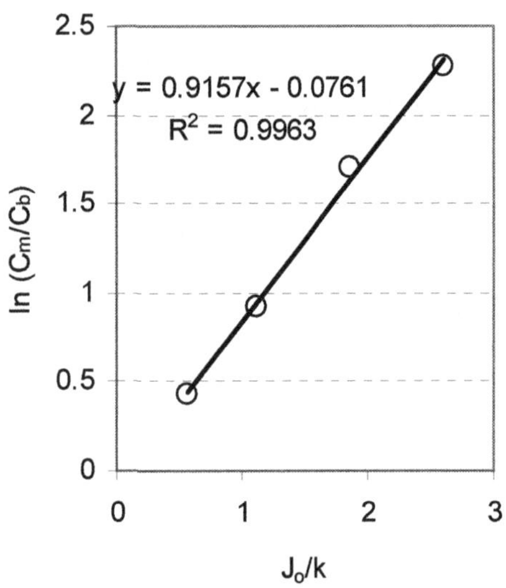

(iii) ESNA

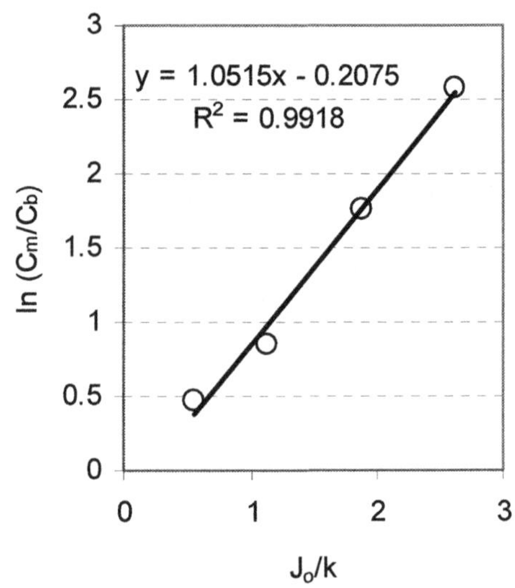

(iv) TFC-S

(j) Constant (K) of membranes for ClO_4^-

Figure 8.19 (Continued)

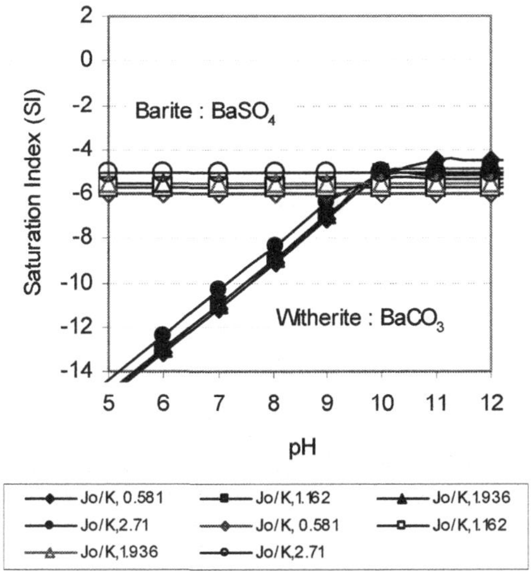

(a) Barium Compounds

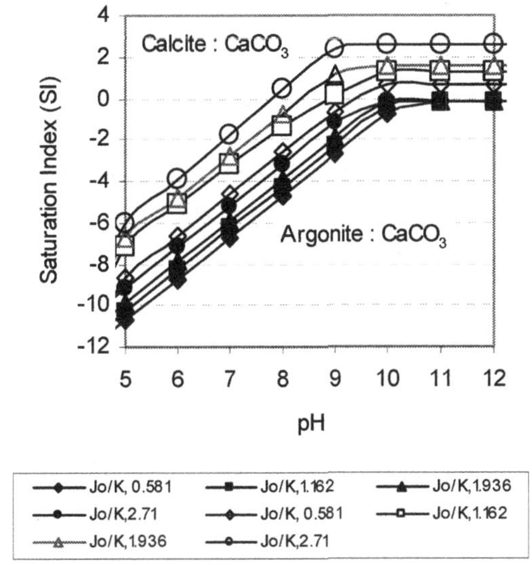

(b) Calcium-carbonate Compounds

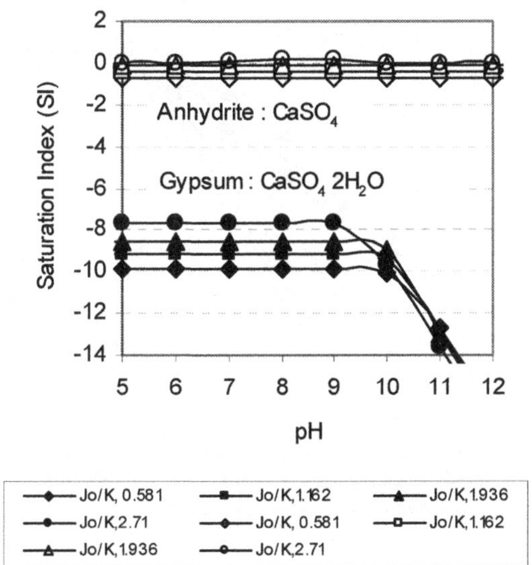

(c) Calcium-sulfate Compounds

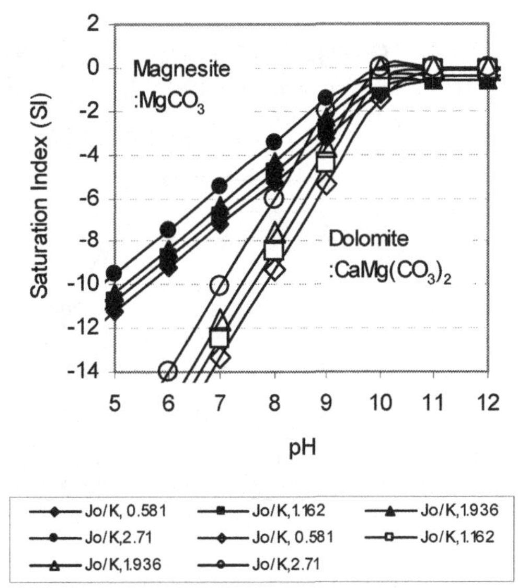

(d) Magnesium Compounds

Figure 8.20 Saturation index (SI) on LFC-1 membrane surface at different pH and J_o/K ratio

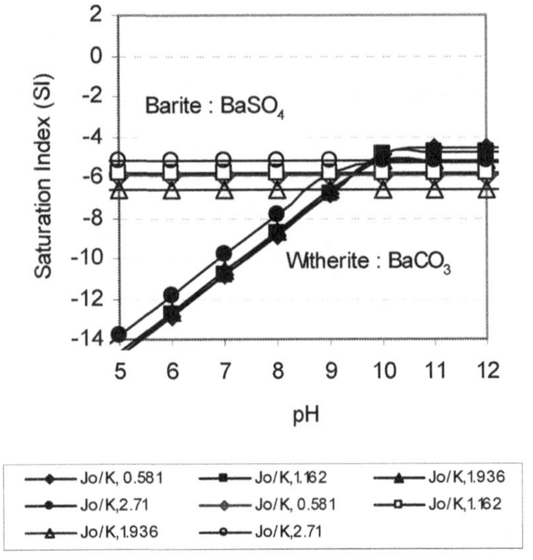

(a) Barium Compounds

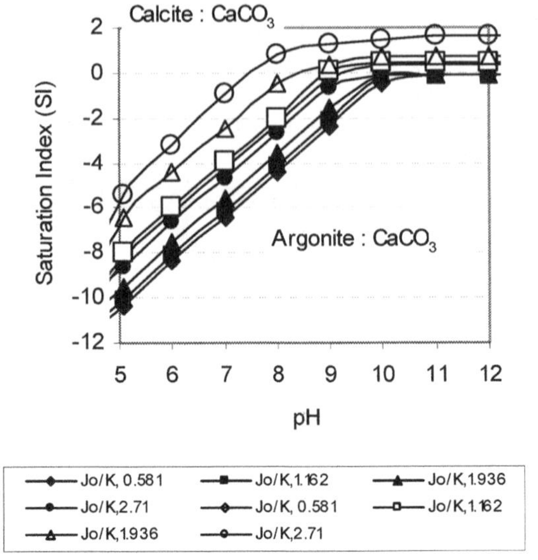

(b) Calcium-carbonate Compounds

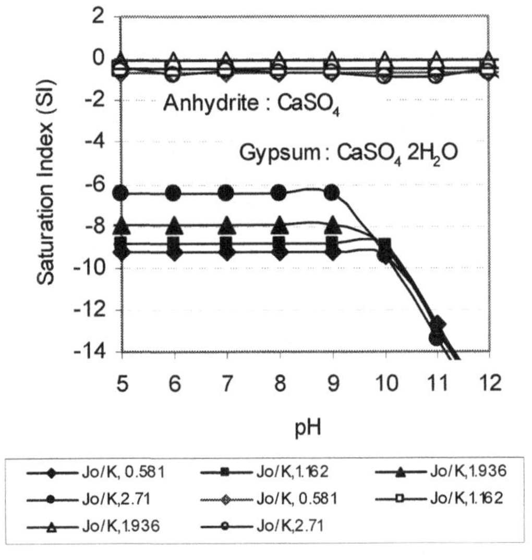

(c) Calcium-sulfate Compounds

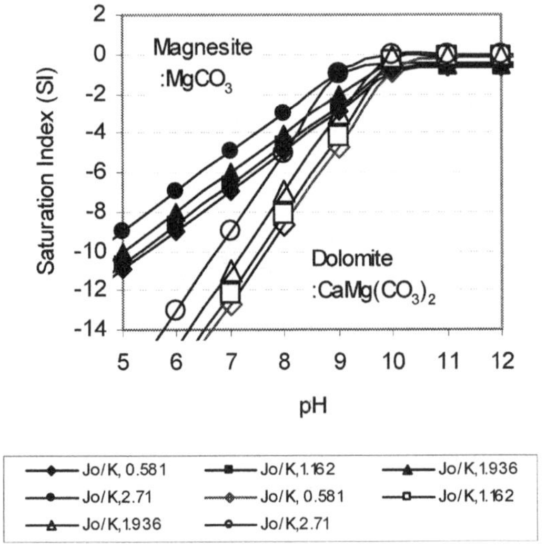

(d) Magnesium Compounds

Figure 8.21 Saturation index (SI) on TFC-ULP membrane surface at different pH and J_o/K ratio

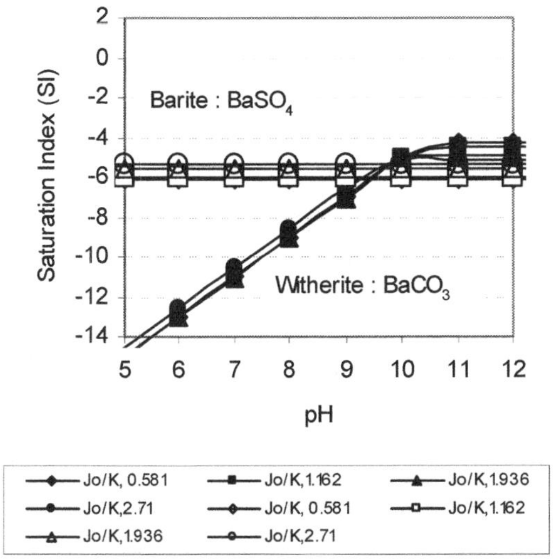

(a) Barium Compounds

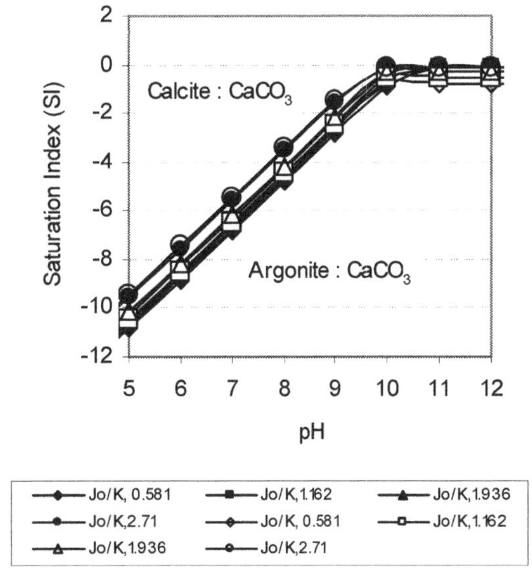

(b) Calcium-carbonate Compounds

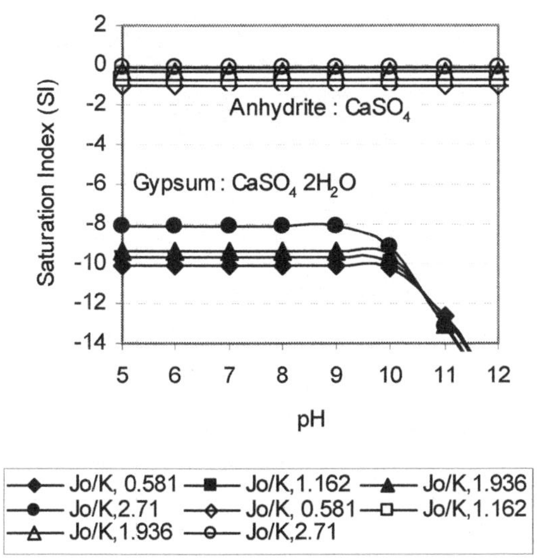

(c) Calcium-sulfate Compounds

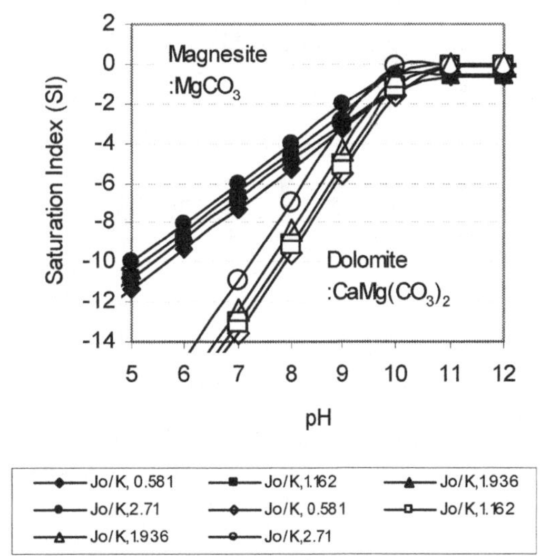

(d) Magnesium Compounds

Figure 8.22 Saturation index (SI) on ESNA membrane surface at different pH and J_o/K ratio

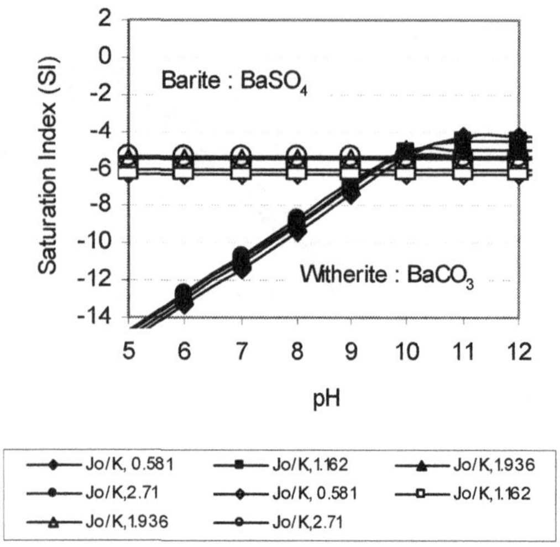

(a) Barium Compounds

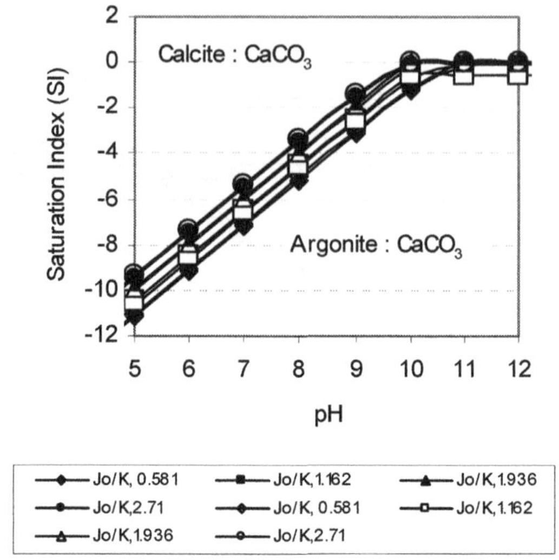

(b) Calcium-carbonate Compounds

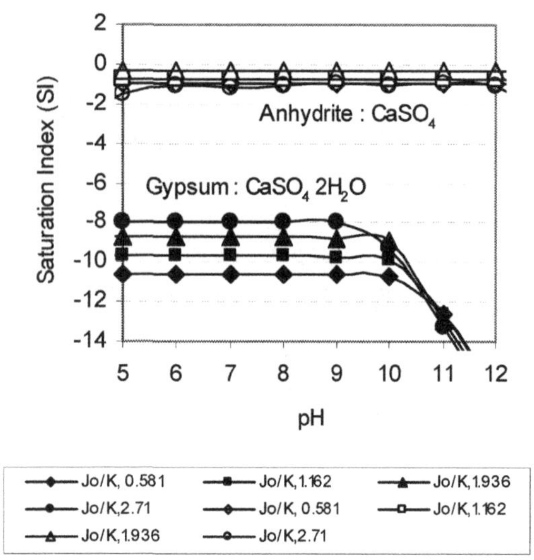

(c) Calcium-sulfate Compounds

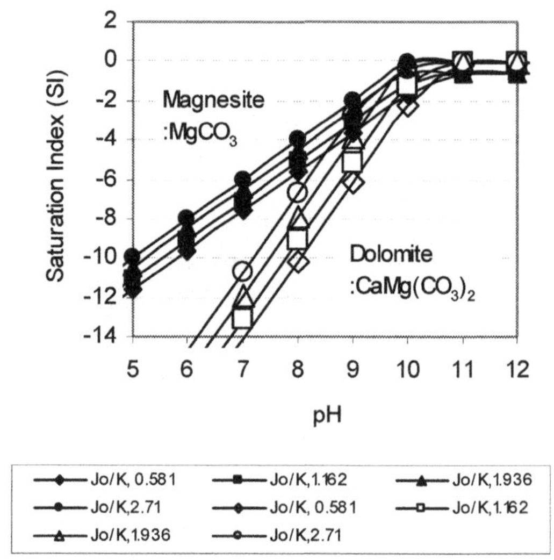

(d) Magnesium Compounds

Figure 8.23 Saturation index (SI) on TFC-S membrane surface at different pH and J_o/K ratio

CHAPTER 9
A RATIONAL BASIS OF RECONCILING TRADEOFFS IN THE CHOICE BETWEEN FULL-STREAM NF vs. SPLIT-STREAM RO TREATMENT

INTRODUCTION

The economics of water treatment using RO and NF technology has been continuously improving with a reduction of product water cost as a result of lower investment costs and decreased power consumption. The objective of this evaluation was to determine the economic feasibility of operating a membrane treatment system at higher recovery based on prevailing economic parameters, and to also address the economic tradeoffs between split-stream RO versus total-stream NF treatment.

In this study, we have employed an RO design software to: i) evaluate product water quality and corrosivity after membrane treatment, ii) compare full-stream treatment with NF to split-stream treatment with RO to reduce the corrosivity while meeting perchlorate removal goal, and iii) determine the effect of major design operational parameters on water cost, including the blending ratio of feed to product water, maintenance cost, scale inhibitor dose, interest rate, power cost, plant life, and membrane life. For the cost evaluation, we have adopted cost parameters from the optimized system design of a water facility installation: a \$2,500,000 (1.0 MGD) investment for a RO water plant at Marco Island, Florida.

PARAMETERS FOR EVALUATION

The evaluation was conducted based on a blend of Colorado River water and State Project water (CRW/SPW). We applied CRW/SPW quality to the water treatment facility, as previously defined (Table 8.1).

Table 9.1 summarizes the equipment data based on published cost estimations and other communications (Wilf and Klinko, 1994, 1997a,b). The product water cost was evaluated based on these parameters.

Table 9.1
Summary of parameters for cost evaluation

Parameters	Value	Reference
Plant capacity, gpd	1,000,000	Marco Island, FL, water plant
Investment, \$	2,500,000	Marco Island, FL, water plant
Interest rate, %	8	Wilf and Klinko, 1997b
Plant factor, %	95	Hydronautics Co.
Power Cost, \$/kwhr	0.07	X-cel Energy, CO
Avg. Plant life, yr	20	Wilf and Klinko, 1997b
Avg. Membrane life, yr	5	Wilf and Klinko, 1997b
Membrane cost, \$/element	700	Wilf and Klinko, 1997b
Number of elements	166	Marco Island, FL, water plant
Inhibitor cost, \$/lb	0.5	Solutia Inc. (Dequest)

Correlation Between Feed Pressure and Power Requirement

The power requirement is directly related to feed pressure and feed water flow. A higher recovery rate requires a higher feed pressure to overcome increasing osmotic pressure.

The power requirement includes the electricity consumed by the pretreatment system and the high pressure feed pumps.

The power requirement (P_w) is a function of the flow (Q) and the total head (H) and can be represented by equation 9.1 (Lee and Lin, 1999).

$$P_W = jQH \tag{9.1}$$

PW: Power requirement, KW (m· m^3/min) or hp (ft· gal/min)
Q: discharge, m^3/min or gpm
j: constant at 20 °C
0.163 (for SI units), 2.525×10^{-4} (for British units)
H: total head, m or ft

Total head, H, is related to the Bernoulli equation (Equation 9.2), which can be applied to determine the total dynamic head on the pump. The energy equation expressing the head between the suction and discharge line of a pump is as follows:

$$H = \frac{P_d}{\gamma} + \frac{V_d^2}{2g} + z_d - \left(\frac{P_S}{\gamma} + \frac{V_S^2}{2g} + z_S \right) \tag{9.2}$$

H: total dynamic head, m or ft
P_d, P_s: gage pressure at discharge and suction, respectively, N/m^2 or lb/in^2
γ: specific weight of water, N/m^3 or lb/ft^2
V_d, V_s: velocity in discharge and suction nozzles, respectively, m/s or ft/s
g: gravitational acceleration, 9.81 m/s^2 or 32.2 ft/s^2
z_d, z_s: elevation of discharge and suction gage above the datum, m or ft

Figure 9.1 shows the relation of power requirement and feed pressure for two different membranes at various recoveries. A higher feed pressure is directly related to the power required for increased recovery.

Total Water Cost and Water Quality As a Function of Blending Ratio for Corrosion Control and Product-Water Perchlorate From NF Full vs. Split RO Treatment

The use of RO and, to a lesser extent, NF membranes will provide a significant rejection of hardness (e.g. Ca^{2+}) and carbonate alkalinity (HCO_3^- and CO_3^{2-}), rendering the product water corrosive and potentially problematic in complying with the lead and copper rule. When considering the selection of RO vs. NF, important factors include target rejection (e.g. perchlorate) and pressure requirements related to specific flux. One possible application for RO is a situation where very high (> 90%) removals of cations and anions are required. In situations where somewhat lower removals are acceptable, NF membranes have some advantages over RO because of their permeabilities, greater selectivity, and being less prone to form scale. However, another possibility involves tradeoffs between full feed stream by NF versus partial split-stream treatment by RO. In some cases, RO split-stream treatment may be more cost-effective and provide better water quality than NF full-stream treatment.

Tables 9.2 and 9.3 summarize the water quality and the total water cost as a function of blending ratio of feed water versus permeate for the LFC-1 (RO) and ESNA (NF) at a recovery of 70%. In both tables, the water quality was varied by the given blending conditions between feed flow and permeate flow providing the total product water. With increasing feed flow and decreasing permeate flow, concentrations of cations and anions increased, however, the total water cost decreased. Comparing the total water cost of the LFC-1 and ESNA, the cost of the RO full-stream treatment is $1.26/Kgal, higher than NF full-stream treatment. However, the water quality by the NF treatment at a high recovery was lower than the RO split-stream treatment as well as the RO full-stream treatment.

Besides the treatment by full RO vs. full NF, comparing RO split treatment to NF full treatment, RO split-stream treatment showed better performance in terms of water quality and cost than NF full treatment. As seen in Tables 9.2 and 9.3, split ratios of feed (10% to 30%) and permeate (90% to 70%) in RO treatment showed higher rejections of all parameters including perchlorate than NF full treatment, a RO split ratio of feed 50% and permeate 50% showed slightly higher performance than NF full treatment. However, with respect to the rejection of perchlorate, split ratios of RO treatment (feed 10% to 70% and permeate 90% to 30%) showed better rejection than NF full treatment. For lower recoveries of NF full treatment (Figure 8.9(c)), the rejection of perchlorate was higher than that of 70% recovery, but perchlorate rejection decreased with increasing recovery. For the split ratios of feed 10% to 30% and permeate 90% to 70% of RO, total water cost was similar or less (0~22%) than NF full treatment, and water quality was better and potential corrosivity was reduced. Also, considering provisional action level (California Department of Health Service, CDHS) of 4 μg/L, the appropriate rejection of perchlorate can be obtained by RO split treatment at 70% recovery with a split ratio of feed 0% to 10% and permeate 100% to 90% (U.S.EPA, 1998). However, a higher split ratio can be employed with a lower recovery of < 70%.

It should be recognized that this exercise described in this chapter is based on one specific RO and one specific NF membrane, and that the relative results would change for analysis of different membranes (e.g a higher pressure NF and/or a lower pressure RO). Nevertheless, the framework presented here is applicable to evaluating tradeoffs between full-stream NF versus split-stream RO treatment.

Figure 9.2 presents the alkalinity and pH variations as a function of the blending ratio. As seen in Figure 9.2 (a) and (b), pH and alkalinity are important factors affecting the corrosion potential. In both figures, the dotted line is the marginal guide to provide corrosion control of copper and lead. According to Edwards et al, the corrosion potential is higher below pH 6.5 and 25 mg/L as $CaCO_3$ of alkalinity (Edwards et al., 1996 and Dodrill and Edwards, 1995). At these ranges, releases of copper and lead are much greater than at high pH and high alkalinity.

In Figure 9.2 (a), RO full treatment shows that corrosion potential is high because the alkalinity is lower than 25 mg/L and the pH range is also around 6.5. However, as the blending ratio increases, the corrosion potential decreases. However, in case of the NF full treatment, there are no corrosion effects.

Table 9.2
Water quality and total water cost as a function of blending ratio of feed water vs. permeate of LFC-1 (RO) at recovery of 70%

Parameters	Avg. Feed Water Quality	Water Quality and Total Water Cost as a Function of Blending Ratio of Feed vs. Permeate at Recovery 70%					
		Feed* (0%) : Permeate (100%)	Feed* (10%) : Permeate (90%)	Feed* (30%) : Permeate (70%)	Feed* (50%) : Permeate (50%)	Feed* (70%) : Permeate (30%)	Feed* (90%) : Permeate (10%)
pH	7.93	6.5	6.77	7.11	7.35	7.57	7.8
Condutivity, mS/m	91	6.9	15.4	32.5	49.6	66.7	83.8
TDS, mg/L	499	41	87	178	270	361	453
Langlier saturation index	0.1	-3.5	-2.5	-1.6	-1.0	-0.5	-0.1
Calcium, mg/L	55.1	2.3	7.6	18.1	28.7	39.3	49.8
Magnesium, mg/L	21.4	0.5	2.6	6.8	11	15.1	19.3
Sodium, mg/L	63.1	9.4	14.8	25.5	36.3	47	57.7
Potassium, mg/L	3.4	0.4	0.7	1.3	1.9	2.5	3.1
Barium, mg/L	0.073	0.001	0.008	0.023	0.037	0.051	0.066
Carbonate, mg/L	0.4	0	0	0.1	0.2	0.3	0.4
Bicarbonate, mg/L	101.4	12.2	21.1	39	56.8	74.6	92.5
Sulfate, mg/L	190	8.3	26.5	62.8	99.2	135.5	171.8
Chloride, mg/L	60.5	7.3	12.6	23.3	33.9	44.5	55.2
Nitrate, mg/L	0.3	0	0	0.1	0.2	0.2	0.3
Silica, mg/L	3.2	0.3	0.6	1.2	1.8	2.3	2.9
Perchlorate, μg/L	100	15	21	39	58	75	92
Total water cost, $/Kgal	—	1.26	1.12	0.882	0.63	0.378	0.126

* Feed ( ): The percentage of feed flow at the total product water
Permeate ( ): The percentage of permeate flow at the total product water

Table 9.3

Water quality and total water cost as a function of blending ratio of feed water vs. permeate of ESNA (NF) at recovery of 70%

Parameters	Avg. Feed Water Quality	Water Quality and Total Water Cost as a Function of Blending Ratio of Feed vs. Permeate at Recovery 70%					
		Feed* (0%) : Permeate (100%)	Feed* (10%) : Permeate (90%)	Feed* (30%) : Permeate (70%)	Feed* (50%) : Permeate (50%)	Feed* (70%) : Permeate (30%)	Feed* (90%) : Permeate (10%)
pH	7.93	7.78	7.8	7.83	7.86	7.89	7.92
Condutivity, mS/m	91	43.5	48.3	58.1	67.9	77.7	87.4
TDS, mg/L	499	248	273	323	374	424	474
Langlier saturation index	0.1	-0.6	-0.5	-0.4	-0.2	-0.1	0
Calcium, mg/L	55.1	20	23.5	30.5	37.6	44.6	51.6
Magnesium, mg/L	21.4	9	10.2	12.7	15.2	17.7	20.2
Sodium, mg/L	63.1	41.1	43.3	47.7	52.1	56.5	60.9
Potassium, mg/L	3.4	1.7	1.9	2.2	2.6	2.9	3.2
Barium, mg/L	0.073	0.02	0.025	0.036	0.047	0.057	0.068
Carbonate, mg/L	0.4	0.2	0.2	0.3	0.3	0.3	0.4
Bicarbonate, mg/L	101.4	66.7	70.2	77.1	84.1	91	97.9
Sulfate, mg/L	190	74.8	86.3	109.4	132.4	155.4	178.5
Chloride, mg/L	60.5	32.1	34.9	40.6	46.3	52	57.7
Nitrate, mg/L	0.3	0.2	0.2	0.2	0.3	0.3	0.3
Silica, mg/L	3.2	1.9	2	2.3	2.6	2.8	3.1
Perchlorate, μg/L	100	88	89	92	94	96	99
Total water cost, $/Kgal	—	1.134	1.017	0.791	0.565	0.339	0.113

* Feed ( ): The percentage of feed flow at the total product water
Permeate ( ): The percentage of permeate flow at the total product water

Effect of Maintenance Costs

The maintenance costs of a membrane facility affect the total water cost. If maintenance is assumed to be 3% of capital, effects can be simulated based on previous work (Wilf and Klinko, 1997b and Taylor et al., 1989).

Figure 9.3 shows the effect of maintenance costs (% of capital) on total water cost and maintenance cost for the LFC-1 and ESNA at a recovery of 70%. As seen in Figure 9.3, the water cost increases with increasing maintenance percentage. Under conditions of no blending, the water cost by NF treatment is approximately 10% less than by RO treatment.

Effect of Interest Rate

The effect of interest rate is based on a range of values from 1% to 15%. Most previous studies have employed an interest rate of 7 to 8% (Ericsson and Hallmans, 1991, Taylor et al., 1989, and Wilf and Klinko, 1997b).

Figure 9.4 shows the effect of interest rate on total water cost and capital cost. In Figure 9.4 (a) and (b), comparing interest rates from 1% to 15%, various total water costs ($/Kgal) and capital costs ($/Kgal) are shown for both membrane treatments.

Effect of Power Cost

The calculations relating power cost ($/Kwhr) to total water cost are presented in Figure 9.5. The calculations were based on a power rate of $0.04/KWhr to $0.14/KWhr. According to the selected electricity provider, power costs ($/KWhr) vary according to locale and industrial classification. However, a representative range is reflected by $0.06/KWhr or $0.07/KWhr in Colorado and $0.14/KWhr in California. Figure 9.5 shows the values of unit power cost ($/Kgal) and total water cost ($/KWhr) for $0.04/KWhr to $0.14/KWhr for two different membrane treatments. As expected, comparing total water cost of Colorado and California, the water costs of Colorado area are not only 15% less than California, but power costs ($/Kgal) are 55% lower.

Effect of Inhibitor Dose

As mentioned in Chapter 8, antiscalants are used to inhibit inorganic precipitative scale on membrane surfaces. In the flux decline tests, inhibition tests of precipitation under various conditions (e.g. pH, conductivity, types and doses of antiscalants), and comparisons of membrane characterization such as XRD, FTIR, SEM, NAA with/without antiscalant, revealed that antiscalants were effective in the inhibition of inorganic fouling. Based on the results of these experiments, Dequest 2000 was more effective than other antiscalants. Thus, we simulated the water costs and chemical costs based on this antiscalant dose. According to the manufacturer (Solutia Inc.), the cost of Dequest 2000 is $0.50/lb cost.

Figure 9.6 shows the effect of inhibitor dose on chemical cost and total water cost. Based on the previous tests, 3 ppm of Dequest 2000 was an adequate concentration to inhibit inorganic fouling; we evaluated the chemical cost and water cost over an inhibitor dose range of 0 to 20 ppm. In the simulation results, the inhibitor cost accounts for 7% (20 ppm) and 0.8% (3 ppm) of total water cost, however, considering an adequate dose (3 ppm) and higher water production, inhibitor addition is cost effective in membrane treatment.

Effects of Plant Life and Membrane Life

Both plant life and membrane life affect total water cost. In many studies, the average membrane plant life and membrane replacement rate has been to be assumed 20 years and 20% per year, respectively (Ericsson and Hallmans, 1991, Darwish et al 1989, and Wilf and Klinko, 1997b).

Figure 9.7 presents the effect of plant life on total water cost and capital cost. Comparing water cost and capital cost ($/Kgal) over 10 years or 30 years versus a baseline of 20 years, water cost and capital cost (10 years) of the RO plant increase 22% and 73%, and those of the NF plant increase 32% and 40%, respectively. However, water cost and capital cost (30 years) of the RO plant decrease 8% and 13%, and those of the NF plant decrease 9% and 13%, respectively.

Figure 9.8 shows the effect of membrane life on total water cost. As seen in Figures 9.7 and 9.8, a decrease in membrane life results in a higher total water cost, however, total water cost is almost constant beyond a replacement period of 5 years.

SUMMARY

In this study, we have evaluated the costs and water quality of full vs. split stream treatment with respect to corrosion potential and perchlorate rejection, and the effects on total water cost associated with major design and operational parameters. The results of this evaluation are as follows:

- As recovery increases, the membrane system requires a higher feed pressure, which is directly related to a higher power requirement.
- Comparing results of treatment of RO-split stream treatment vs. NF full treatment, lower water treatment costs and better water quality can be obtained by RO split-stream treatment than NF full treatment.
- To control the corrosion potential of RO permeate, we investigated the effects on water quality associated with blending of full vs. split treatment. At 70% recovery, RO split stream treatment reduced the corrosion potential of RO full stream treatment, providing a cost-effective alternative to, and better water quality than, NF full stream treatment.
- Considering design and operational parameters such as maintenance cost, interest rate, power cost, chemical cost, plant life and membrane replacement, each parameter affected the product water cost.

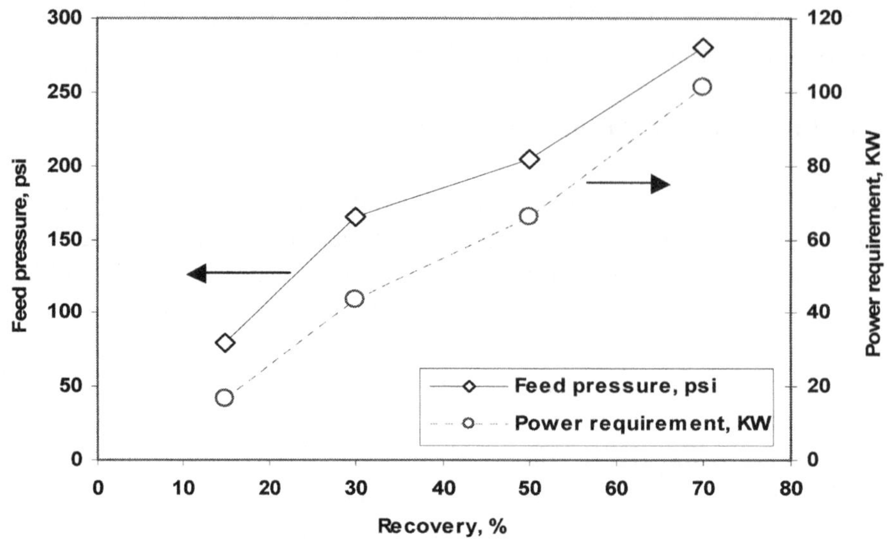

(a) LFC-1 (RO)

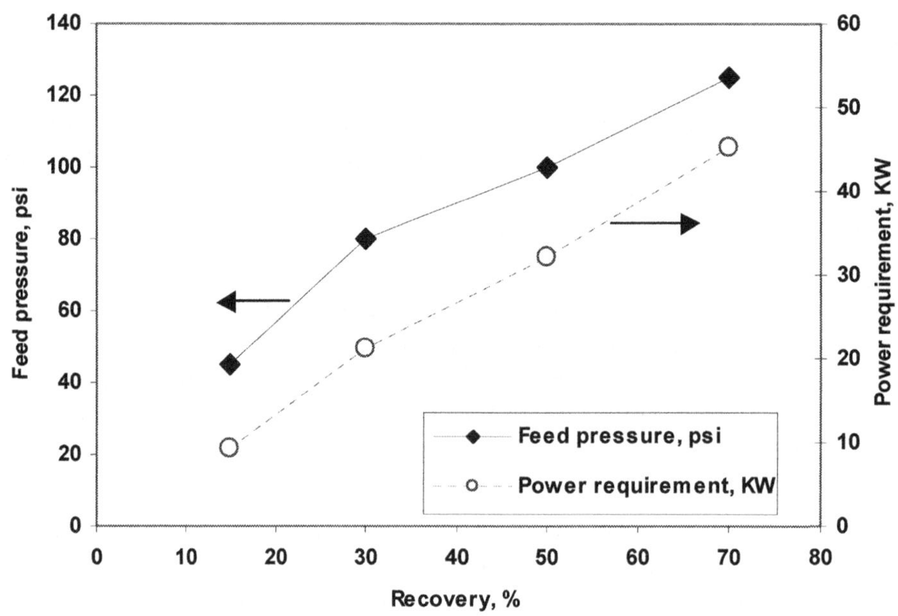

(b) ESNA (NF)

Figure 9.1 Feed pressure and power requirement at various recoveries

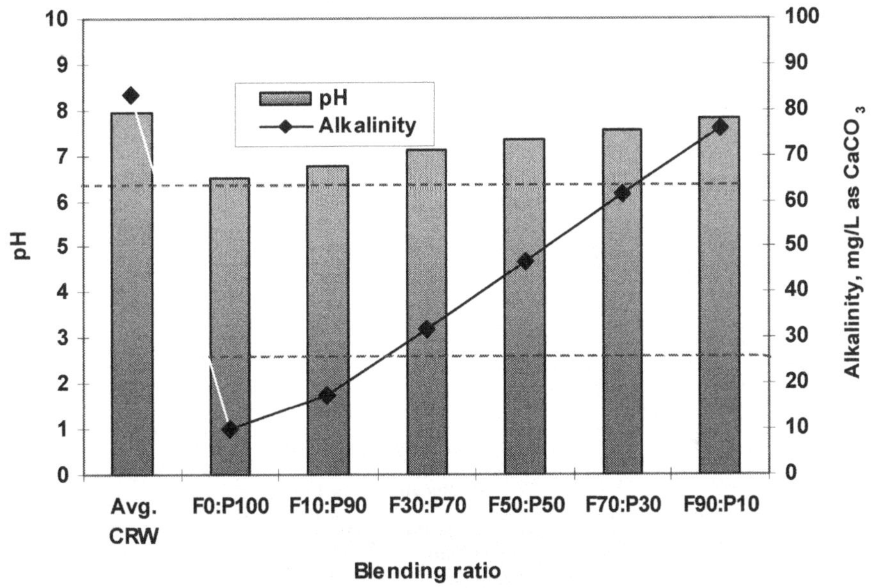

(a) LFC-1 (RO)

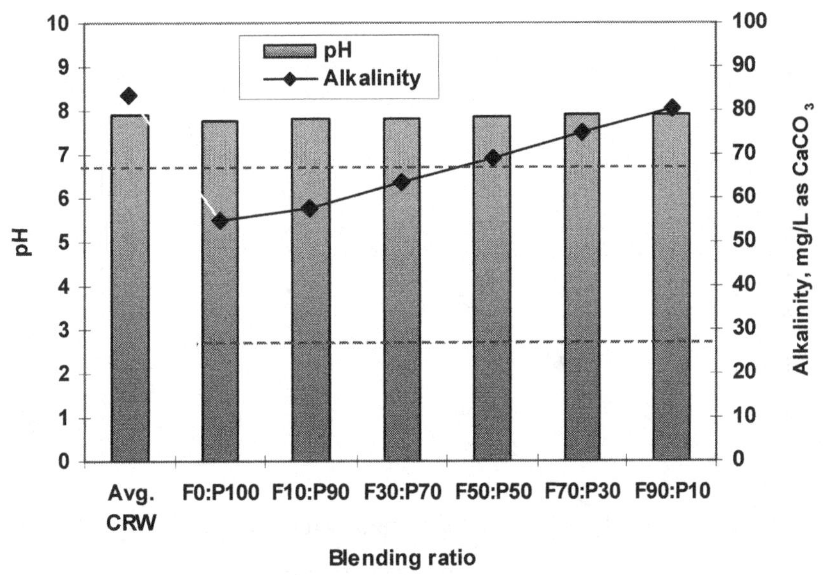

(b) ESNA (NF)

Figure 9.2 Alkalinity and pH variation as a function of blending ratio between CRW feed (F) and permeate (P) at recovery 70% (average perchlorate concentration 100 μg/L spiked in CRW)

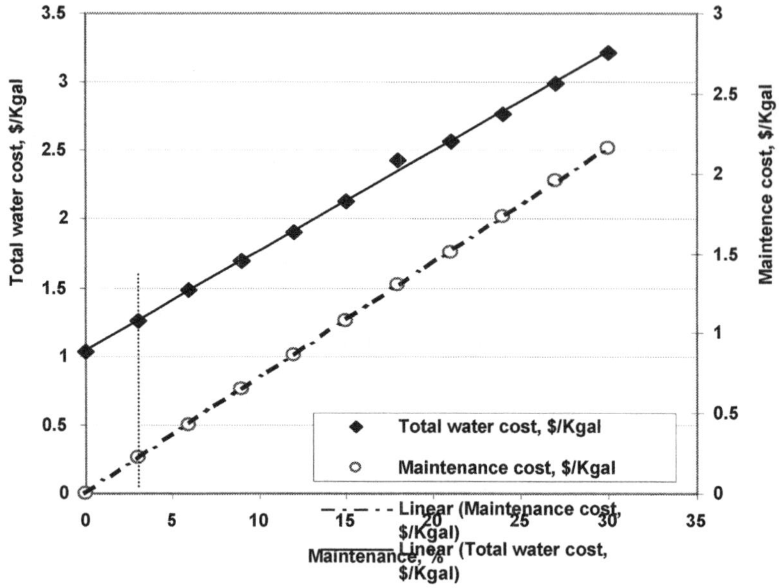

(a) LFC-1 (RO)

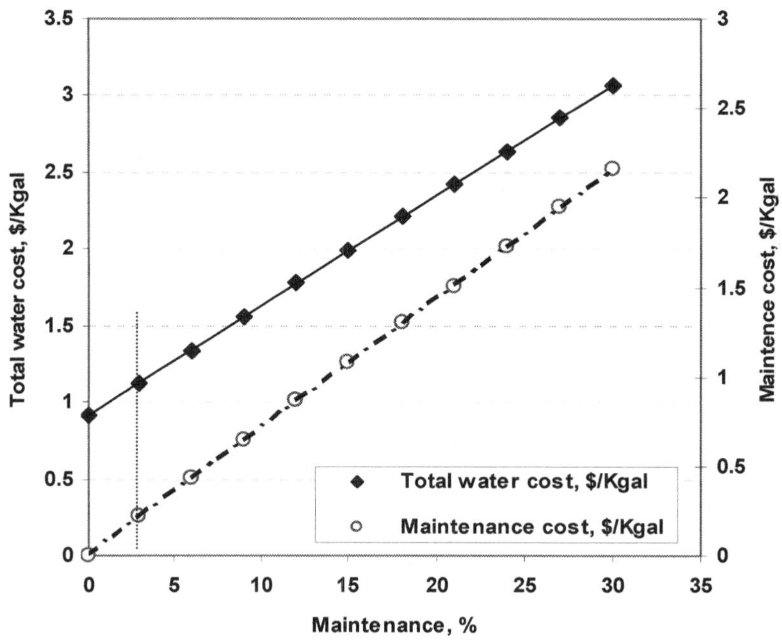

(b) ESNA (NF)

Figure 9.3 Effect of maintenance (% of capital) on total water cost and maintenance cost at a recovery of 70% and with no blending

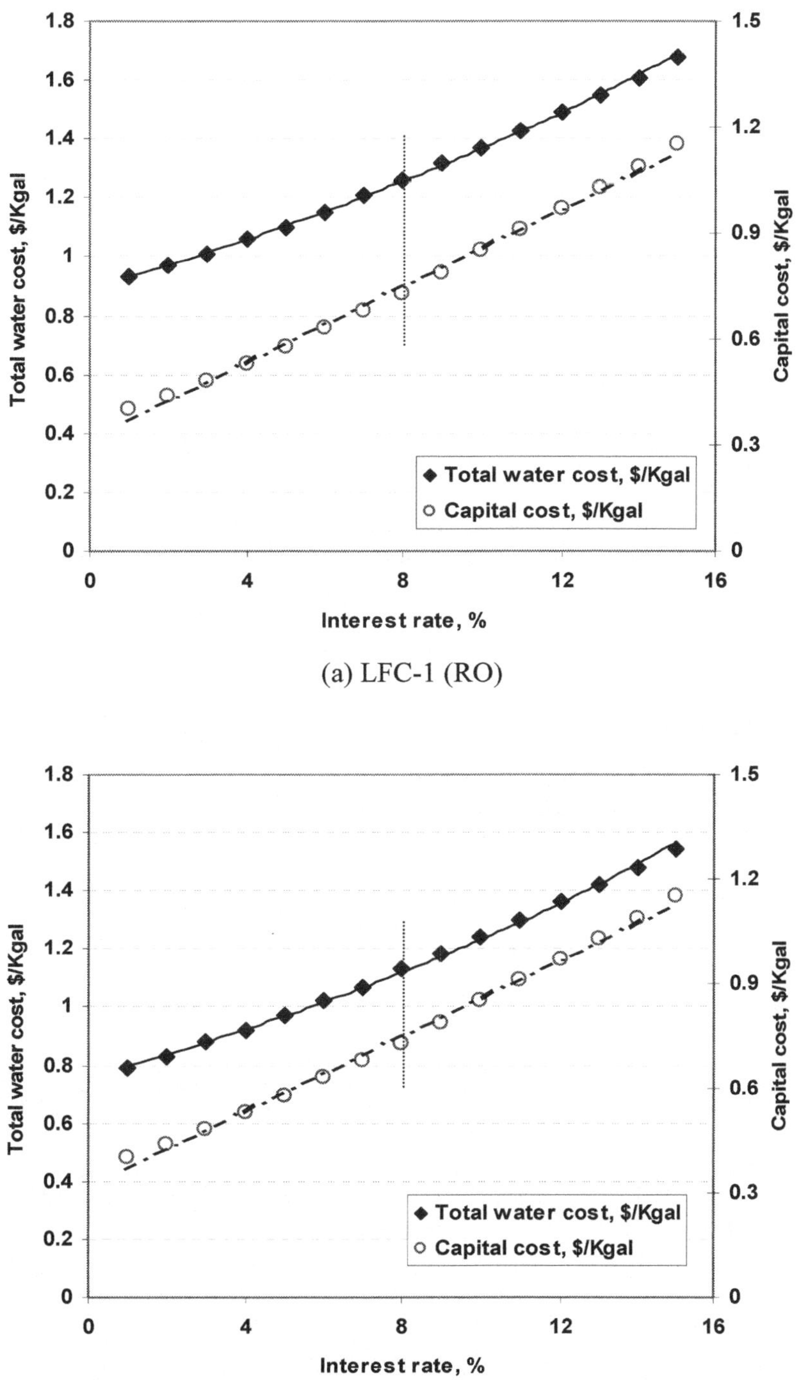

(a) LFC-1 (RO)

(b) ESNA (NF)

Figure 9.4 Effect of interest rate on total water cost and capital cost at a recovery of 70% and with no blending

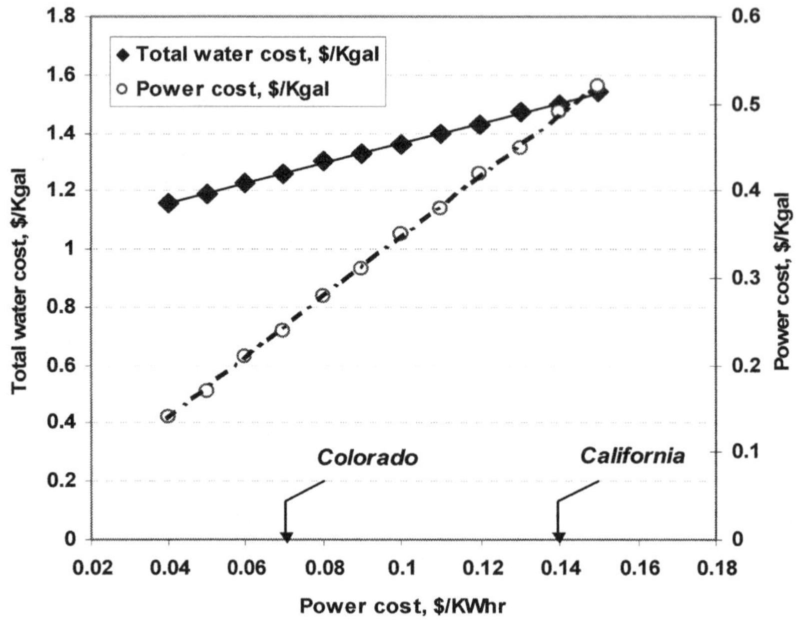

(a) LFC-1 (RO)

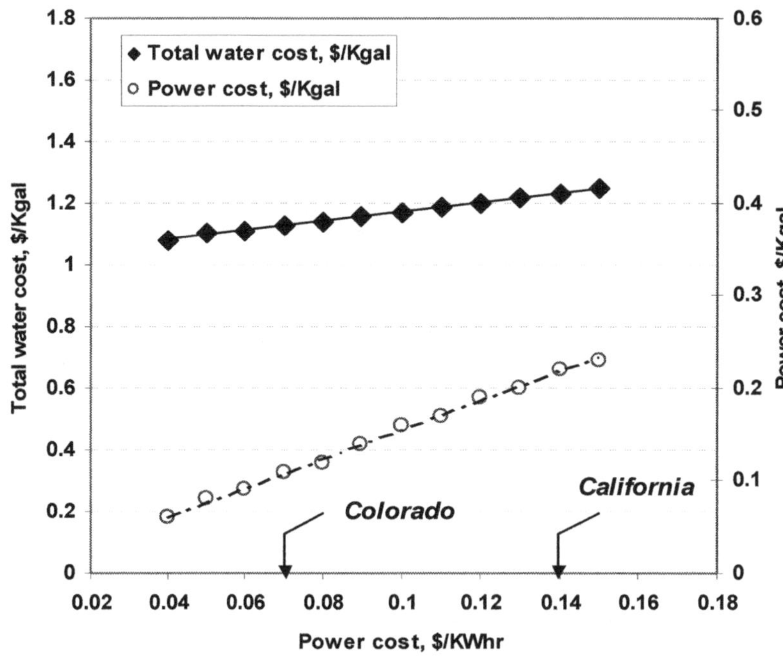

(b) ESNA (NF)

Figure 9.5 Effect of power cost ($/KWhr) on total water cost and unit power cost ($/Kgal) at a recovery of 70%

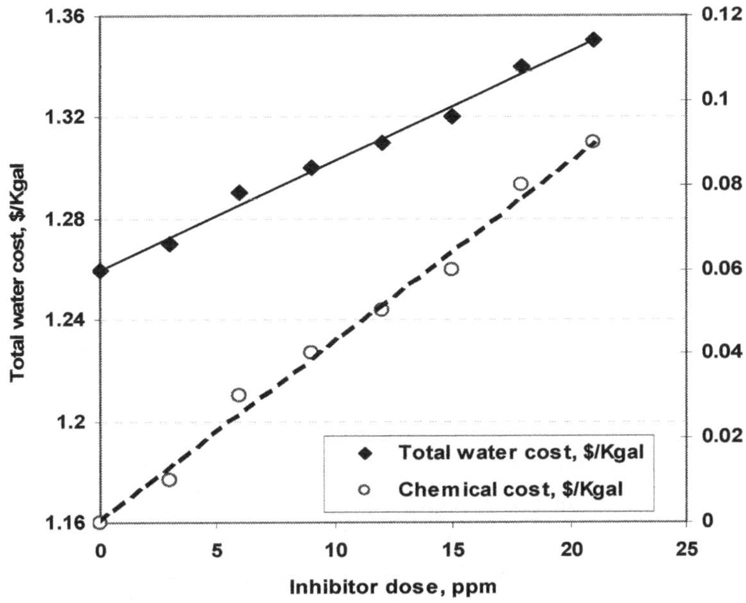

(a) LFC-1 (RO)

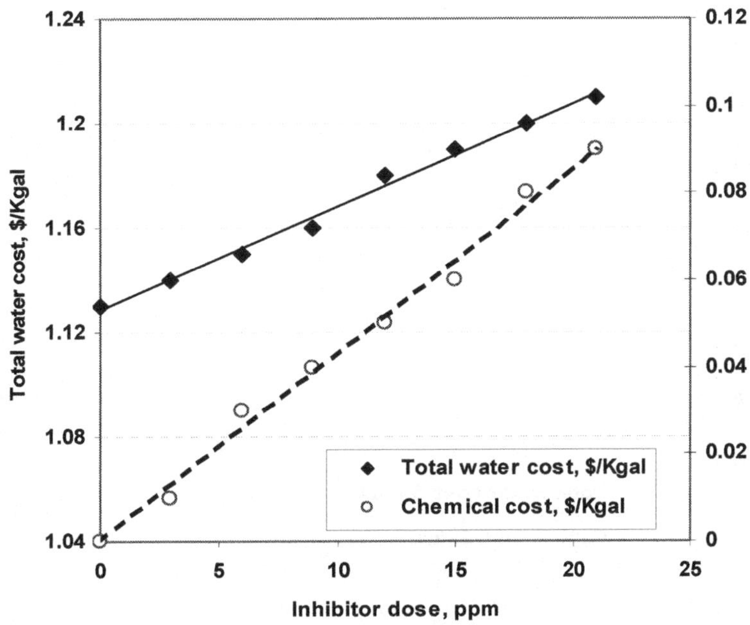

(b) ESNA (NF)

Figure 9.6 Effect of inihibitor dose on chemical cost and total water cost at a recovery of 70% (no blending, cost of Deguest 2000: $0.50/lb)

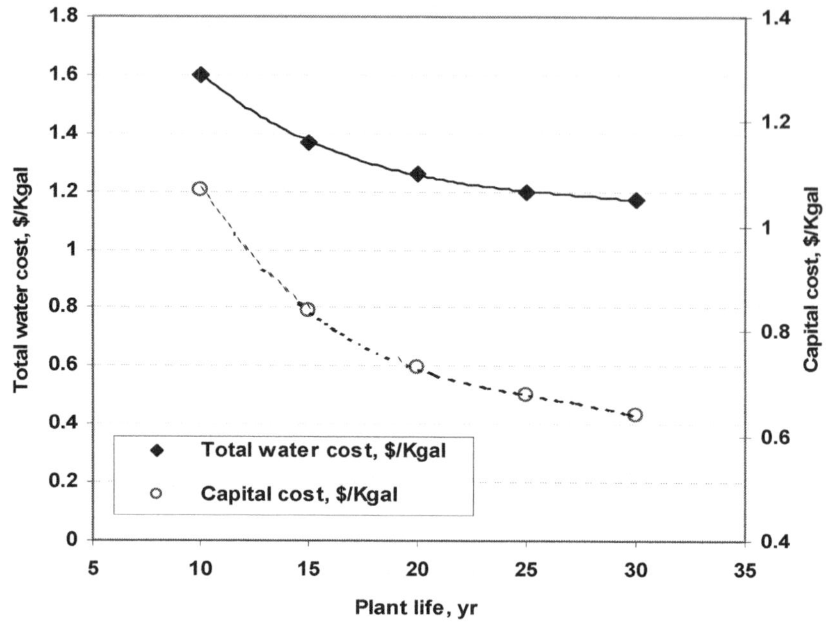

(a) LFC-1 (RO)

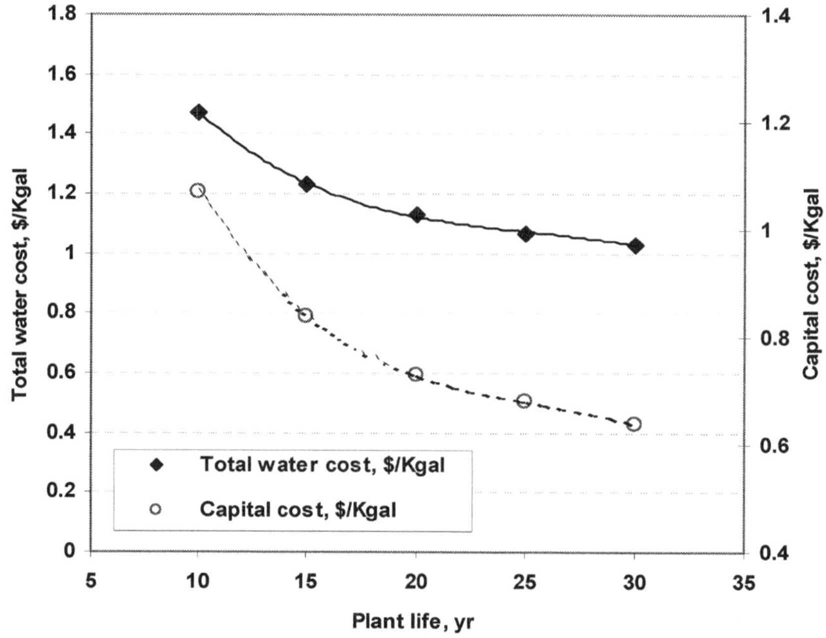

(b) ESNA (NF)

Figure 9.7 Effect of plant life on total water cost and capital cost at a recovery of 70%

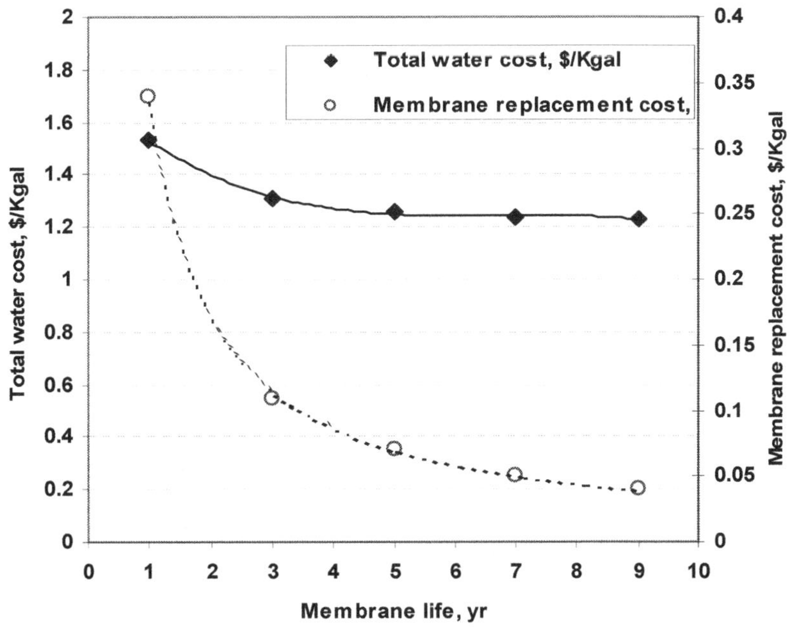

(a) LFC-1 (RO)

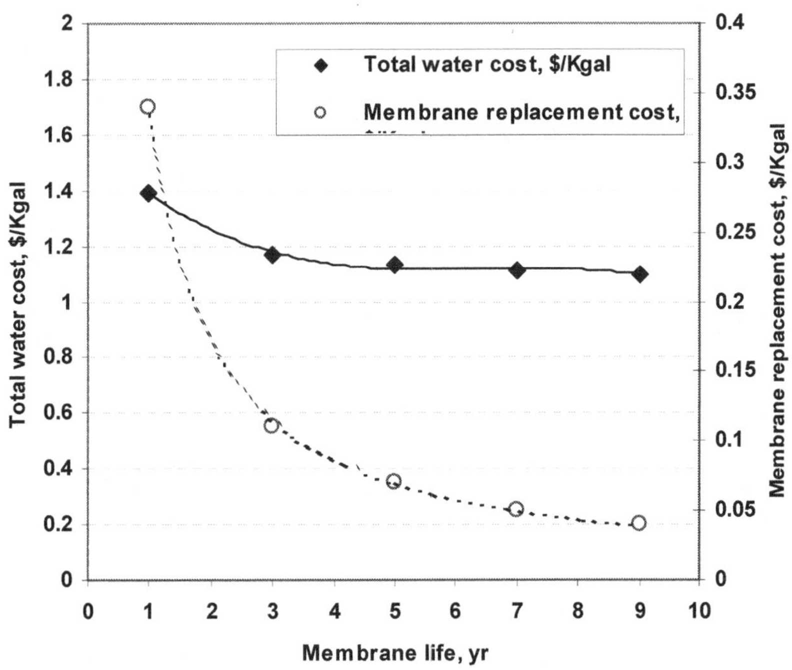

(b) ESNA (NF)

Figure 9.8 Effect of membrane life on total water cost and membrane replacement costs at a recovery of 70%

CHAPTER 10
SPECIAL STUDIES: (I) USE OF SURFACTANT MODIFIED ULTRAFILTRATION FOR PERCHLORATE REMOVAL AND (II) PERCHLORATE OXIDATION EFFECTS ON MEMBRANES

SPECIAL STUDY I: USE OF SURFACTANT MODIFIED ULTRAFILTRATION FOR PERCHLORATE

INTRODUCTION

Ultrafiltration (UF) alone is not very effective in ClO_4^- rejection; because of the pore size of UF membranes, observed rejections can be attributed to electrostatic exclusion.

In this work, a cationic or anionic surfactant was added to synthetic and natural waters at a concentration less than its critical micellar concentration (CMC) to enhance UF rejection of ClO_4^-. Perchlorate is an anion with a molecular mass of 100 and an ionic radius of 1.37×10^{-10} m (calculated from Stokes Einstein Equation based on its diffusion coefficient).

In this study, we have used existing theories; Donnan equilibrium and extended Nernst-Planck models; with bench-scale measurements to predict the transport of target ions (in this case ClO_4^-) through a negatively charged membrane in the presence of a cationic or anionic surfactant. We have also investigated how the presence of a surfactant affects electrostatic interaction (charge repulsion) and steric (size) exclusion mechanisms between a negatively charged porous UF membrane and the perchlorate anion. When a dilute solution containing ions is brought into contact with a membrane possessing a fixed charge, the passage of ions possessing the same charge as the membrane (co-ions) can be inhibited. This condition is termed Donnan Exclusion (Strathmann H., 1992). The use of the extended Nernst-Planck model in conjunction with the Donnan equilibrium condition provides the possibility of characterizing the effective membrane pore size and effective charge density (Bowen and Mukhtar, 1996). These relationships can be used to predict the separation of mixtures of electrolytes at the membrane/solution interface. Effective membrane charge depends on the bulk co-ion concentration associated with mono-valent and divalent co-ions if solution concentrations are described in equivalents (Bowen and Mukhtar, 1996).

BACKGROUND

Morel suggested that small concentrations of cationic surfactant significantly modified cellulose acetate membrane properties concerning nitrate ion rejection (Morel *et al.*, 1997); the surfactant addition produced a strong adsorption of surfactant molecules, which acted on the Donnan potential difference. Cho performed a rejection test with an uncharged model species, polyethylene glycols (PEGs), to determine the effects of mono-valent and di-valent ions, Cl^-, Na^+, and Ca^{2+}, on membrane pore size (Cho *et al.*, 2000). Cho found that an ultrafiltration (UF) membrane, whose nominal molecular mass cut-off (MMCO) was 8,000, rejected smaller molecular mass PEGs in deionized water than in NaCl or $CaCl_2$ solutions (Cho *et al.*, 2000). Electrostatic exclusion is known to be one of the dominant mechanisms between a negatively charged solute and a negatively charged UF membrane.

Membrane surface charge can be determined by zeta potential, calculated from measurement of streaming potential. A zeta potential is defined as the potential across the fluid sheer plan at the membrane surface. Childress and Elimelech studied the effects of anionic/cationic surfactants on the zeta potential of membrane (one reverse osmosis and two nanofiltration (NF)) surfaces in the presence of 0.01 M NaCl (Childress and Elimelech, 1996). In their research, the negatively charged sulfate functional groups of the anionic surfactant molecule caused the membrane to become more negatively charged. Additionally, the effect of the adsorbed cationic surfactant was opposite to that of the adsorbed anionic surfactant in that the membrane became less negatively charged. Surface charge also depends on the functional groups associated with different membrane materials.

An attenuated total reflection-Fourier transform infrared (ATR-FTIR) spectrum can be used to qualitatively determine clean/fouled membrane functional groups. Cho compared the FTIR spectra of UF (GM) and NF (ESNA) membranes (Cho *et al.*, 1998). Infrared spectroscopy was also used to identify organic compounds that are readily distinguished from the absorption pattern of other compounds of an NF membrane (NF 200) (Her *et al.*, 2000). Contact angle is an index of the hydrophobicity of the membrane surface, which can be measured by captive bubble and sessile drop methods (Zhang *et al.*, 1990; Gekas *et al.*, 1992).

A hydrodynamic operating parameter, the ratio of the initial pure water flux (J_o) to the estimated back-diffusional mass transfer coefficient (k) through the boundary layer, has been developed to facilitate comparison of results from different membranes in bench-scale cross-flow flat-sheet tests (Cho *et al.*, 1999).

METHODS AND ANALYSES

Membrane and Testing Unit

A UF membrane, GM (Desal), made of sulfonated polyethersulfone coated with an ultrathin polyimide was used for the perchlorate rejection tests. This membrane is characterized by ionizable functional groups such as carboxylic acids based on ZP, and its nominal MMCO as provided by the manufacturer. The surface charge was estimated by measuring the streaming potential of the membrane. A commercial electrokinetic analyzer (EKA) measurement apparatus was employed to estimate zeta potential of the membrane with the measured streaming potential.

Table 4.1 describes the characteristics of the membrane based on zeta potential (surface charge), material (functional group), nominal MMCO (pore size), and permeability (pore size and/or pore density). Figure 4.1 and Table 4.4 show a schematic of the experimental system and operating conditions used for bench-scale cross-flow flat-sheet tests.

Surfactants and Source Waters

A cationic surfactant and an anionic surfactant were used in this study. The cationic surfactant was tetradecyl trimethyl ammonium bromide (TTAB) and the anionic surfactant was sodium dodecyl sulfate (SDS). Table 10.1 shows that the CMC of the TTAB and SDS is 3.4 mM and 8.5 mM, respectively. Three different concentrations of the TTAB (0.17 mM, 0.34 mM, and 1.7 mM) that are lower than the CMC were used in experiments. Two different SDS concentrations, 2 mM and 4 mM, were also employed. All of the synthetic surfactant and perchlorate anion solutions (present at concentrations of 100 μg/L of $KClO_4$) were prepared from pure water at pH 7.6, in binary mixtures with other salts (KCl and/or $CaCl_2$), and at varying solution conductivities (30 mS/m and/or 115 mS/m). The synthetic waters were prepared with pure water, which was deionized followed by 0.2 μm-filter filtration. Potassium hydroxide or hydrochloric acid was added to pure water to adjust solution pH. Also, a blend of Colorado River water and State Project water (CRW/SPW) from the Metropolitan Water District in California was used to perform bench-scale cross-flow flat-sheet membrane tests. The CRW/SPW contains approximately 4 μg/L of ClO_4^-. The CRW/SPW was prefiltered with a 0.45 μm filter for all membrane tests.

Table 10.1
Surfactant characteristics

Surfactant	Formula	CMC (mM), 25 °C	Surfactant Concentration (mM)
Tetradecyl trimethyl ammonium bromide (TTAB)	$CH_3(CH_2)_{13}N(CH_3)_3Br$	3.4	0.17, 0.34, and 1.7
Sodium dodecyl sulfate (SDS)	$CH_3(CH_2)_{10}CH_2OSO_3Na$	8.5	2 and 4

Tables 4.2 and 4.3 describe the characteristics of source (feed) waters used in this study. Several analyses were performed, including dissolved organic carbon (DOC), ultraviolet adsorption at 254 nm (UVA_{254}), conductivity, total dissolved solids (TDS), and pH of the CRW/SPW. Anions and cations in the CRW/SPW were measured to predict how mono- and divalent co- and counter ions, Cl^-, SO_4^{2-}, K^+, and Ca^{2+}, influenced perchlorate rejection in the absence/presence of cationic and anionic surfactants.

Estimation of the Membrane Pore Size Change by PEG Rejection Tests

Rejection tests were performed with a range of PEGs (relative molecular mass 2,000 - 10,000) by using a cross-flow flat-sheet filtration unit for clean and surfactant modified GM membranes. A PEG concentration of approximately 5 mg/L as DOC was used for each of the four PEG rejection measurements. The PEG was introduced to the clean and surfactant modified membranes as a pure component (pH 7.6 and varying conductivity < 3.0 mS/m), and in binary mixtures with other salts (KCl and/or $CaCl_2$) and varying solution conductivity (30 mS/m and/or 115 mS/m). KOH and HCl were added to adjust the solution pH. Each bulk water was filtered and wasted for 30 minutes for membrane pore stabilization, and a permeate sample of 30 mL was collected for analyzing PEG rejection based on DOC rejection. PEG rejection, based on initial feed concentration, was calculated by using Equation 10.1:

$$R_j(initial\ feed),\% = \frac{C_{if} - C_{pp}}{C_{if}} \times 100\,\% \qquad (10.1)$$

where, C_{if} is the PEG concentration in the initial feed fluid and C_{pp} is the PEG concentration in the permeate. These measurements were performed not to estimate the MMCO of the membrane, but to determine the effects of surfactant conditioning on membrane pore size, which influences perchlorate transport by steric (size) exclusion due to the change of membrane pore size.

Streaming Potential Measurements

Streaming potential was measured with an electrokinetic analyzer apparatus (EKA). The same procedures and the same EKA as those used by Wilbert were used to measure streaming potential (Wilbert *et al.*, 1999). For the streaming potential measurements, membrane samples were cut to fit the measurement cell and then wetted in KCl and $CaCl_2$ solution in the presence of anionic/cationic surfactants at a sudden pH 7.6 and stored in a refrigerator for the soaking time specified in the experimental design. ZP measurements in the absence/presence of anionic/cationic surfactants were performed to estimate how the changes of the membrane surface charge affect perchlorate transport by electrostatic repulsion.

ZP was calculated from the measured streaming potential using the method described by Fairbrother and Mastin, which is based on The Helmholtz-Smoluchowski relationship (Equations 4.3 and 4.4)

Measurements of Contact Angle and FTIR

A gomiometer (Rame-Hat inc. NRL, CA) was used to measure the contact angle of the membrane. The contact angle is a measure of the hydrophobicity of the membrane and the energy of interaction of the membrane with other materials (Gekas *et al.*, 1992). Contact angles of the membrane were measured before and after surfactant modification to indicate surfactant adsorption onto/into the membrane. The contact angle was measured by a sessile drop method, based on measuring the contact angles between the water droplet and the membrane surface.

ATR-FTIR spectrum was measured using a Nicolect 752 spectroscope with a 45° ZnSe flat plate crystal, to determine functional groups associated with the membrane surface materials and surfactants. FTIR analyses were performed on membrane specimens before and after cationic/anionic surfactant conditioning. A potassium bromide pellet was prepared to perform FTIR analysis for a solid sample of the cationic surfactant.

RESULTS AND DISCUSSION

Prediction of Perchlorate Transport Based on Surfactant Modified Membrane Characterization: PEG Rejection and Zeta Potential

PEG rejections are plotted versus average molecular mass of PEG for the GM membrane at pH 7.6 and two solution conductivities, as established with $CaCl_2$ (see Figure 10.1). Figure 10.1 describes the results of PEG rejection trends with the cationic/anioinc surfactant modified GM membrane. The objective of these PEG rejection measurements is not only to determine the changes of the surfactant modified membrane pore size but also to estimate perchlorate transport by the surfactant modified membrane in the presence of co- and counter ions. PEG rejection increased with increasing the cationic surfactant, TTAB, and Ca^{2+} concentrations at pH 7.6. From these results, an assumption is made that perchlorate rejection can increase due to a decrease of the membrane pore size when the TTAB is adsorbed onto/into the membrane surface/pores, especially in the presence of divalent ion, Ca^{2+}. However, there was no significant difference based on PEG rejection trends between the clean and anionic surfactant modified membranes. It is concluded that the anionic surfactant modified membrane does not significantly affect perchlorate rejection due to steric exclusion in the presence of Ca^{2+} (see Figure 10.1-c).

Figure 10.2 shows the effect of cationic/anionic surfactant conditioning on zeta potential. All streaming potentials were measured at pH 7.6 with KCl or $CaCl_2$ background solution (30 mS/m or 115 mS/m) in the presence of varying amount of cationic/anionic surfactants. The ZP of the clean GM membrane in the absence of surfactant with KCl of 30 mS/m is -9.6 mV. The ZP decreases toward the isoelectric point with increasing TTAB and Ca^{2+} concentration. However, ZP was not significantly affected by the SDS anionic surfactant conditioning, which also proves that Donnan potential was not significantly affected by the anionic surfactant conditioning. From these results, two observations are made in that perchlorate rejection can decrease due to the membrane surface charge becoming less negative with the cationic surfactant conditioning, and that perchlorate rejection may not be significantly affected by electrostatic interaction due to anionic surfactant modification.

Determination of Cationic/Anionic Surfactant Adsorption: Contact Angle and FTIR

Figure 10.3 illustrates the membrane surface of the cationic/anionic surfactant modified GM membrane. The surfactants have both a hydrophobic tail and a hydrophilic head. The hydrophilic head of the surfactant has a negative/positive charge depending on cationic or anionic surfactant type. Figure 10.3 also illustrates the adsorption mechanism of cationic/anionic surfactants on the negatively charged membrane surface. Because the surface of the GM membrane is negatively charged at pH 7.6, the head of the TTAB cationic surfactant attaches to the negative charged surface of the membrane. Here, the tails of the cationic surfactant are hydrophobic, thus the hydrophobicity increases on the surface, and a relatively large contact angle is shown (see Figure 10.3-a). Also, in the case of high concentration of TTAB adsorption, tails of the cationic surfactant stand opposite to each other (Childress and Elimelech, 1996). Thus, the heads of the cationic surfactant are inclined to the upper side, and the membrane surface shows relatively large hyrophilicity (Rosa and De Phino, 1997). For this reason, contact angle is smaller than that of a low concentration TTAB modified membrane (see Figure 10.3-b).

Figure 10.3-c also illustrates the adsorption of anionic surfactant on the membrane surface. Hydrophilic heads of the anionic surfactant are inclined to the upper side, and the contact angle of the membrane is relatively smaller than that of the hydrophobic surface as shown in Figure 10.3-c. However, if the pH value is lower, the surface charge of the GM membrane will eventually approach the isoelectric point and the surface may become more positive at relatively low pH. Under this condition, the hydrophobic tails of the anionic surfactant are inclined to the upper side and the contact angle will be not shown in Figure 10.3 (Childress and Elimelech, 1996). Also a competition between anions such as perchlorate and chloride and the negatively charged membrane inhibited SDS adsorption, which influenced contact angle. Table 10.2 summarizes the results of contact angle measurements at varying surfactant concentrations and conductivities. From Table 10.2, it is generalized that contact angle increased at relatively low TTAB concentration. Contact angle was not significantly affected at relatively high TTAB concentration and by SDS conditioning. These results are consistent with previous studies (Morel *et al.*, 1997; Childress and Elimelech, 1996).

The FTIR spectra of a clean GM membrane and a GM membrane with adsorbed TTAB in the presence of Ca^{2+} and Cl^{-} are compared in Figure 10.4. The clean GM membrane exhibited IR peaks of aromatic double bonded carbons (around 1,500 and/or 1,600 cm^{-1}) and carboxylic groups (around 1,250 cm^{-1}) (see Figure 10.4-a). The cationic surfactant, TTAB, $CH_3(CH_2)_{13}N(CH_3)_3Br$ (molecular mass. 336.41), adsorbed onto the GM membrane, exhibits

Table 10.2
Summary of contact angle measurements at varying cationic/anionic surfactant concentrations and conductivities

	Conc. (mM)	pH	Conductivity (mS/m)	Contact angle (°)
Clean GM membrane				45
TTAB modified GM membrane	1.7	8	30 w/ $CaCl_2$	47
	1.7	8	115 w/ $CaCl_2$	48
	0.34	8	30 w/ $CaCl_2$	61
	0.34	8	30 w/ KCl	63
	0.34	8	115 w/ $CaCl_2$	61
	0.17	8	30 w/ KCl	56
	0.17	8	30 w/ $CaCl_2$	56
SDS modified GM membrane	4	8	115 w/ $CaCl_2$	49
	4	8	30 w/ $CaCl_2$	42

relatively dominant IR peaks of methyl groups (around 2750-2950 cm^{-1}) in Figure 10.4-b. The IR peaks shown in Figure 10.4-b for the GM membrane conditioned with the cationic surfactant, TTAB, verified that the cationic surfactant, TTAB, was absorbed onto/into the surface/pores of the negatively charged membrane in the presence of co- and counter ions, Ca^{2+} and Cl^-. Figure 10.4-b also verifies that methyl groups which are relatively dominant in TTAB were adsorbed onto and/or into the GM membrane. Figure 10.4-c shows that the cationic surfactant, TTAB, was also adsorbed onto/into the GM membrane under the conditions with various anions, cations, and NOM in the CRW/SPW. However, Figure 10.4-d verifies that the anionic surfactant, SDS, was not significantly adsorbed onto/into the GM membrane in the presence of Ca^{2+} and Cl^- due to possibly electrostatic repulsion occurring between the negatively charged GM membrane and the anionic SDS surfactant.

Perchlorate Rejection and Flux-Decline With Pure Water and CRW/SPW Without Surfactant Conditioning

Perchlorate rejection by the GM membrane was monitored over time for 22 hours with pure water (at pH 7.6 and conductivity < 3.0 mS/m) and CRW/SPW spiked with 100 μg/L (into pure water) and 96 μg/L (into CRW) ClO_4^- since the CRW/SPW included 4 μg/L ClO_4^-. Flux decline trends were also monitored at the same time. In Tables 4.2, 4.3, and 4.4, the feed water qualities of CRW/SPW, ions in CRW/SPW, and experimental operating conditions of cross-flow flat-sheet filtrations with the membrane, respectively, are presented. In Figure 10.5, the membrane shows a relatively high perchlorate rejection (80%) with pure water in the presence of a minimal amount of co- and counter ions. However, the GM membrane consistently provides less than 5% perchlorate rejection with approximately 8% flux-decline over time with CRW/SPW. In this work, perchlorate rejection did not appear to be affected by flux decline during short-term bench-scale filtration tests without surfactant conditioning.

From the results shown in Figure 10.5, an assumption is made that electrostatic exclusion significantly contributes to perchlorate rejection from the relatively low ionic strength (conductivity) synthetic water for the porous membrane even though its pore size is much larger than perchlorate ion. However, perchlorate rejection by this membrane is significantly reduced from CRW/SPW compared to relatively low ionic-composition synthetic water. Yoon suggested that ions in CRW caused the ZP of the membrane to significantly become less negative; thus, electrostatic repulsion decreased between the perchlorate ion and the GM membrane. In that study, the ZP of the GM membrane became less negative with increasing divalent cation, Ca^{2+}, due to ion binding (Yoon, 2001).

From the results shown in Figure 10.5, an assumption was made that perchlorate transport can be affected by the changes of membrane surface charge and membrane pore size due to anionic/cationic surfactant conditioning. In the previous sections, cationic surfactant conditioning shows significant effects on membrane surface charge and pore size, which can affect perchlorate transport; perchlorate rejection can increase by the decrease of the membrane pore size and decrease by the membrane surface charge becoming less negative due to the cationic surfactant conditioning.

Perchlorate Rejection and Flux Decrease at Varying Surfactant Concentrations and Conductivities

The effects on perchlorate rejection by the GM membrane from changes in solution conductivity and surfactant concentration were determined with synthetic and CRW/SPW waters. Perchlorate rejection was higher in the presence of a divalent counter ion, Ca^{2+}, than in the presence of a mono-valent counter ion, K^{+}, when a TTAB concentration of 0.17 mM was added into a synthetic water at concentration of 96 μg/L of ClO_4^- at pH 7.6. (See Figure 10.6-a). This is possibly because more steric (size) exclusion occurred in the presence of the divalent cation, Ca^{2+}, than in the presence of the mono-valent, K^{+}, under the conditions of cationic surfactant conditioning, which influences the membrane pore size based on PEG rejection and flux-decrease tests as shown in Figures 10.1 and 10.9. Figure 10.6-b shows that a relatively large amount of perchlorate was rejected when more divalent counter ion, Ca^{2+}, was introduced to the membrane surface in the presence of a TTAB concentration of 1.7 mM. This result was because of the relatively high decrease of membrane pore size with increasing the Ca^{2+} concentration. However, perchlorate rejection was consistently less than 5% with anionic surfactant conditioning even in the presence of Ca^{2+} and Cl^- ions (115 mS/m) at both SDS concentrations of 2 mM and 4 mM (see Figure 10.7). This result was consistent with the assumption, based on the results of PEG rejections and ZP measurements with the anionic surfactant conditioned membrane.

Higher perchlorate rejection was obtained at a higher cationic surfactant presence with CRW/SPW (See Figure 10.8-a). In Figure 10.8-a, a relatively large TTAB concentration of 1.7 mM shows a relatively high perchlorate rejection of 80% with the CRW/SPW. On the contrary, the anionic surfactant conditioned membrane with two SDS concentrations of 2 mM and 4 mM shows consistently less than 5% of perchlorate rejection even with CRW/SPW (see Figure 10.8-c).

Membrane permeate flow in each experiment was monitored over time to determine the effects of surfactant adsorption on perchlorate rejection associated with flux-decrease. In Figure 10.9-a, membrane flux decreased in the presence of a divalent counter ion, Ca^{2+}, compared to that in the presence of a mono-valent counter ion, K^{+}, with surfactant conditioning for the GM membrane. A relatively high flux-decrease occurred in response to conditioning with a relatively high level of TTAB, 1.7 mM, in the presence of divalent counter ion, Ca^{2+} (see Figure 10.9-b). These flux-decrease trends were consistent with an assumption that ClO_4^- rejection increases with decreasing membrane pore size due to cationic surfactant adsorption onto/into the surface/pores of the negatively charged membrane. In Figure 10.9-b, approximately 85% flux decrease occurred with cationic surfactant at a TTAB concentration of 1.7 mM, with surfactant conditioning over time for the GM membrane. However, relatively low flux-decrease occurred in response to conditioning with two SDS levels, 2 mM and 4 mM, in the presence of Ca^{2+} (see Figure 10.10).

Figure 10.11 shows the effect of cationic/anionic surfactant, TTAB and SDS, conditioning on flux-decrease with CRW/SPW for the GM membrane when 96 μg/L (9.6×10^{-7} M) of ClO_4^- was spiked into CRW/SPW. The purpose here was to evaluate the effects of natural organic matter (NOM) and other co- and counter ions in CRW/SPW on flux-decrease in conjunction with surfactant conditioning. Flux-decrease trends with the CRW/SPW were greater than those with synthetic water having only Ca^{2+} and Cl^- (see Figures 10.9-a and 10.11-b) with cationic surfactant conditioning. These results describe that NOM and other ions also significantly affected the flux-decrease trends for the cationic/anionic surfactant conditioning.

Table 10.3 summarizes overall perchlorate rejection and flux-decrease results performed with various concentrations of surfactants and conductivities (some of these results were not shown graphically in figures). From Table 10.3, it is generalized that membrane flux decreased in the present of a divalent counter ion, Ca^{2+}, compared to that in the presence of a mono-valent ion, K^+, with surfactant conditioning of the GM membrane. Relatively high flux decrease occurred in response to conditioning with a relatively high level of TTAB, 1.7 mM, in the presence of divalent ion, Ca^{2+}. However, a relatively small amount of flux-decrease occurred with the anionic surfactant conditioning for both synthetic and CRW/SPW waters.

Table 10.3
Overall comparison of perchlorate rejection and flux decrease at varying surfactant concentrations and conductivities for a UF membrane

Surfactant/ Source Water	Surfactant conc. (mM)	Conductivity (mS/m)	pH	ClO_4^- rejection (%)	Flux decrease (%)
TTAB (synthetic)	1.7	30 w/ $CaCl_2$	8	50	79
	1.7	115 w/ $CaCl_2$	8	65	88
	0.34	30 w/ $CaCl_2$	8	23	46
	0.34	30 w/ KCl	8	15	27
	0.34	115 w/ $CaCl_2$	8	15	48
	0.17	30 w/ KCl	8	15	18
	0.17	30 w/ $CaCl_2$	8	22	30
SDS (synthetic)	4	115 w/ $CaCl_2$	8	<3	<5
	2	115 w/ $CaCl_2$	8	<3	<5
TTAB (CRW/SPW)	1.7	75.7	7.6	83	89
	0.34	75.7	7.6	27	78
SDS (CRW/SPW)	4	75.7	7.6	<5	<5
	2	75.7	7.6	<5	<5

SUMMARY

The objective of this study was to surfactant-modify a negatively charged UF membrane in terms of the fundamental mechanisms, steric/size exclusion and electrostatic exclusion, to enhance perchlorate rejection from with synthetic water and CRW/SPW. CRW/SPW was selected as a model source (feed) water for this study because CRW contaminated with perchlorate has been widely used as a drinking source water. Since CRW/SPW has relatively high levels of ions, it can also be used as a representative source (feed) water to study perchlorate transport through a negatively charged membrane.

ClO_4^- rejection by a UF membrane modified with cationic surfactant was greater than expected, based mostly on steric/size exclusion as a result of a decrease of the membrane pore size. However, a limitation in use of a cationic surfactant to modify a negatively charged UF membrane for perchlorate rejection was that a significant amount of flux-decrease also occurred. As expected, perchlorate transport was not affected by anionic surfactant conditioning, which did not significantly influence membrane pore size or membrane surface charge.

Two understandings were obtained from this study. This study was performed not only to modify a negatively charged membrane with cationic/anionic surfactants for ClO_4^- rejection but also to understand perchlorate transport phenomena by changes in steric (size) exclusion and electrostatic exclusion, which can be significantly influenced by cationic surfactant conditioning.

SPECIAL STUDY II: PERCHLORATE OXIDATION EFFECTS ON MEMBRANES

Membrane rejection is affected by MMCO, membrane material and charge, pH, ionic strength, and feed water. Perchlorate-containing water can possibly affect membrane rejection because perchlorate has an oxidizing potential that can change the membrane material and alter flux decline and rejection. Membrane selection is being based on membrane material, MMCO, and degree of surface charge, making it necessary to compare perchlorate oxidation effects on different membranes. In this part of the study, we have evaluated perchlorate oxidation effects on four different types of membranes including RO, NF, and UF membranes. Membrane flux and perchlorate rejection have been compared before/after perchlorate oxidation. We focused on RO, NF, and UF membranes using a dead-end stirred-cell unit for perchlorate oxidation of membrane material. We focused on four different membranes (TFC-ULP, ESNA, MX07, and GM). The membranes are characterized in Table 4.1.

These experiments were performed with 50 µg/L of perchlorate spiked into pure water, before and after soaking with 1,000 µg/L of perchlorate solution for 24 hrs, to examine perchlorate oxidation effects on membrane perchlorate rejection. Each test used 4L of reservoir feed connected to the dead-end stirred–cell unit. Membrane fluxes were monitored for 24 hours and compared before/after soaking with 1,000 µg/L of perchlorate solution. Perchlorate rejection tests were performed using the dead-end stirred-cell filtration unit (effective surface area: 28.8 cm^2). Perchlorate rejections (50 µg/L) were also compared before/after soaking with 1,000 µg/L of perchlorate solution.

RESULTS AND DISCUSSION

Dead-End Stirred-Cell Filtration

The TFC-ULP membrane is a RO membrane. Figure 10.12 shows flux trends before and after perchlorate oxidation for the membrane. Even with the low pressure, the TFC-ULP membrane exhibits relatively high water flux, as much as the MX07 membrane. We studied possible oxidation effects of perchlorate on flux trends with a low feed concentration of perchlorate (50 μg/L) before and after soaking in 1,000 μg/L of perchlorate for 24 hrs. This test was performed at a relatively low pressure for a RO membrane (483 kPa (70 psi)) due to the maximum pressure (75 psi) of the dead-end stirred-cell unit. The membrane was not affected by perchlorate oxidation based on membrane flux before/after soaking with 1,000 μg/L.

The ESNA membrane is an NF membrane made of polyamide. Polyamide membranes are known for their good chemical stability. Flux decline test results are shown in Figure 10.13; the applied pressure was 483 kPa (70 psi). The membrane showed constant flux and perchlorate rejection trends before/after with 1,000 μg/L of perchlorate solution. The ENSA membrane showed no oxidation effects in its flux and perchlorate rejection trends.

Aromatic polyamide membranes are often preferred as membrane materials because of their outstanding mechanical, thermal, chemical, and hydrolytic stability, as well as their permeate selective properties, particularly in reverse osmosis. A pressure of 483 kPa (70 psi) was applied to the MX07 membrane. Figure 10.14 shows that perchlorate did not change the aromatic polyamide membrane's flux trends. We have tested sulfonated polyimide dominated membranes, GM (UF). Figure 10.15 shows the results of flux trends for the GM membrane. No changes in flux trends were observed before and after soaking with 1,000 μg/L of perchlorate for 24 hrs. The GM membrane demonstrated good chemical stability upon exposure to perchlorate.

Perchlorate Rejection Tests

Perchlorate rejections were investigated before and after perchlorate oxidation (exposure to 1,000 μg/L for 24 hours). The results from subsequent rejection tests with 50 μg/L of perchlorate are shown in Figure 10.16. As shown in Figure 10.16, the RO (TFC-ULP) and NF (ESNA and MX07) membranes provided high perchlorate rejections. The UF (GM) membrane still provided rejections of over 80%. All of the tested membranes showed differences in perchlorate rejection percentage of less than 3%, comparing conditions before and after perchlorate oxidation. As a result of this test, perchlorate rejection is not significantly affected by perchlorate oxidation for the TFC-ULP, ESNA, MX07, and GM, membranes.

SUMMARY

- All of the tested membranes generally exhibited, good chemical resistance to perchlorate oxidation.
- Perchlorate rejection percentages were over 95, 90, and 80 percent by RO, NF, and UF membranes, respectively. There was little difference in ClO_4^- rejection after exposure to 1,000 μg/L of perchlorate for 24 hours.

- Since only short-term rejection tests were performed, long-term performance should be indentified for future research.

The flux trends of all the tested membrane were not affected by perchlorate in 50μg/L rejection tests run after soaking with 1000μg/L of perchlorate for 1 day

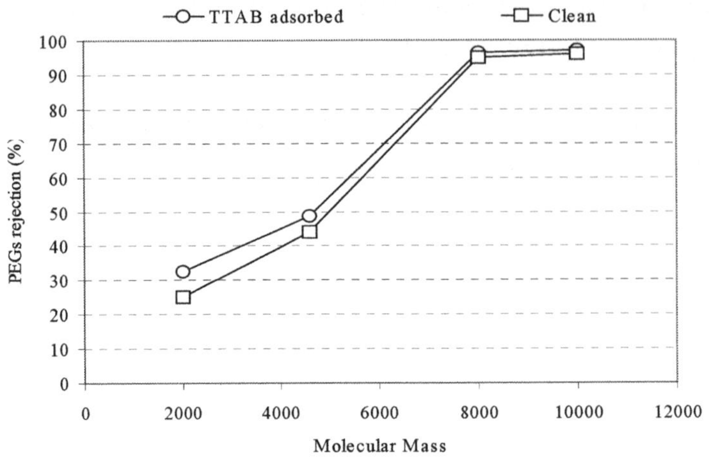

(a) TTAB 0.17 mM and 30 mS/m $CaCl_2$

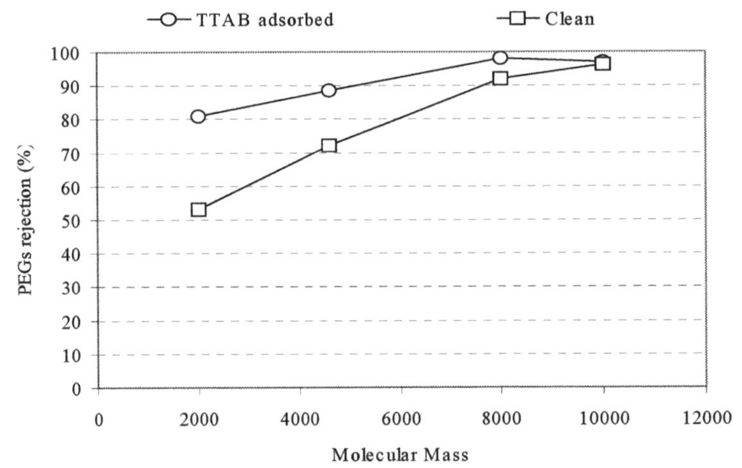

(b) TTAB 1.7 mM and 115 mS/m $CaCl_2$

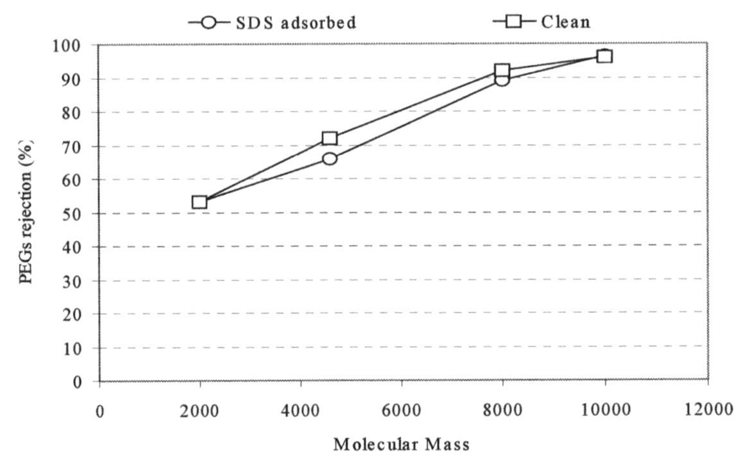

(c) SDS 2 mM and 115 mS/m $CaCl_2$

Figure 10.1 Comparison of PEG rejections by surfactant modified membranes

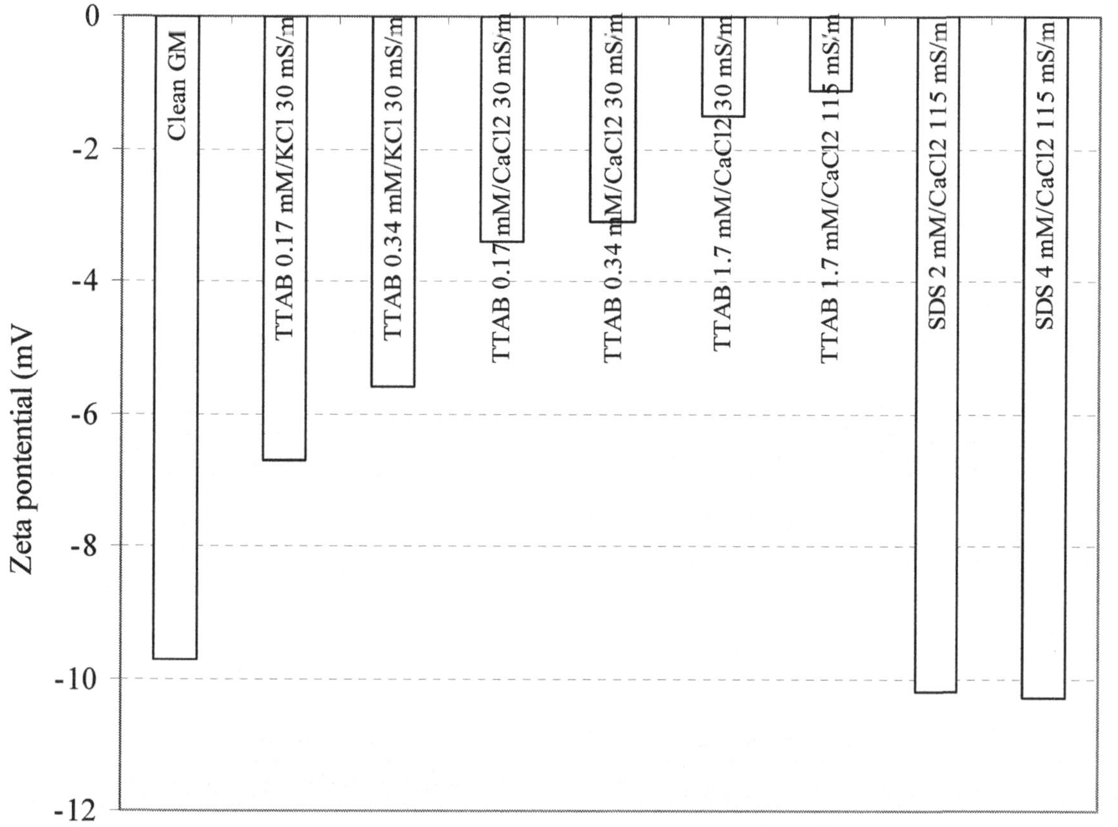

Figure 10.2 Comparison of zeta potentials of cationic/anionic surfactant conditioned UF membranes with synthetic water

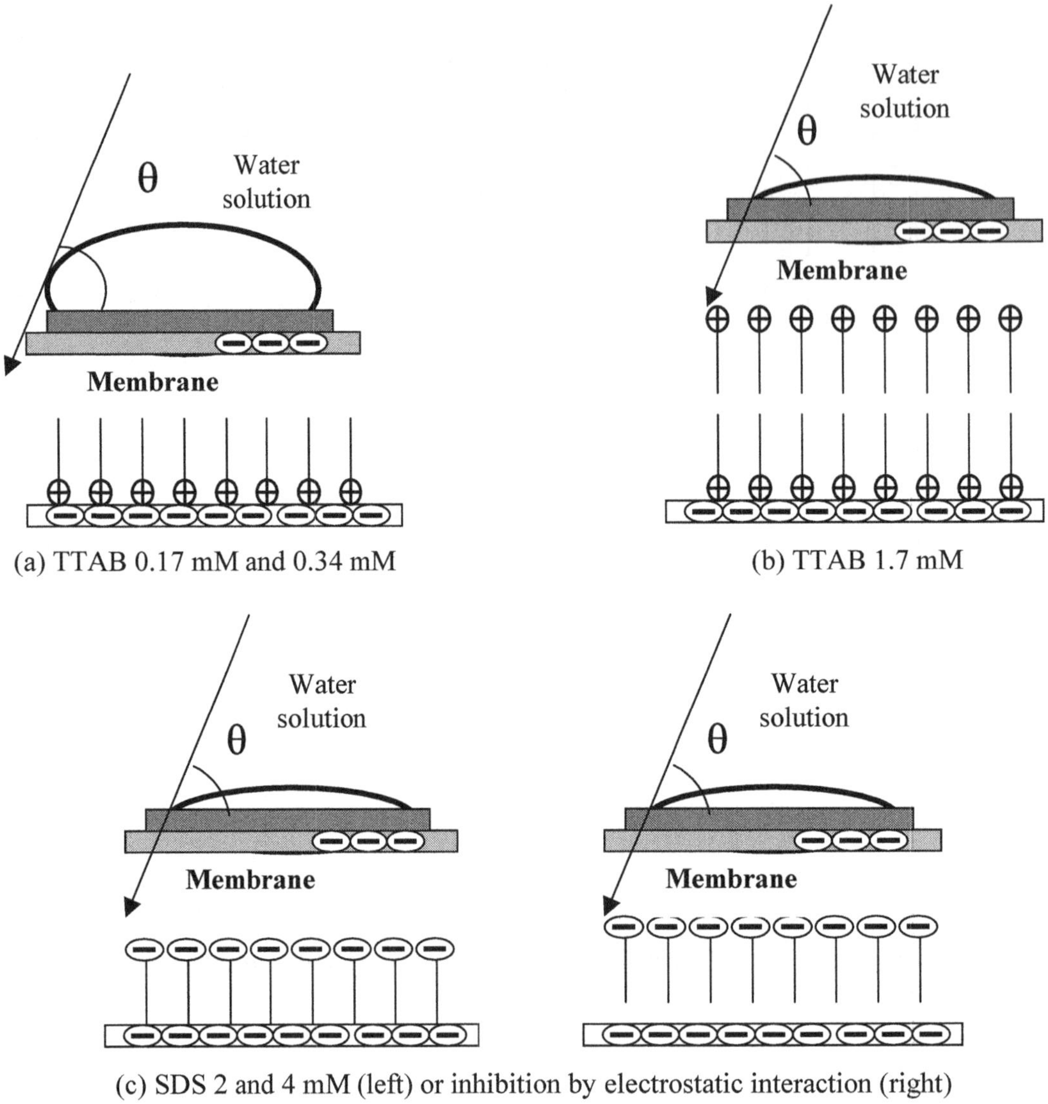

Figure 10.3 Illustration of the relationship between cationic/anionic surfactants on a negatively charged UF membrane and contact angle of water drop

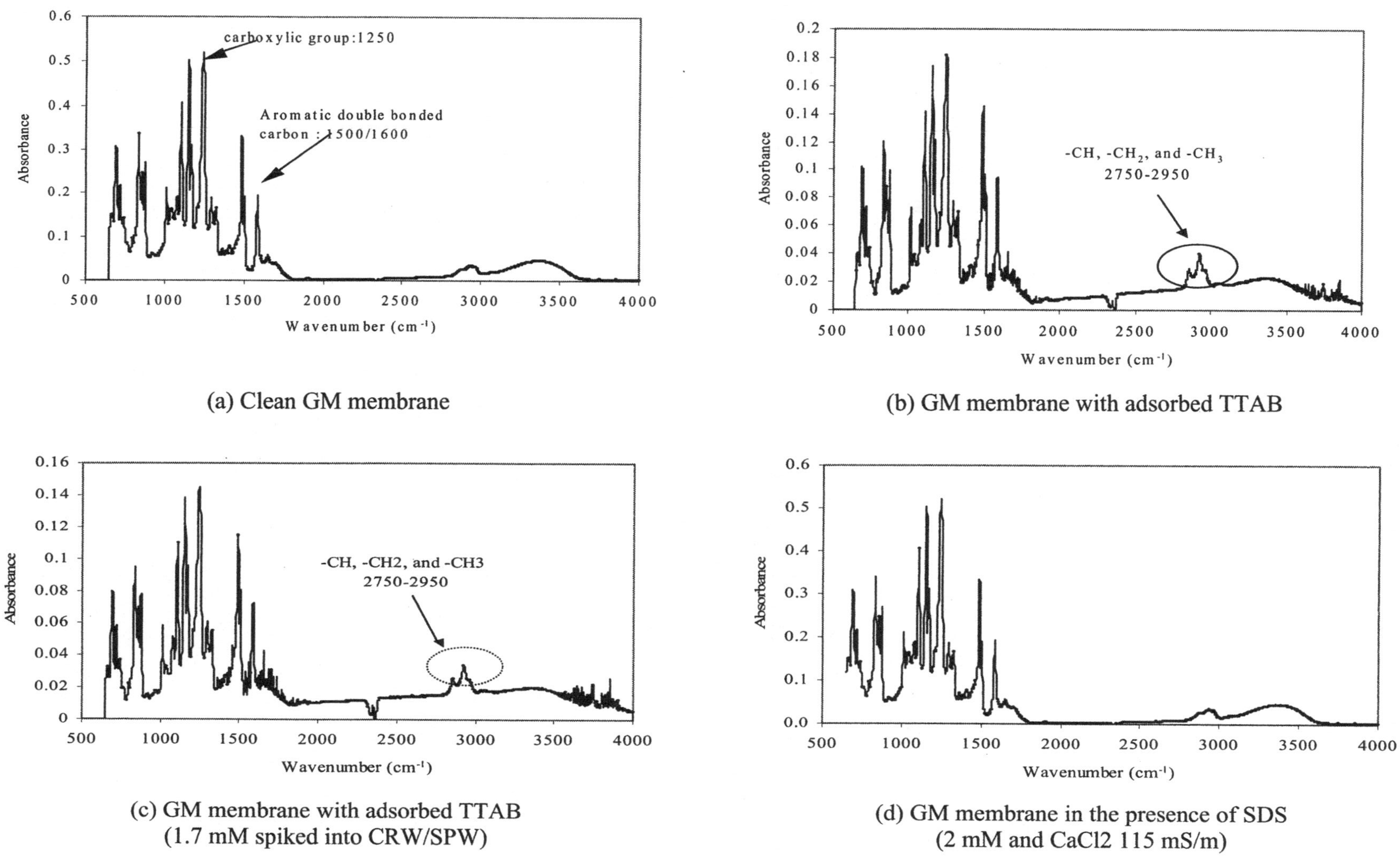

(a) Clean GM membrane

(b) GM membrane with adsorbed TTAB

(c) GM membrane with adsorbed TTAB
(1.7 mM spiked into CRW/SPW)

(d) GM membrane in the presence of SDS
(2 mM and CaCl2 115 mS/m)

Figure 10.4 Comparison of FTIR spectra of clean and cationic/anionic surfactant modified UF membrane with synthetic and CRW/SPW waters

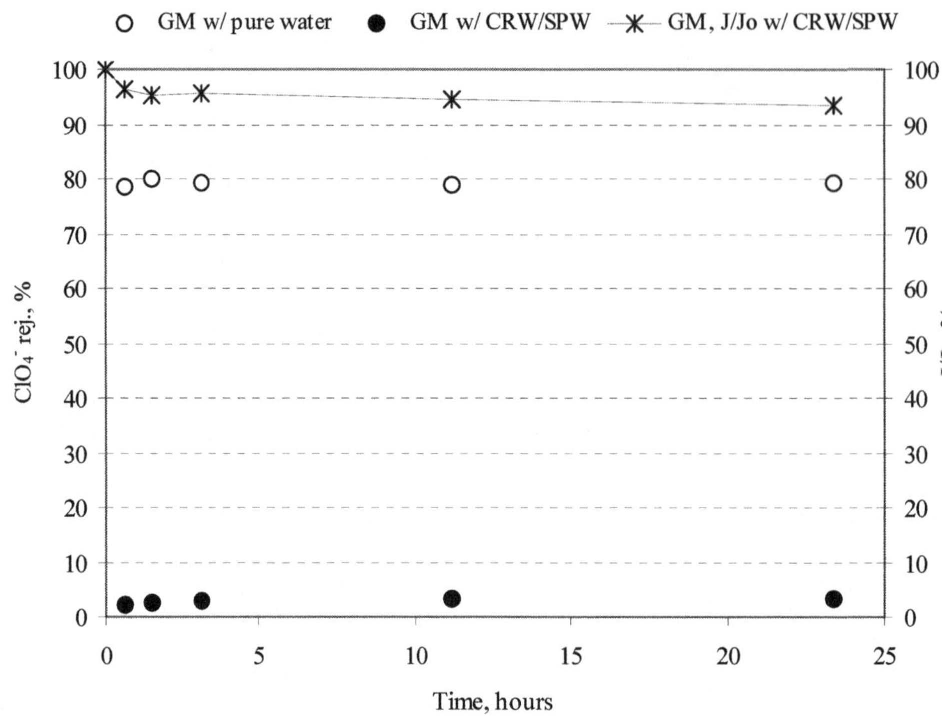

Figure 10.5 Comparison of perchlorate rejection and flux-decline trend with CRW/SPW for a UF membrane

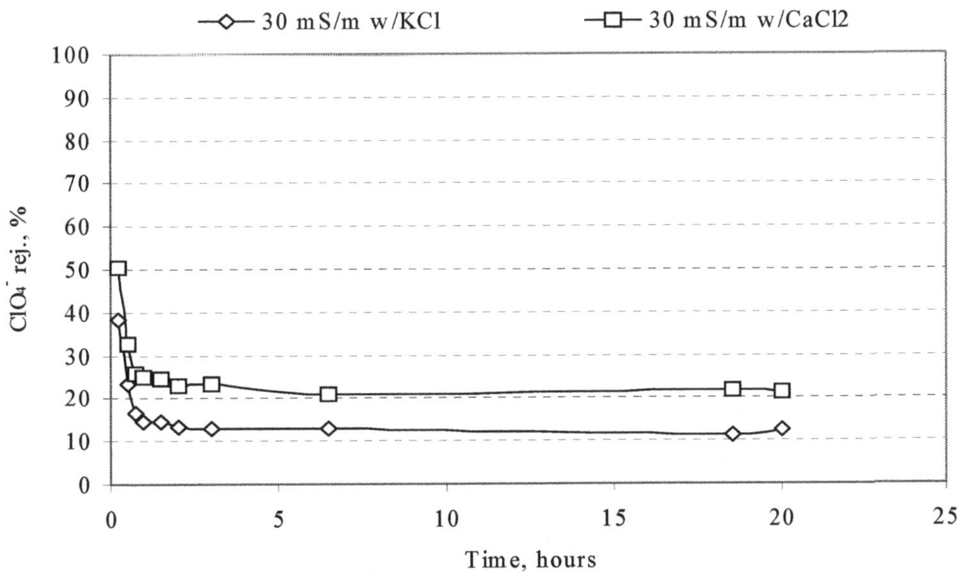

(a) TTAB 0.17 mM

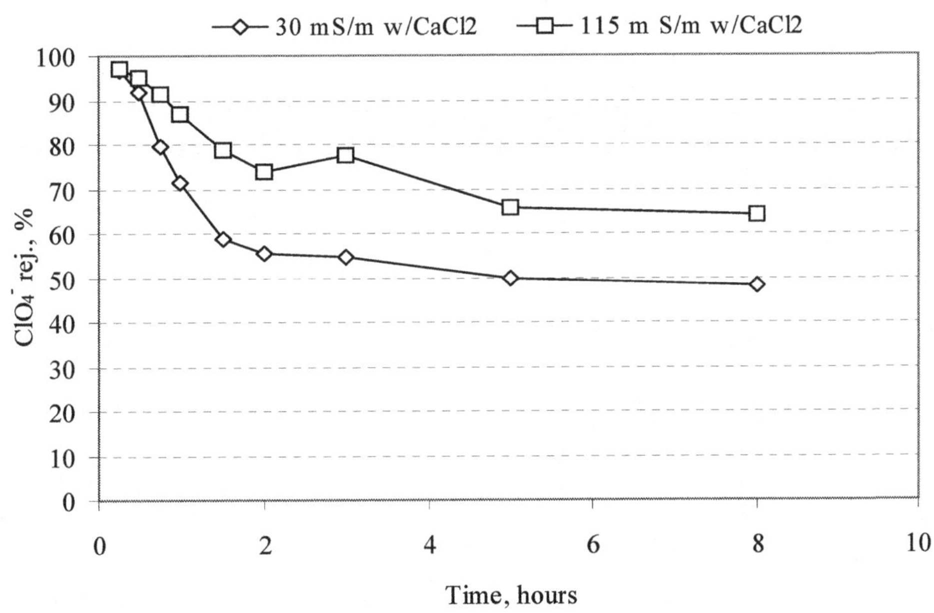

(b) TTAB 1.7 m M

Figure 10.6 Effect of cationic surfactant, TTAB, conditioning on perchlorate rejection in the presence of co- and counter ions with synthetic water for a UF membrane

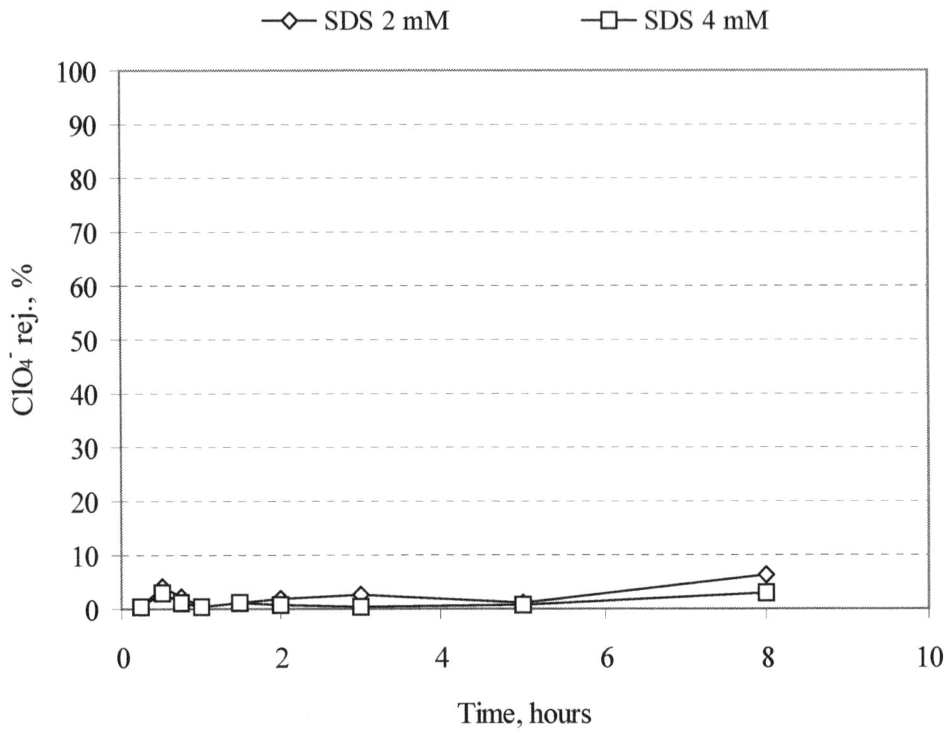

Figure 10.7 Effect of anionic surfactant, SDS, conditioning on perchlorate rejection in the presence of co- and counter ions (115 mS/m $CaCl_2$) with synthetic water for a UF membrane

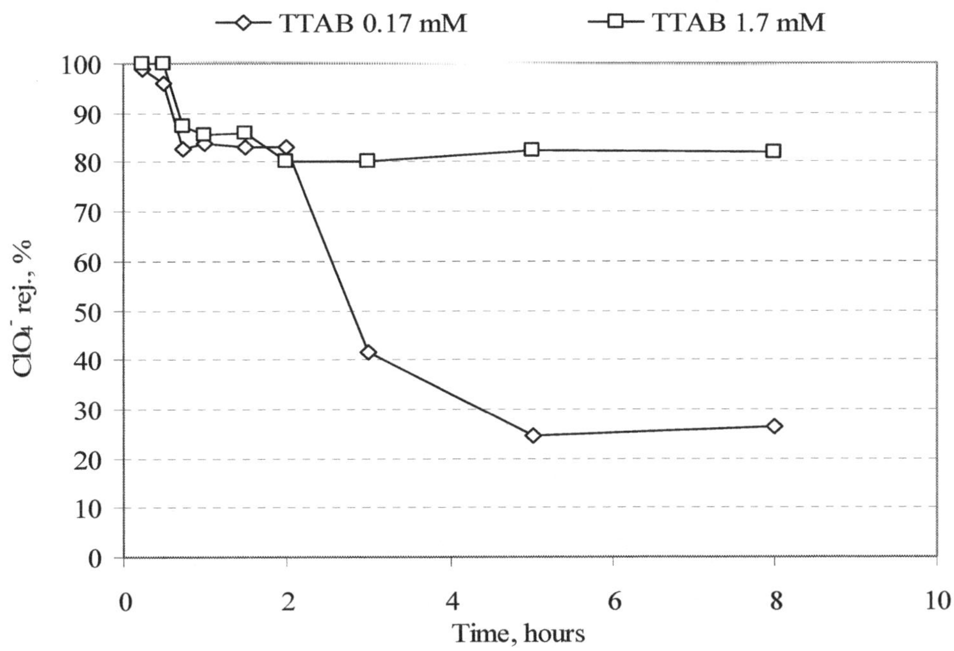

(a) CRW/SPW

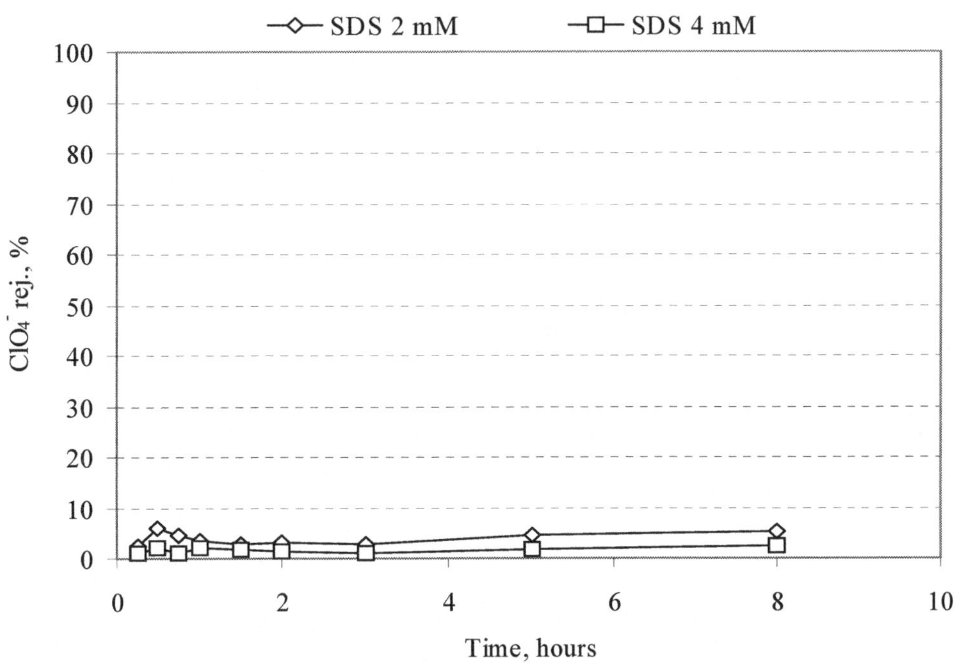

(b) CRW/SPW

Figure 10.8 Effect of cationic/anionic surfactants, TTAB and SDS, conditioning on perchlorate rejection with CRW/SPW for a UF membrane

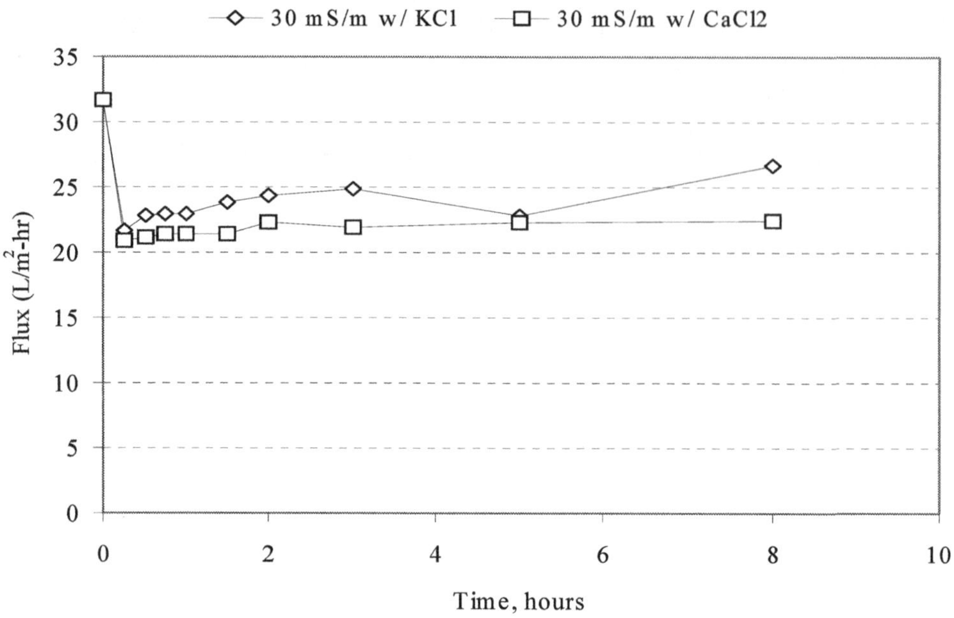

(a) TTAB 0.17 mM

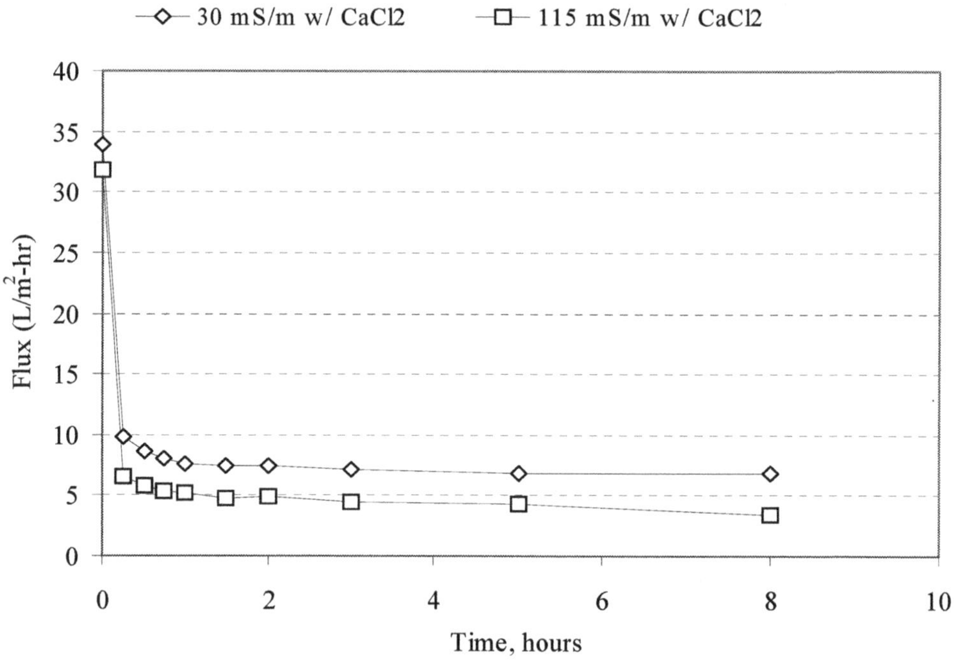

(b) TTAB 1.7 mM

Figure 10.9 Effect of cationic surfactant, TTAB, conditioning on flux decrease with synthetic water for a UF membrane

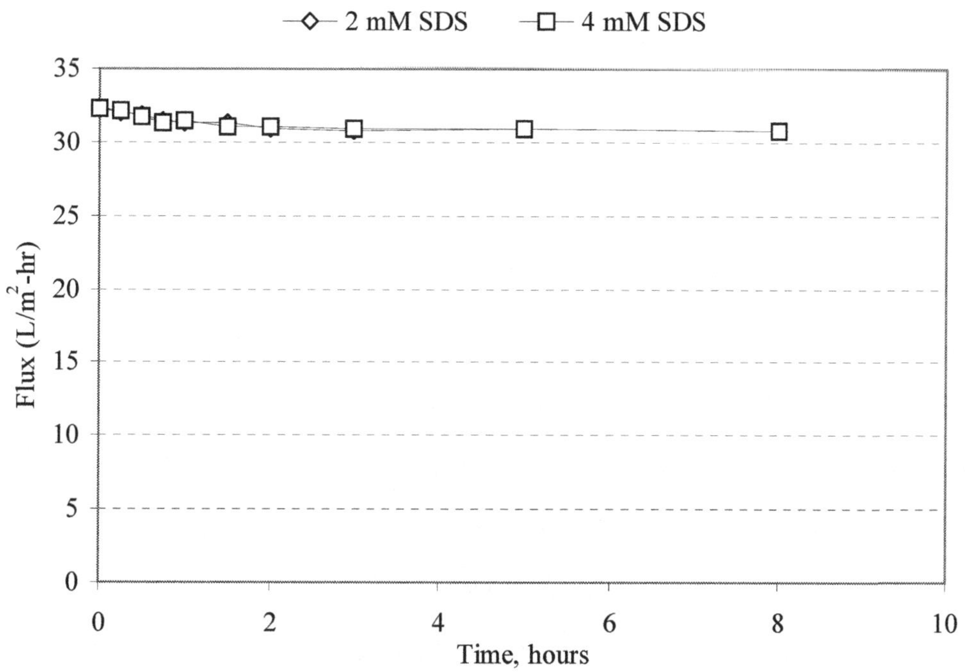

Figure 10.10 Effect of anionic surfactant, SDS, conditioning on flux decrease with synthetic water (115 mS/m $CaCl_2$) for a UF membrane

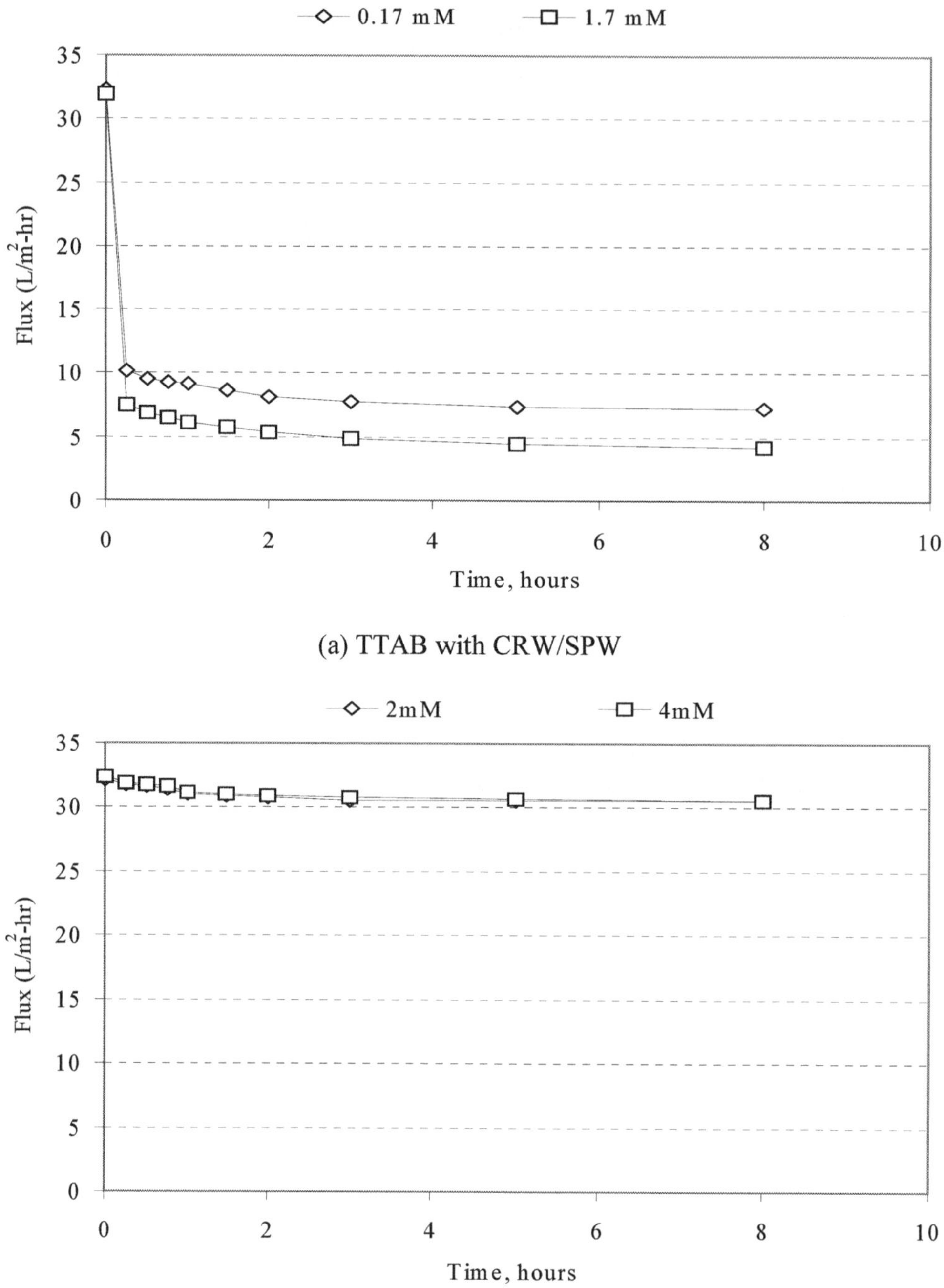

(a) TTAB with CRW/SPW

(b) SDS with CRW/SPW

Figure 10.11 Effect of anionic surfactant, SDS, conditioning on flux decrease with CRW/SPW for a UF membrane

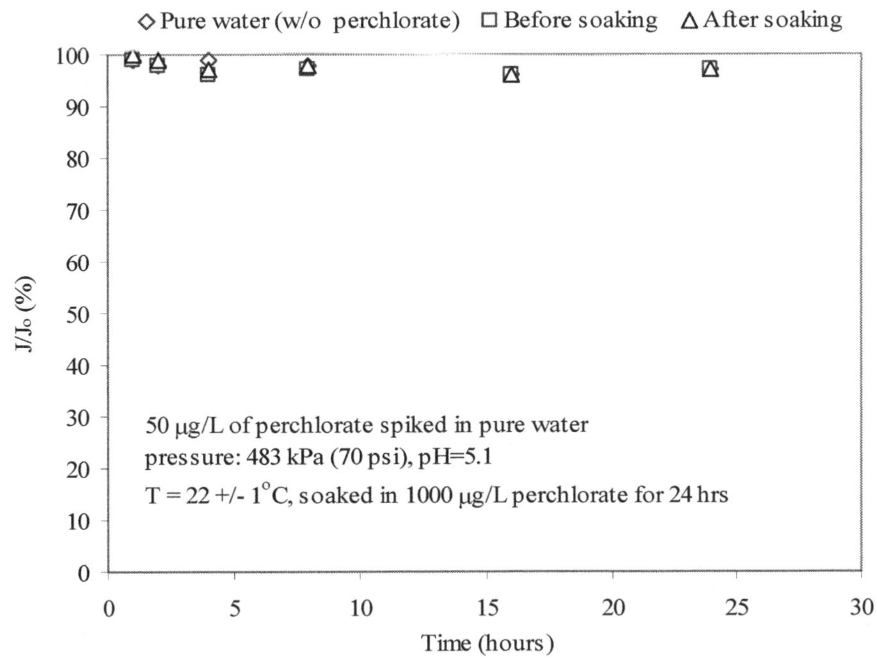

Figure 10.12 Comparison of flux before/after perchlorate oxidation for the TFC-ULP membrane

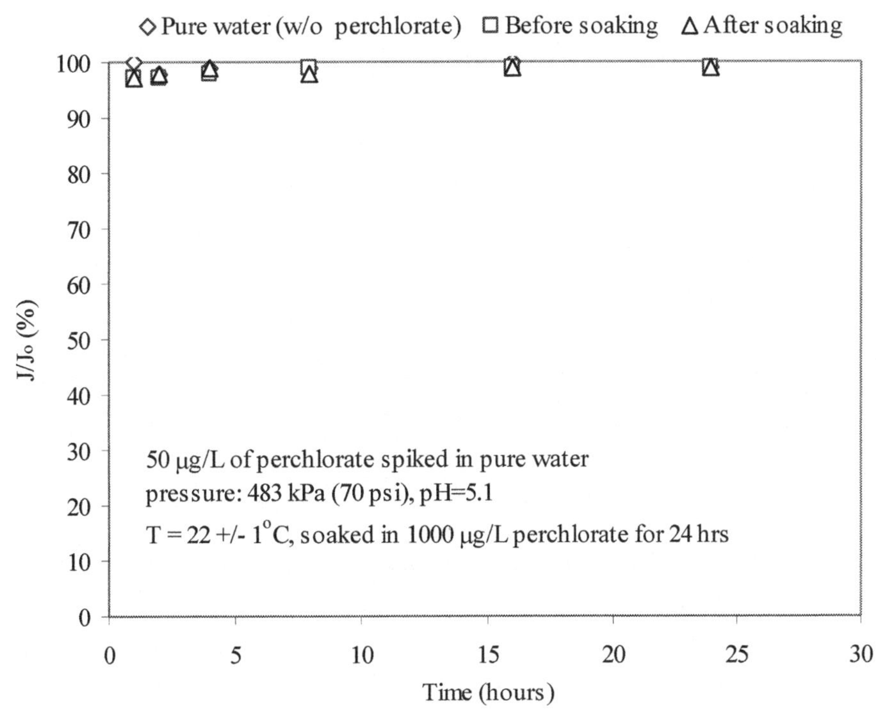

Figure 10.13 Comparison of flux before/after perchlorate oxidation for the ESNA membrane

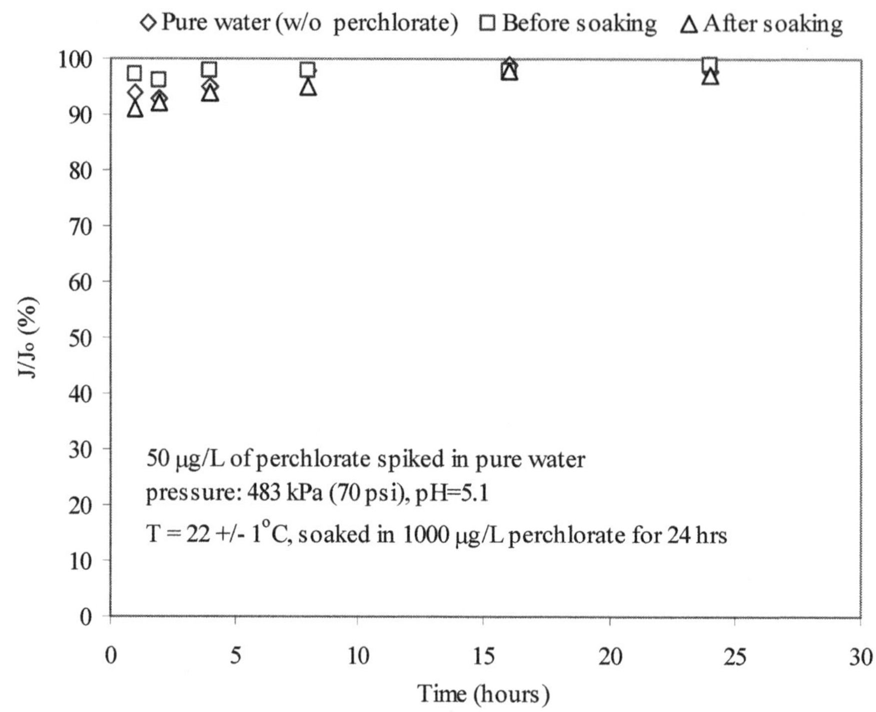

Figure 10.14 Comparison of flux before/after perchlorate oxidation for the MX07 membrane

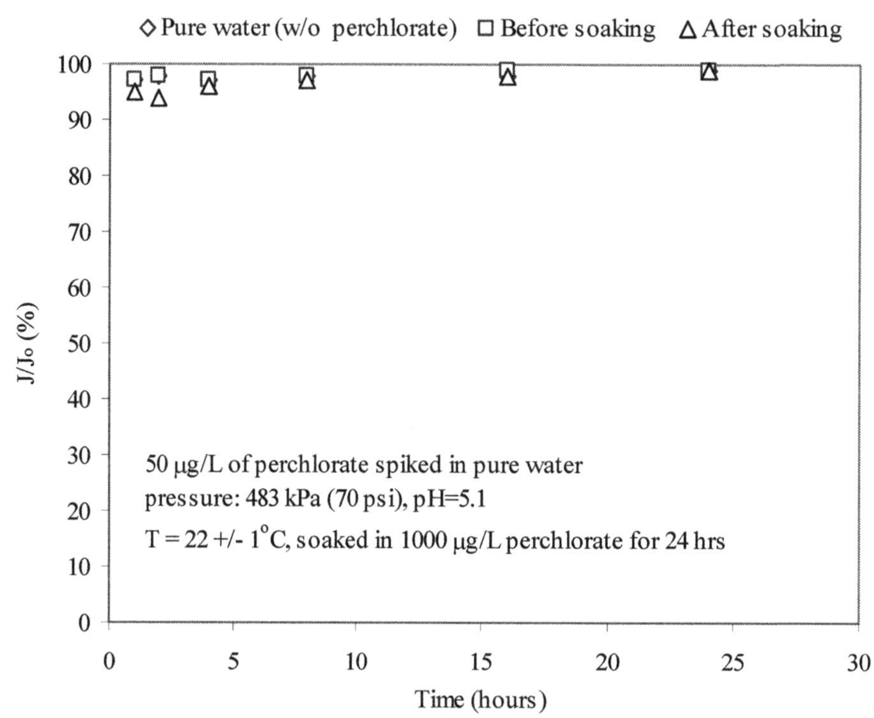

Figure 10.15 Comparison of flux before/after perchlorate oxidation for the GM membrane

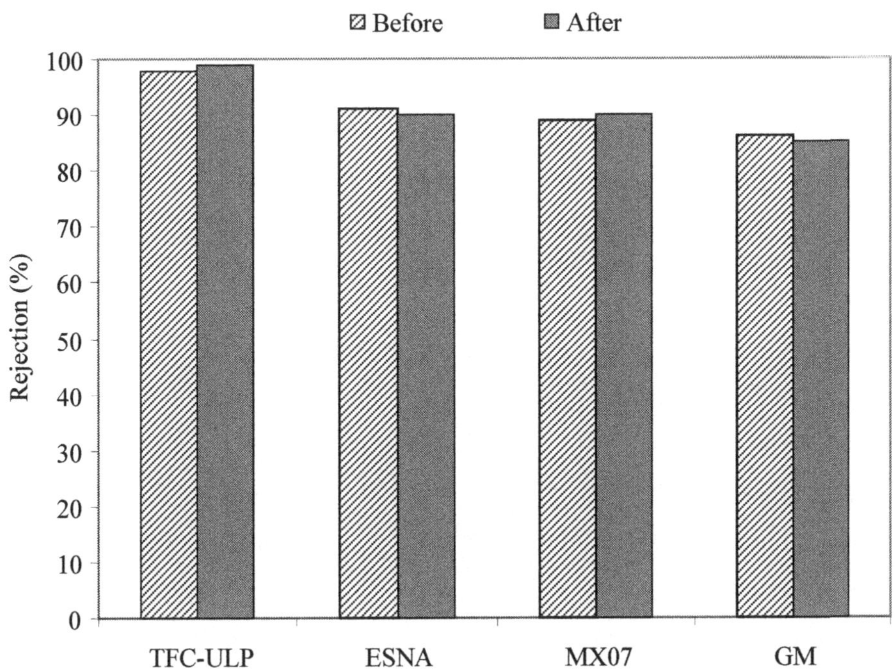

Figure 10.16 Comparison of perchlorate rejection (50 μg/L feed) before/after oxidation (exposure to 1,000 μg/L for 24 hours)

CHAPTER 11
SINGLE ELEMENT TESTS FOR PERCHLORATE REJECTION

INTRODUCTION

Single element tests were conducted in a two-phase study at Metropolitan's La Verne Pilot Plant (La Verne, CA). During Phase I of the tests, the single element membrane unit was used to confirm the findings from the bench-scale cross-flow unit. One reverse osmosis (RO) membrane (LFC-1, Hydranautics, Oceanside, CA), two nanofiltration (NF) membranes (ESNA, Hydranautics, CA, and TFC-S, Koch Membrane Systems, Inc. Wilmington, MA) and one ultrafiltration (UF) membrane (GM Desal, Osmonics, Vista, CA) were selected based on the bench-scale cross-flow tests done at the University of Colorado, and these single elements were evaluated in a series of 6-hour tests with either 100 or 2,000 μg/L of spiked perchlorate. Tests were conducted with no recycle flow to simulate the lead stage, and with a 50-percent recycle (i.e., rejected concentrate constituted 50 percent of the feed flow) to simulate a later stage in an array. The feed was blend of Colorado River water (CRW) and State Project water (SPW) that had been treated in a conventional treatment plant and dechlorinated with sodium sulfite. For the UF membrane, a cationic surfactant (myristyltrimethylammonium bromide) was injected upstream of the membrane, in the predetermined amounts, in order to condition the membrane for higher rejection of perchlorate.

In Phase II, two of the membranes tested in Phase I (an RO, and a NF membrane) were selected for evaluation of long-term perchlorate rejection trends. Each of the membranes was operated for approximately 30 days (or until it fouled) in both the recycle and non-recycle modes.

OBJECTIVES

The objective of this study was to determine the removal/rejection of perchlorate by pressure-driven membranes, including RO, NF, and tight UF. The following goals were pursued during the course of the subject study:

- Evaluate the perchlorate removal performance of production-sized (4-in. diameter) RO, NF, and UF membranes in a natural water.
- Evaluate the effects of perchlorate feed concentration on perchlorate rejections over a range of concentrations possible in the contaminated drinking water sources.
- Evaluate the effect of recycling on perchlorate rejections.
- Evaluate the long term rejection of perchlorate and potential fouling of a membrane.
- Evaluate the effect of charge and size of ions by monitoring the background water constituents.

EXPERIMENTAL DESIGN

Membrane Pilot Plant Description

The single element membrane system was a skid-mounted unit consisting of a pump, pressure vessel, sample ports, and the instrumentation required to monitor pressure and adjust/monitor flow in each segment of the system. The pressure vessel accommodated a single 4-in.-diameter by 40-in.-long spiral membrane. The unit was capable of operating at pressures of up to 250 pounds per square inch (psig) and at feed flow rate of up to 20 gpm. The membrane test unit could be operated in either a non-recycle or a recycle mode. In the non-recycle mode, the reject (concentrate) stream was discharged to waste. Conversely, in the recycle mode, a portion (50 percent) of the reject stream was returned to, and blended with, the feed stream.

Other supplementary setups included a dechlorinating system using a flow through sodium sulfite unit (ELTECH International Corp., Tablet Feeder, Model 200), dechlorinated feed water holding tank, multistage booster pump (Model 2PC44, TEEL/Dayton, Niles, IL 60714), 5 micron pre-filter (Hytrex II Cartridge, Nimbus Water Systems, San Marcos, CA), perchlorate feed pump (Series J, Milton Roy, Ivyland, PA), perchlorate stock solution tank, and concentrate treatment tank with ion exchange resin (ASB1 Anion Resin Type 1, C1 form, Syberon Chemicals, Birmingham, NJ 07011). Throughout this study, a blend of CRW and SPW treated with conventional treatment at the F.E. Weymouth plant (La Verne, CA) was used after dechlorinating. The schematic of the membrane-testing unit is shown in Figure 11.1.

Operations

Each membrane was run for 4 hours prior to each test to ensure that the preservative solution and any other foreign material had been flushed from the membrane. At the completion of the initial rinse cycle a particle sample was taken to prove that the membrane was operating correctly. Once the membrane was placed in service, the system was checked and sampled twice daily to monitor operation of the test platform. The data collected allowed operators to observe the quality and trends of the membranes operation.

During and at the completion of the 30-day run, a complete set of samples was taken and the membrane was washed. The system was then adjusted to operate in a 50% recycle mode for another 30-day test. At the completion of the recycle run, the membrane was washed and placed in the storage solution (2% solution, AM-88 Membrane Preservative, Applied Membranes, Inc., San Marcos, CA 92069).

Short Term Tests

Membranes selected for the short term series (LFC-1, TFC-S, ESNA, and GM) were run for a total of 8 hours per test. The first two hours were used to ensure that the system had reached steady state with the perchlorate system spiking at the appropriate rate of 100 or 2000 μg/L. Prior to each test, the membranes were equilibrated for two hours, and pressure and flowrate at the feed, concentration, and permeate were monitored throughout the run. Samples/analyses including particles >2 microns, electroconductivity (EC), pH, temperature, and turbidity were taken every hour at the feed, permeate, and concentrate. This was done to ensure the quality of the run as well as to provide accurate trending for the membranes ability to remove perchlorate from the feed solution. At the completion of the run a full sample set including cations and anions was taken. A detailed description of the samples taken is found in the sub-section entitled Sampling in this section.

After the final sample set was taken, the perchlorate feed system was shut down, and the system was allowed to flush with straight feed water for one hour.

Long Term Tests

Based on the single element tests of the four membranes, two membranes (TFC-S and LFC-1) were selected for the 30-day phase II series of the tests to study the long-term rejection of perchlorate and fouling of the membrane. Each membrane was run for 4 hours prior to be placed in-service for the 30-day test. At the completion of the initial rinse cycle, a particle sample was taken to prove that the membrane was operating correctly.

The membrane was subsequently checked twice daily and sampled to monitor operation of the test platform. The operating parameter data collected allowed operators to adjust the pressure so that the membrane can be run at a constant feed flow rate.

At the completion of the 30-day run, a complete set of samples was taken and the membrane was washed. Depending on the membrane performance, a complete set of the samples was taken prior to 30 days. EC was monitored closely for early signs of possible decrease in perchlorate rejection. Prior to the 50 percent recycle test, the membrane was washed with the cleaning procedure discussed in the next section. The system was then adjusted to operate in a 50 percent recycle mode for another 30-day test. At the completion of the recycle run, the membrane was washed and placed in storage.

Membrane Cleaning

Prior to the start of each phase of the project or storage, the membranes were cleaned with both acidic (pH 2.5 citric acid solution) solution to remove scale and caustic (pH 10.5 solution with equal parts sodium tripolyphosphate, trisodium phosphate, and Na-EDTA) solution to remove organic foulants and/or biofilm (see Table 11.1). Each wash cycle consisted of permeate collection and a 1 hour premix of the cleaning solution. The acid cleaning cycle was followed by a caustic cleaning cycle with rinse cycle after each solution. Each cleaning cycle took approximately 2 hrs including acid cleaning, rinse, and caustic cleaning. The cleaning solution was then drained, and the unit was rinsed with feed water for 2 hours. Additionally, the membranes were cleaned when the normalized flux decreased 15 percent or a significant increase in salt passage was observed. The membranes were cleaned per the RO membrane manufacturer's guidelines.

For certain cases, the membranes were cleaned after each fouling episode using a commercially-avaiable acidic membrane cleaner (AM-11; Applied Membranes, Inc., San Marcos, CA). The cleaning agent was selected to control the predicted foulants, barium sulfate ($BaSO_4$) and calcium carbonate ($CaCO_3$). A 2 percent solution of AM-11 was dissolved into permeate water and the pH was adjusted to below 3.0. The system was flushed with the cleaning solution at 50 psi for 10 min to loosen the foulants from the membranes. The flow rate was less than 3 gpm, per the manufacturer's recommendation. After this time period, the flow rate was increased for 20 min to clear any foulants from the system. Finally, the system was flushed with permeate water until the pH returned to normal. The water production (WPC) always increased immediately following a cleaning cycle and exhibited a gradual decline over time.

At the completion of each cycle, flux performance was measured to ensure that the membrane was operating within specifications. This was done to ensure the quality of each run and to provide each test with optimum operating conditions for the membranes.

Table 11.1
Composition of cleaning solution

Caustic:	EDTA disodium	1.9 lbs.	25 Gallons
	Tripolyphosphate	1.9 lbs.	
	Tribasic sodium phosphate	1.9 lbs.	
Acid:	Citric acid	1.9 lbs.	25 Gallons

SAMPLING

Samples were collected from feed, permeate, and concentrate sample ports as shown in Figure 11.1. Operational conditions monitored included feed, permeate, and concentrate flow rates and pressures. The membrane feed water was monitored to ensure the absence of chlorine. Field measurements included temperature, conductivity, pH, and turbidity; particle counts of the feed, permeate, and concentrate streams; and the silt density index of the feed stream. A complete set of samples taken at the beginning and end of the test included measurements for perchlorate, chloride, sulfate, fluoride, total dissolved solids, bromide, TOC, aluminum, manganese, strontium, barium, silica, calcium, barium, sodium, potassium, total alkalinity, and UV absorbance.

Analytical Methods

All inorganic and organic analyses were performed at Metropolitan's Water Quality Laboratory. Parameters monitored include perchlorate (ClO_4^-) and other anions (SO_4^{2-}, Cl^-, F^-, and Br^-) by ion chromatography (IC), carbonate (HCO_3^- and CO_3^{2-}) alkalinity by titration, cations (Ca^{2+}, Mg^{2+}, Ba^{2+}, Na^+, K^+, Al^{3+}, Sr^{2+}) by ICP, and silica (SiO_2). These measurements of dissolved inorganic species were supplemented by parameters to characterize natural organic matter (NOM) (total organic carbon (TOC) and UV absorbance (UVA_{254})) and colloids/particles (turbidity and particle counting). All analyses were conducted in accordance with the procedures described in *Standard Methods* (APHA, AWWA, and WEF, 1998), except as noted in the following sections:

- ***Perchlorate.*** Perchlorate samples were analyzed using an ion chromatograph (model DX300; Dionex Corp., Sunnyvale, Calif.) modified with a 200-µL sample loop. An ion-chromatography (IC) analytical column (model AG11, Dionex), an anion micromembrane suppressor, and a conductivity detector were used. The method detection limit was 1.4 µg/L, and the reporting limit was 4.0 µg/L.
- ***Bromide.*** Bromide analyses were conducted on an ion chromatograph (model 2010, Dionex Corp.) with a 20- or 50-µL sample loop. An IC analytical column (model AS4A, Dionex), an anion micromembrane suppressor, and a conductivity detector were used.

RESULTS AND DISCUSSION

Feed Water Quality

The pertinent water quality parameters of the feed water treated in the membrane studies are listed in Table 11.2.

Table 11.2
Feed water quality (dechlorinated Weymouth plant effluent)

Parameter	Feed to membrane average (range)
Turbidity (NTU)	0.14 (0.08–0.24)
Total dissolved solids, TDS (mg/L)	460 (293–680)
Total alkalinity (mg/L as $CaCO_3$)	105 (73–145)
Total hardness (mg/L as $CaCO_3$)	202 (109–274)
Total organic carbon, TOC (mg/L)	3.46 (2.47–4.54)
Calcium (mg/L)	58 (52-68)
Magnesium (mg/L)	24.5 (21.5-27)
Bromide (mg/L)	0.11 (0.06–0.21)
UVA_{254}(1/cm)	0.054 (0.029–0.085)
Conductivity (μmho/cm)	746 (465–1,009)
Chloride (mg/L)	79 (60–112)
Silica (mg/L)	13 (10–17)
Sulfate (mg/L)	150 (56–257)
Fluoride (mg/L)	0.21 (0.11–0.28)
Aluminum (μg/L)	71 (35–117)
Strontium (μg/L)	673 (217–1,050)
Barium (μg/L)	73 (34–106)
Perchlorate (μg/L)*	158 (79–273) for 100 μg/L run 1514 (1165–1960) for 2,000 μg/L run 4084 (3238–5552) for 2,000 μg/L recycle run

* with spike

Membrane Characteristics

Characteristics of the membranes used for the study are summarized in Table 11.3. All of the membranes evaluated were negatively charged, with zeta potentials ranging from –45.4 mV (highly charged) to –29.8 mV (less charged). The contact angles listed provide an index of the hydrophobicity of the membrane surface, the higher the contact angle, the more hydrophobic the membrane. For example, the ESNA (NF), with a contact angle of 60.3°, was more hydrophobic than the GM-Desal (UF), which had a contact angle of 45.5°. The TFC-S (NF) membrane had a relatively high pure-water permeability (PWP), whereas the ESNA (NF) and LFC-1 (RO) membranes had comparatively low permeabilities. As expected, because of its relatively large nominal molecular weight cut-off (MWCO), the GM-Desal (UF) exhibited the greatest permeability among the membranes tested. The four membranes were made of different materials (Table 11.3). The LFC-1 (RO) and TFC-S (NF) membranes were polyamide thin-film composite (TFC) membranes, while the ESNA (NF) and GM-Desal (UF) membranes were aromatic polyamide and sulfonated polyimide membranes, respectively. The surfaces of these membranes were made of a combination of aromatic and aliphatic polymers and had ionizable functional groups such as carboxylic (COO^-) and sulfonic (SO_3^-) groups.

Table 11.3
Membrane characterization: physical-chemical properties and perchlorate rejection

Membrane	LFC1	TFC-S	ESNA	GM
Type	RO	NF	NF	UF
MMCO* (from manufacturer)	n/a	200	200	8,000
MMCO* (estimated with PEG)	n/a	n/a	560	8,000
ZP (mV) at pH 8 and 115 mS/m	-45.4	-42.6	-43.2	-29.8
Contact angle (°)	n/a	n/a	60.3	45.5
Surface area (ft^2)	85	79	85	90
PWP† (L/m2-day-kPa)	0.75	1.99	1.05	2.81
ClO_4^- rej.at 2.1 mS/m‡	99	88	93	87
ClO_4^- rej.at 115 mS/m‡	99	79	88	29
Pressure, kPa (psi)	523 (76)	385 (56)	406 (59)	124 (18)
Jo/k (R, % §)	0.60 (3.6)	0.60 (1.6)	0.40 (2.6)	0.40 (1.4)
Manufacturer	Hydranautics	Fluid System	Hydranautics	Desal
Material	Polyamide TFC	Polyamide TFC	Aromatic polyamide	Sulfonated polyimide

n/a—Not available

* Estimated by polyethylene glycols
† Pure-Water Permeability
‡ At pH 8: pH and conductivity were adjusted by adding HCl, NaOH/KOH, and NaCl/KCl.
§ R: recovery

Membrane Performance

The primary measures of membrane performance for drinking water treatment are permeate flux, solute rejection, and hydraulic recovery. Permeate flux (J) is the rate at which a volume of water is filtered per unit area of the membrane and is typically expressed in units of gal/ft^2-day (gfd) or L/m^2-hr (lmh). A related measure of performance is specific flux, which is defined by

$$\text{Specific Flux} = J / \Delta P \tag{11.1}$$

where ΔP is the net pressure applied to the membrane. Observed rejection is the measure of how well the solute of interest is removed from the feed water by the membrane. Observed rejection is defined by

$$\text{Observed Rejection (\%)} = 100\,(1 - C_p / C_f) \tag{11.2}$$

where C_p is the solute concentration in the permeate stream and C_f is the solute concentration in the feed (bulk) stream. Hydraulic recovery is the yield of product water from the filtration process,

Hydraulic Recovery (%) = 100 (Q_p / Q_f) (11.3)

where Q_p is the flow rate of the permeate stream and Q_f is the flow rate of the feed stream.

The performances of the RO, NF, and UF membranes are discussed below in terms of perchlorate rejection, specific flux, hydraulic recovery, and permeate water quality.

Perchlorate Rejection

The LFC-1 (RO) and TFC-S (NF) membranes both showed excellent removal of perchlorate, ranging from > 94.9 to approaching 100 percent removal under all conditions (Figure 11.2). The ESNA (NF) membrane, with a higher MWCO, was less effective, but it also performed well, removing 86.7 to 92.1 percent of the perchlorate. The GM (UF) membrane proved ineffective for perchlorate removal (0 to 10.3 percent removal). Addition of a cationic surfactant to the GM membrane to assist in perchlorate removal only served to plug the membrane. Overall, the applied perchlorate concentrations (100 μg/L and 2,000 μg/L) and the recycling of 50 percent of the concentrate had little effect on the removal efficiency.

While RO results can be explained by simple solution-diffusion as a rejection mechanism for perchlorate, NF membranes can be explained by two potential mechanisms for perchlorate rejection: steric rejection (i.e., mechanical sieving) and electrostatic rejection (i.e., charge exclusion). Given that perchlorate has a molecular weight of 100 daltons and a hydrated radius of 3.38 angstroms (Å), effective perchlorate rejection can be achieved by both steric rejection and electrostatic rejection when using negatively charged NF membranes with low MWCOs, and by electrostatic rejection alone when using NF membranes with higher MWCOs. The ineffective rejection of perchlorate by the UF membrane indicates that steric rejection is the major mechanism of perchlorate separation by the UF membrane.

Specific Flux

The operating conditions and performance of the membranes tested are given in Table 11.4. At feed pressures of 86–88 psig for the TFC-S (NF) membrane and 85–109 psig for the ESNA (NF) membrane, the average flux was approximately 15 gfd. At feed pressures of 99 to 105 psig, the LFC-1 (RO) membrane produced an average flux of approximately 3.64 gfd. The feed pressure of 53 psig for GM-Desal (UF) without recycle concentrate flow and 89 psig for the GM-Desal (UF) with 50-percent recycle concentrate flow produced an average flux of approximately 13.9 gfd. Specific flux is a direct indicator of the membranes' water mass transfer rate, which accounts for the energy (water pressure) and ionic strength of the solution on the membrane. Higher specific flux values indicate greater membrane efficiency for water production and lower operating (energy) costs. Both NF membranes, the TFC-S and the ESNA, averaged specific flux rates of 0.17 gfd/psig, while the RO membrane has a lower specific flux of 0.04 gfd/psig. The specific fluxes of GM-Desal (UF) are 0.27 and 0.15 gfd/psig without and with recycle concentrate flow, respectively.

Table 11.4
Membrane operating conditions for single-element tests

Type	Target ClO_4^- (μg/L)	Recycle Ratio (%)	Feed Pressure (psig)	Feed Flow (gpm)	Reco-very (%)	Perm. Flux (gfd)	Specific Flux (gfd/psig)	Conc. Flow (gpm)	Conc. Pressure (psi)
LFC-1 (RO)	100	0	105	1	20	3.39	0.03	0.8	120
	100	0	115	4.3	14	10.2	0.088	3.7	110
	100	50	99	1.05	21	3.73	0.04	0.31	115
	2,000	0	100	1.02	21.6	3.73	0.04	0.8	112
LFC-1 (RO)	2,000	0	125	4.3	14	10.2	0.081	3.7	120
	2,000	50	100	1.03	21.4	3.73	0.04	0.3	115
TFC-S (NF)	100	0	88	5.15	16.5	15.5	0.18	4.3	88
	100	50	87	4.92	16.7	14.9	0.17	1.65	88
	2,000	0	87	4.96	17.3	15.7	0.18	4.1	87
	2,000	50	86	5.02	17.1	15.7	0.18	1.62	86
ESNA (NF)	100	0	85	5.02	17.3	14.7	0.17	4.15	87–90
	100	50	88–94	5.02	17.3	14.7	0.16–0.17	1.6	90–100
	2,000	0	88–89	5.02	17.3	14.7	0.16–0.17	4.15	89–90
	2,000	50	105–109	5.02	17.3	14.7	0.13–0.14	1.6	117–119
GM-Desal (UF)	100	0	52–53	5.02	17.3	13.9	0.27	4.15	50
	100	50	89	3.56	24.4	13.9	0.16	0.9	90
	2,000	0	54	5.02	17.3	13.9	0.26	4.15	49–50
	2,000	50	90	3.56	24.4	13.9	0.15	0.9	100

Hydraulic Recovery

Hydraulic recovery (R) was maintained at approximately 20 percent for the RO membrane (with and without recycle) and between 16 and 17 percent for the NF membranes. In order to maintain a 50-percent recycle and a permeate flow rate of 0.87 gpm through the UF membrane, the feed pressure had to be increased from 50 to 90 psig and the feed flow decreased from 5.02 to 3.56 gpm. This resulted in an increase in the hydraulic recovery from 17 percent, without recycle flow, to 25 percent when recycling the concentrate. It is apparent from the increased feed pressure and decreased feed flow that, at the conditions tested, the UF membrane was more sensitive to higher dissolved solids in the membrane feed influent than either the RO or NF membrane. There was no significant increase in perchlorate removal associated with the increase in hydraulic recovery.

Permeate Water Quality

Figures 11.3 through 11.5 show the percent removal of EC, bromide, and TOC, respectively. EC reduction is a surrogate for the removal of ions from the filtered water. The removal efficiencies of the membranes were very similar for EC and perchlorate. The LFC-1 (RO) and TFC-S (NF) membranes removed 98.2 to 99.1 percent and 97.6 to 98.0 percent of the EC, respectively, while the ESNA (NF) removed 93.4 to 94.3 percent of the EC (Figure 11.3). The GM (UF) membrane removed only 7.8 to 9.4 percent of the EC. These data indicate that EC may provide a simple surrogate measure of the membrane's ability to remove perchlorate. Permeate conductivity ranges were 7 to 8 (LFC-1, RO), 15 to 18 (TFC-S, NF), 32 to 38 (ESNA, NF), and 548 to 595 μmho/cm (GM-Desal, UF) when the system was operated without concentrate recycling. When the system was operated with 50-percent concentrate recycling, the permeate conductivity increased slightly to 9 to 17 (LFC-1, RO), 20 to 23 (TFC-S, NF), 43 to 58 (ESNA, NF), and 436 to 654 μmho/cm (GM-Desal, UF).

Removal of bromide ion was less definitive, as there was none detected in the permeate in most of the RO and NF evaluations (Figure 11.4). Bromide removal ranged from >66.7 percent to approaching 100 percent in the RO and NF evaluations. The GM UF membrane removed 12.2 to 72.1 percent of the bromide. TOC removal was similar in the TFC-S (NF), LFC-1 (RO), and ESNA (NF) evaluations, ranging from 85.4 to 96.6 percent removal (Figure 11.5). The NF and RO membranes appeared to be equally effective in TOC removal, reducing TOC concentrations to between 0.13 and 0.47 mg/L. In contrast, the GM (UF) achieved only 48.8 to 58.7 percent removal, reducing the TOC to concentrations ranging from 1.41 to 1.85 mg/L in the permeate. Decreased removal of low-molecular-weight fractions of natural organic matter (NOM), and thus TOC, by UF membranes, as compared to NF and RO membranes, has been previously observed in other studies (J. Mallevialle et al., 1996)

Examination of UVA_{254}, turbidity, and particle-count data indicated that the LFC-1 (RO), TFC-S (NF), ESNA (NF), and GM-Desal (UF) membranes produced permeates with acceptable water quality (Table 11.5). The LFC-1 (RO), TFC-S (NF), and ESNA (NF) membranes removed UVA_{254}-absorbing constituents to levels ranging from the detection limit of the UV spectrophotometer (0.004 1/cm) to 0.009 1/cm. The GM-Desal (UF) membrane removal of UV_{254}-absorbing constituents averaged 59 percent, reducing the UVA_{254} levels to 0.024 to 0.028 1/cm. The turbidity and particle counts were also effectively reduced by the LFC-1 (RO), TFC-S (NF), ESNA (NF), and GM-Desal (UF) membranes. The average turbidity in the permeate ranged from 0.05 to 0.09 NTU, while the average permeate particle count ranged from 3 to 28 particles/mL.

Effect of Ion Charge and Size

NF membranes differ from RO membranes on the molecular level. RO membranes are made of three amide groups surrounding the benzene molecule. NF membranes contain two amide groups and one carboxyl group. The carboxyl group causes the NF lattice to be weaker, thus yielding a larger pore size than RO. UF membranes differ from NF and RO membranes in material. Polysulfone is one of the most widely used UF membrane materials, with a wide range of pore sizes (from 1 to 20 nm).

Mechanisms of ion separation by RO, NF, and UF are different. Unlike the mechanism of physical sieving that dominates the rejection behavior of UF membranes, rejection of ions by RO and NF membranes depends in a complex fashion on the chemistry of the solute-membrane interactions, resulting in separation of ions by solution/diffusion and exclusion (e.g., ion charge or ion size).

Table 11.5
Permeate water quality summary

Parameter	RO		NF				UF	
	LFC-1		TFC-S		ESNA		GM-Desal	
	No Recycle	50% Recycle	No Recycle	50% Recycle	No Recycle	50% Recycle	No Recycle	50% Recycle
Turbidity (NTU)	0.06	0.07	0.08	0.06–0.07	0.08	0.05–0.07	0.06	0.06–0.09
Total dissolved solids (mg/L)	4-8	3–8	2–29	6–135	24–32	30–34	326–358	257–401
Total alkalinity (mg/L as $CaCO_3$)	3–4	4	4	4–9	6	7	85–88	67–84
Total hardness (mg/L as $CaCO_3$)	1	0	0.6–1	1–2	2	2	135–156	97–178
Total organic carbon (mg/L)	0.1–0.2	0.25–0.45	0.13–0.26	0.12–0.16	0.19–0.47	0.168–0.26	1.41–1.42	1.61–1.85
Calcium (mg/L)	0.1-0.7	N/A	N/A	N/A	N/A	N/A	N/A	N/A
Manganese (mg/L)	<5	<5	<5	<5	<5	<5	<5	<5
Bromide (mg/L)	<0.02	<0.02	<0.02	<0.02	<0.02	<0.02	0.056–0.07	0.072–0.1
UVA_{254}, (1/cm)	ND–0.005	ND–0.004	ND	0.004–0.006	<0.004–0.009	<0.004	0.023–0.024	0.028–0.032
Conductivity (µmho/cm)	7–11	9–17	15–18	20–23	32–38	43–58	548–595	436–654
Chloride (mg/L)	2.0–2.3	2.2	3.2–3.5	4.0–4.5	5.1–5.8	7.6–8	68–69	68–75
Silica (mg/L)	<1	<1	<1	<1	1.6–1.8	1.7	12.2–12.9	11.6–14.5
Sulfate (mg/L)	<2-2.7	<2-2.8	2.3-3.1	3.2-3.2	<4	<4-4.4	85-101	48-124
Fluoride (mg/L)	ND	<0.02-0.03	<0.02-0.02	<0.02-0.27	<0.02-0.02	<0.02-0.03	0.16-0.17	0.09-0.20
Boron (mg/L)	N/A	N/A	N/A	N/A	N/A	N/A	N/A	N/A
Aluminum (mg/L)	<5	<5-6	<5	<5	<5	<5	<5	<5
Strontium (mg/L)	<20	<20	<20	<20	<20	<20	370-513	187-592
Barium (mg/L)	<5	<5	<5	<5	<5	<5	43-56	29-65
Particle count (#/mL)	4-17	11-13	3-4	2-11	5	10	12-28	7-15
Perchlorate (µg/L) - 100 µg/L run	<4	<4	<4	6	11	27	85	210-220
Perchlorate (µg/L) - 2000 µg/L run	11-16	52	29	135	171	387-736	1760	3910

Note: For the test runs at both 1 gpm and 4.3 gpm feed rates with 20% and 14% recovery respectively.

Ion Charge: Mechanism for PerchlorateRrejection by RO and NF

A unique characteristic of NF and RO membranes is the charge the membrane carries. This charge is in most cases negative, but positively charged membranes do exist. The interaction between the charge on the membrane and the charge of the ions is the driving force behind small ion rejection. Ion rejection was examined as a function of ion charge to determine the effects of highly negative NF and RO membrane charge. In general, the rejection of cations by NF (TFC-S) and RO (LFC-1) membranes with their respective charges obeys the following order (Table 11.6):

$Si^{+4} \geq Al^{+3} \geq$ (Mg^{+2}, Ca^{+2} indicated as total hardness) $\geq Sr^{+2} \geq Ba^{+2}$

Similarly, anion rejection should occur in the following order:

$SO_4^{-2} \geq Cl^- \geq ClO_4^- \geq F^- \geq Br^-$

It was observed that ions that were strongly negative (-2) were rejected the most by RO and NF membranes, as were those ions carrying a strong positive charge (+2 or greater). Ions charged with −1 or +1 were not rejected well. Thus, it appears that RO and NF membranes tended to reject particles that were highly charged.

Ion Size: Mechanism for Perchlorate Rejection by UF

In addition to ion rejection of inorganic solutes as a result of negative ion charge on the membrane, sieving-type rejection—such as rejection of ions with a specific hydrated radius— is also an important mechanism for inorganic solute removal. Ions such as silica, aluminum, magnesium and calcium (total hardness), strontium, barium, sulfate, chloride, bromide, and fluoride were examined. There was no apparent correlation between ion size and ion rejection by the LFC-1 (RO), TFC-S (NF), and ESNA (NF) membranes. Such results would indicate that rather than steric rejection, electrostatic rejection is the likely mechanism of rejection of ions for these membranes.

A UF membrane can be thought of as a barrier allowing the retention of particles larger than the largest pore diameter on its surface. However, ion size is very difficult to define accurately in terms of appropriate parameters such as the effective radii of solvated ions because of the nature of the ion-solvent interactions, thermodynamics, and transport processes. Approximations of ion size, described by ion radius and hydrated radius, were used in evaluating the effects of ion size [3,4]. In general, ion rejection percentage increases with increasing ionic size or hydrated radius (see Table 11.7). The GM-Desal (UF) membrane removed 0–3.28 and 0–2.8 percent of silica and chloride (smaller ions), respectively, while greater rejection was seen with the larger ions such as sulfate (17.2–27.6 percent) and aluminum (12.8–23.0 percent).

Membrane Performance – Long Term Tests

Phase II of the single element membrane study was done to investigate the long-term operational impacts on the perchlorate rejection. One reverse osmosis (RO) membrane (LFC-1) and one nanofiltration (NF) membranes (TFC-S) were tested. Each membrane was tested with a perchlorate spike of 2,000 μg/L, both with and without a 50-percent recycle.

Table 11.6
Ion rejection of reverse-osmosis and nanofiltration membranes

Ions	Ionic Charge	LFC-1 (RO) Percent Removal	TFC-S (NF) Percent Removal	ESNA (NF) Percent Removal
Silica*	+4	~100	~100	85.2–88.3
Aluminum	+3	92.7– ~100	~100	91.3– ~100
Calcium/magnesium (total hardness)	+2/+2	99.5	99.3–99.7	98.3–99.1
Strontium	+2	96.7– ~100	97.5– ~100	95.7– ~100
Barium	+2	92.6– ~100	93.9– ~100	91.2– ~100
Sulfate	-2	98.3–98.9	98.0–98.8	96–97.7
Chloride	-1	97.1–98.0	95.4–96.0	89.2–91.4
Bromide	-1	77.8– ~100	81.5– ~100	66.7– ~100
Perchlorate	-1	94.9–98.8	96.6–97.4	86.7–92.1
Fluoride	-1	90.7– ~100	91.3– ~100	87.6– ~100

* measured as SiO_2

Table 11.7
Ion rejections by GM-Desal UF membrane

Ions	Ionic Radius (Å)	Hydrated Radius (Å)	Percent Removal (%)
Silica	0.4	N/A	0–3.28
Aluminum	0.675	4.75	12.8–23.0
Calcium/magnesium (total hardness)	1.26/0.86	4.12/4.28	11.0–15.6
Strontium	1.45	4.12	4.5–16.7
Barium	1.61	4.04	6.7–17.6
Sulfate	2.44	3.79	17.2–27.6
Chloride	1.67	3.32	0–2.8
Bromide	1.82	3.30	12.2–72.1
Perchlorate	2.26	3.38	0–10.3
Fluoride	1.17	3.52	0–17.3

Results (Long Term Tests)

Operating Parameters and Field Samples

The LFC-1 (RO) membrane and the TFC-S (NF) membrane both showed excellent removals of perchlorate throughout the duration of the test. In general, the permeate particle counts > 2 μm and turbidity were not sensitive indicators of membrane performance change. Since the removal efficiencies of the membranes were very similar for the EC and perchlorate, EC was closely monitored throughout the run for early signs of decline in perchlorate rejection due to possible degradation of the membrane performance. Operating parameters such as flow rate, pressure, and temperature were also monitored to note any changes in the specific flux.

Nanofiltration Membrane

Nanofiltration membrane (TFC-S): without recycle. The specific flux was stable at 0.2 gfd/psi for approximately 15 days of operation for the nanofiltration membrane used as shown in Figure 11.6. After that, the flux started to decline steadily, and the specific flux decreased by 25% by the end of the 30-day run, which indicates the fouling of the nanofiltration membrane. Typically, a 15% drop in specific flux triggers membrane cleaning protocol to prevent a decrease in the cleaning efficiency drop when a membrane is operated beyond the 15% in specific flux.

The EC reduction showed a similar decreasing trend as indicated by the specific flux. Higher than 96% reduction was observed until 15 days of operation (Figure 11.7), which is comparable to the rejection based on the short term test and result. After the membrane started to show the specific flux decline, which is indicative of fouling, the EC removal was reduced to 92%. The perchlorate rejection also decreased to 90% from close to 97% based on the short term results.

The permeate pH showed a general decline, as shown in Figure 11.8, over the duration of the test probably due to the precipitation of $CaCO_{3(s)}$ on the membrane surface. The Langelier Saturation Index (LSI) predicted for this system was approximately + 0.2, indicating that there is a slight potential for scale formation due to the precipitation of calcium carbonate solid. Since the pH was not adjusted for these tests to prevent scaling, it was expected to have some precipitation based on the solubility product of calcium carbonate. However, as the low positive LSI value suggests, the scaling of the membrane is probably not a major factor in the membrane performance degradation compared with fouling of the membrane.

Nanofiltration membrane (TFC-S): with 50% recycle. With 50% recycle, the specific flux was stable at approximately 0.15 gfd/psi for more than 20 days of operation (Figure 11.9). Unlike the previous case without recycle, the specific flux was stable for the duration of 30-day test.

With 50% recycle of the concentrate stream, the EC reduction was approximately 97% as shown in Figure 11.10, which correlates well with the short-term test results. This rejection was observed throughout the duration of the test. The perchlorate rejection showed a small decrease (~95% rejection) compared with the short term rejection results for 50% recycle test with NF (~97%). Overall, good rejection of EC and perchlorate more than 2,000 μg/L initial concentration was observed by the NF membrane. Unlike the previous case where no recycle of the concentrated stream was used, with the recycle of the concentrated stream, the membrane performance exhibited only a slight change in rejection, suggesting that good perchlorate rejection performance can be expected when a multistage membrane array is used in larger scale applications.

The decrease in pH was much greater for this case even considering the fluctuation in the pH readings (Figure 11.11). The permeate pH decreased by more than one pH unit within the first few days of operation. Again the precipitation of calcium carbonate is likely the cause for the decrease in the permeate pH ($Ca^{2+} + HCO_3^- \rightarrow CaCO_3(s) + H^+$), but this precipitation is not expected to be a major component in membrane performance degradation with low LSI value.

Reverse Osmosis (LFC-1)

The low fouling composite (LFC-1) reverse osmosis membrane was tested at a feed flow rate of approximately 1 gpm (100 psi feed pressure) with a recovery of 20%. Although these conditions were outside of the normal operating conditions recommended by the membrane manufacturer, they were used based on the bench scale testing results. Under these conditions with or without the 50% recycle of the concentrate stream, the membrane performance did not change significantly over 30-day period. With or without 50% recycle of the concentrate stream, specific flux and salt rejection exhibited no change during the 30-day run. In order to validate the results obtained for the long term test done with low feed rate and high recovery ratio, a follow up test was also performed using the normal operating conditions recommended by the RO membrane manufacturer. A new membrane was acquired and used for the tests done at the initial feed rate of approximately 4gpm (150 psi feed pressure) with recovery of 14%. Both sets of test data are presented below and show similar trends in terms of the stable specific flux and salt rejection. The general decrease in the pH value was consistent for both 0% and 50% recycle as it was shown for NF.

Reverse osmosis (LFC-1): without recycle at low feed rate. Due to the low feed flow rate used for the RO membrane, the calculated specific flux was much lower than that observed for the NF membrane. Also, the specific flux may have been sensitive to the change in operating conditions, yielding a higher standard deviation as shown in Figure 11.12. That is, the specific flux data points near the middle and the end of the test, may not be due to the actual decrease or increase in the specific flux, but may be attributed to the propagation of errors associated with operating the membrane outside of the recommended conditions.

The EC reduction for this test consistently showed greater than 98% throughout the duration of the test (Figure 11.13), which also support that the change in the specific flux shown in Figure 11.14 might be due to the propagation in errors associated with the operating conditions, and not necessarily an indication of the fouling of the membrane. The perchlorate rejection showed a good correlation with EC reduction data with more than 98% removal of perchlorate at 30 days of operation with ~ 2,000 μg/L of initial perchlorate. Both EC and perchlorate rejections correlates well with the results observed for the short term tests.

As in the previous case, a similar decrease in the permeate pH was observed, but the low amount of scaling, as projected by LSI value, had no significant effect on the reduction of perchlorate removal efficiency.

Reverse osmosis (LFC-1): without recycle at high feed rate. A follow up, long-term test was done with a higher feed flow rate of approximately 4 gpm and a 14% recovery to compare the long term trends for low (1 gpm) and high (4 gpm) feed rates for the RO membrane. The follow up test was done with a new membrane, which was rinsed before the membrane was subjected to the testing, so the fluctuation of the data at the beginning may be that the stable condition has not been reached by the time the test started. After about a week of operation once the membrane performance stabilized, a consistent specific flux of 0.03 gfd/psi (Figure 11.15) was reached. Although the specific flux decreased more than 15% at the beginning of the test, no consistent decrease was observed as a function of time, which would have indicated fouling of the membrane. The direction of specific flux change was not consistent, but rather fluctuated during the test, possibly due to inconsistency in operation or performance of the membrane at the beginning of the test with a new membrane.

For the first few days of operation, EC reduction was unstable, so the test was considered started once stable readings were obtained. After a stable condition was reached, consistently greater than 95% EC reduction and perchlorate rejections were observed. Once again, perchlorate rejection correlated well with the EC reduction data, and after two weeks of operation, approximately 97% reductions of both EC and perchlorate were observed, which is consistent with the short term test results.

Reverse osmosis (LFC-1): with 50% recycle and low feed rate. With 50% recycle of the concentrated stream, more stable specific flux was observed during the 40 days of operation as shown in Figure 11.17 with a 1 gpm feed rate. No significant change in specific flux was observed from the stable value of 0.03 gfd/psi for the duration of the test.

Stable salt rejection was also observed throughout the 40 days of operation. A consistent 98% salt rejection was obtained, and the perchlorate rejection correlated well with the EC rejection data throughout the test. Again the rejections observed were consistent with the short term test results.

Similar to the previous results, the general decreasing trend was observed for the permeate pH. The scattering of the data points observed, especially for the permeate pH, was within the typical standard deviation observed.

Table 11.8 shows the typical feed, concentrate, and permeate concentrations of anions, cations and other parameters for nanofiltration membrane and reverse osmosis filter for both with and without recycle.

SUMMARY

Based on the single element tests, the following conclusions can be made about the treatment options for perchlorate removal:

- The RO (LFC-1) and NF (TFC-S and ESNA) membranes consistently removed more than 86.7 percent of the applied perchlorate.
- RO membranes (LFC-1) rejected perchlorate more effectively than did NF membranes (TFC-S and ESNA).
- The tight UF membrane (GM-Desal) proved ineffective for removing perchlorate.
- Addition of a cationic surfactant to the GM membrane to assist in perchlorate removal only served to plug the membrane.
- Concentrate recycling had no significant effect on the membranes' perchlorate rejection capabilities.
- Perchlorate removal percentages remained relatively constant in this study, regardless of the influent perchlorate concentration.
- UVA_{254}, TOC, conductivity, turbidity, and particle-count data indicated that all membranes tested produced permeates with acceptable water quality (excluding perchlorate levels).
- While RO results can be explained by simple solution-diffusion as a rejection mechanism for ions including perchlorate, the electrostatic rejection promoting perchlorate removal can be significantly related with high pH, lower conductivity, and lower calcium concentrations in the source waters for the NF and UF membranes. The optimum range can be different depending upon water chemistry conditions and membrane types/materials.

Table 11.8
Representative complete data for long term tests

Analyte	RO, no recycle			RO, 50% recycle			NF, no recycle			NF, 50% recycle		
	Feed	Conc	Perm	Feed	Conc	Perm	Feed	Conc	Perm	Feed	Conc	Perm
Chloride (mg/L)	67	86	ND	80	126	ND	66.3	73	7.44	100	130	5
Sulfate (mg/L)	171	221	ND	181	288	ND	161	181	ND	179	233	ND
Fluoride (mg/L)	0.23	0.28	ND	0.3	0.4	ND	0.21	0.22	ND	0.2	0.3	ND
TDS (mg/L)	502	663	12	510	2644	12	464	535	24	567	734	16
Perchlorate (ug/L)	1340	2710	21	2370	4034	41	605	744	59	1290	1924	59
Bromide (mg/L)	0.06	0.08	ND	0.1	0.1	ND	0.05	0.06	ND	0.1	0.2	ND
TOC (mg/L)	2.68	3.42	0.12	3.00	4.60	0.10	2.67	2.97	0.16	3.00	3.80	0.20
Aluminum (ug/L)	120	146	40	149	271	ND	179	211	8	112	143	ND
Iron (ug/L)	ND	ND	39	ND	ND	ND	ND	ND	ND	ND	ND	ND
Manganese (ug/L)	ND	ND	ND	ND	ND	ND	ND	ND	ND	ND	ND	ND
Strontium (ug/L)	679	899	ND	763	1150	ND	729	854	ND	777	979	ND
Barium (ug/L)	78	100	ND	81	122	ND	77	91	ND	87	108	ND
Silica (mg/L)	9.81	12.5	0.48	11	16	0.6	10.1	11.4	1.4	13	17	3.4
Calcium (mg/L)	53	69	0.1	56	90	ND	52	60	0.4	57	75	0.3
Magnesium (mg/L)	21.5	27.5	ND	25	40	ND	21.5	25	0.2	26	33	0.1
Potassium (mg/L)	4.6	5.7	0.1	5.4	9	0.3	3.7	4.3	0.4	5.2	6.6	0.3
Sodium (mg/L)	75	97	2.4	82	129	3.1	69	78	7	88	112	4.8
Total Alk (mg/L)	113	140	4	127	194	5	103	117	8	126	161	6
UV (abs/cm)	0.03	0.04	ND	-	0.1	ND	0.03	0.04	0.01	ND	0.1	ND
Hardness (mg/L)	220	288	1	245	387	ND	213	244	2	245	318	2

- UF results can be explained by simple steric rejection as the rejection mechanism for ions including perchlorate. In general, ion rejection percentage increases with increasing ionic size or hydrated radius.
- For the long term tests, the specific flux declined by 25% for NF without recycle, which indicates fouling of NF membrane.
- Perchlorate rejection decreased from 97% to 92% for NF membrane without recycle after the membrane fouled.
- With 50% recycle, the NF membrane did not foul within 30 days of operation under similar conditions compared without recycle.
- For all long term tests with and without the recycle, and at different feed rate and recoveries, the RO membrane did not show any signs of fouling for all conditions tested, and consistent perchlorate rejection was observed.
- EC reduction was associated with co- and counter ions (e.g. Na^+, Ca^{2+}, Cl^- and SO_4^{2-}) and correlated well with perchlorate rejection, and can be used as a good indicator for perchlorate rejection.

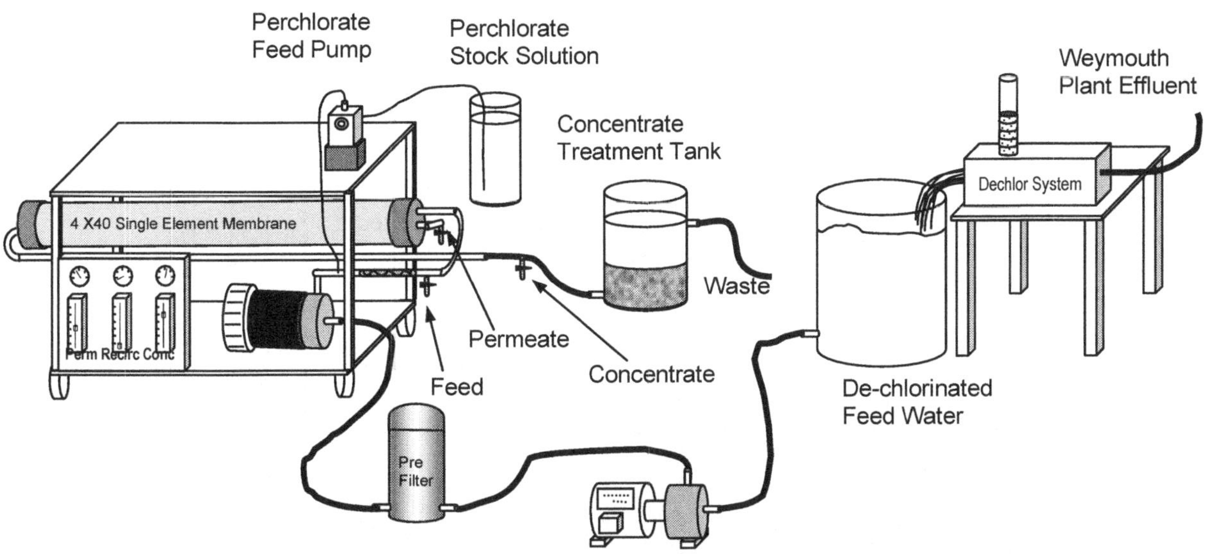

Figure 11.1 Schematic of a single element test platform

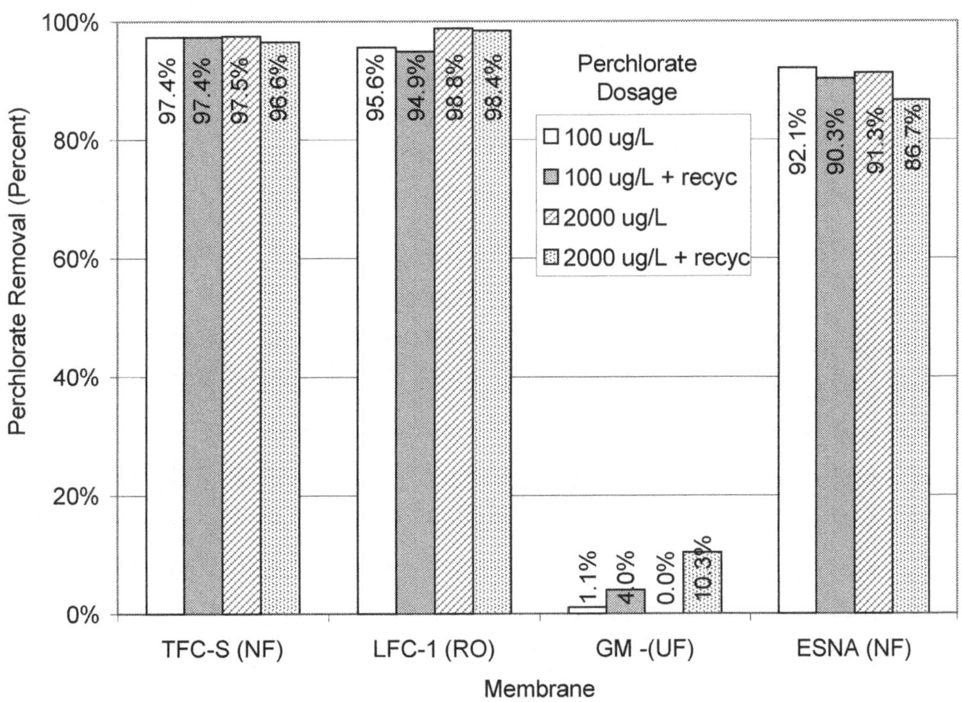

Figure 11.2 Comparison of perchlorate removal

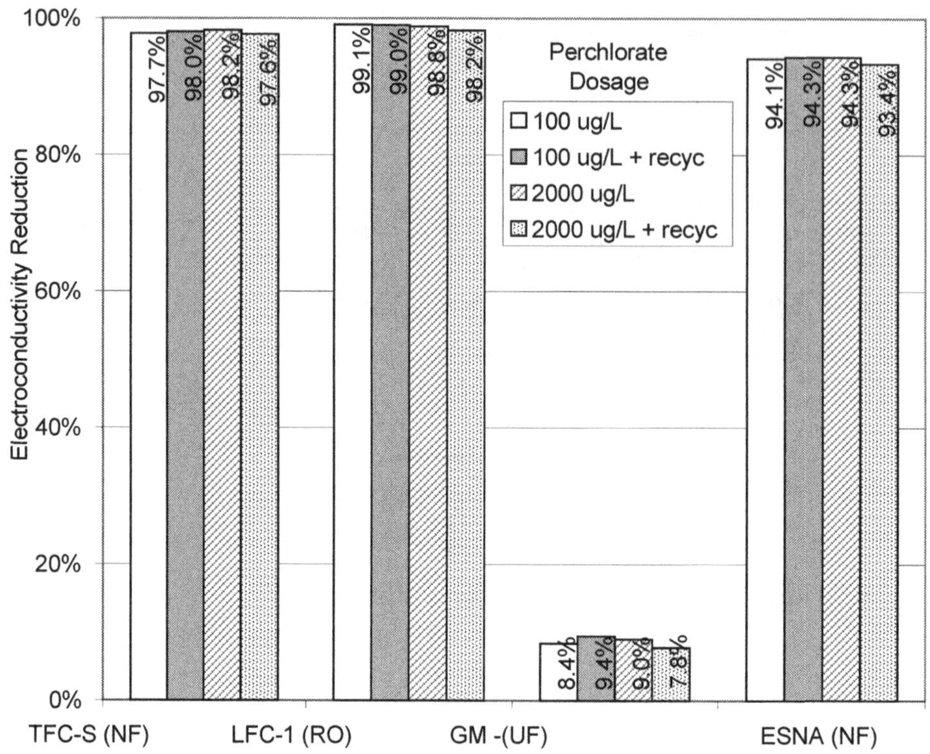

Figure 11.3 Comparison of electroconductivity removal

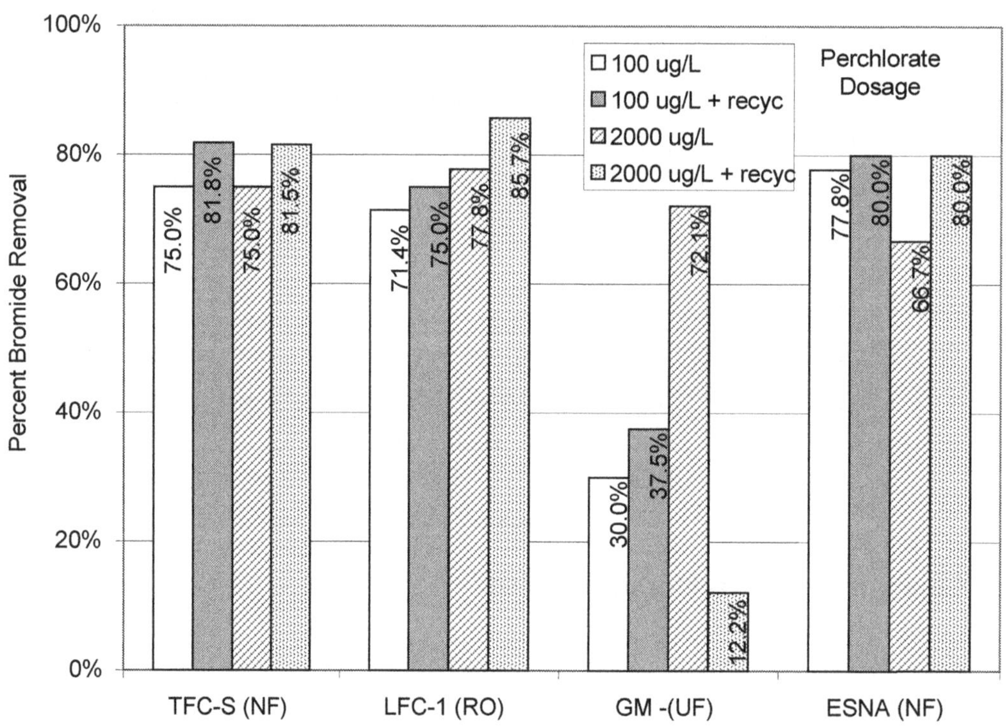

Figure 11.4 Comparison of bromide removal

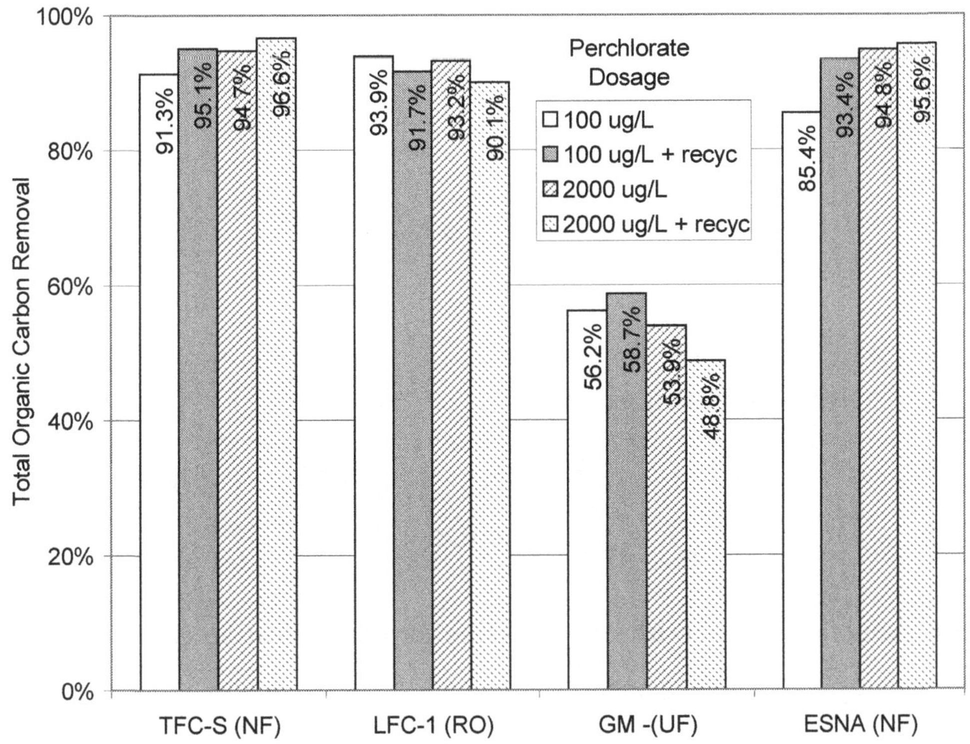

Figure 11.5 Comparison of TOC removal

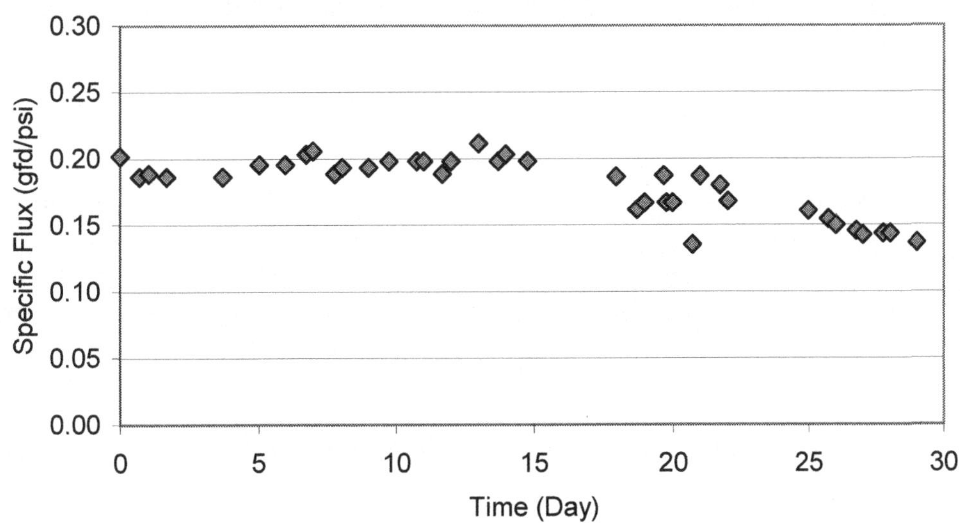

Figure 11.6 Specific flux change for long term test with nanofiltration membrane without recycle

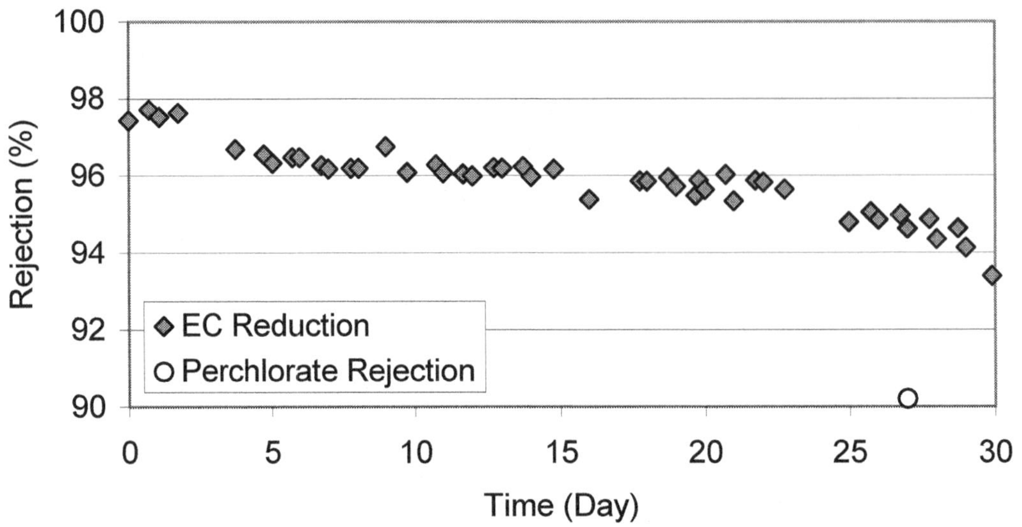

Figure 11.7 EC reduction and perchlorate rejection for nanofiltration membrane

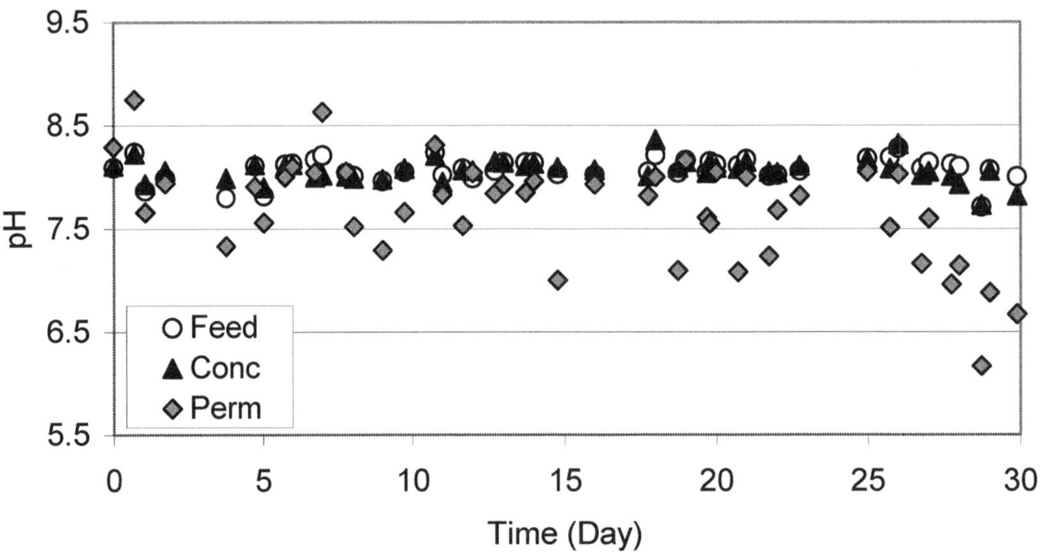

Figure 11.8 pH change for long run test with nanofiltration membrane without recycle

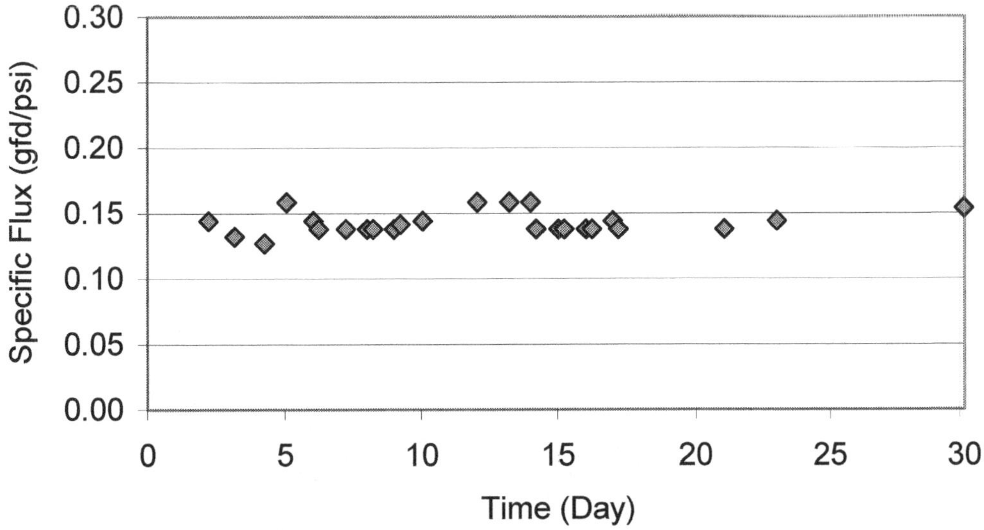

Figure 11.9 Specific flux change for long term test with nanofiltration membrane at 50% recycle

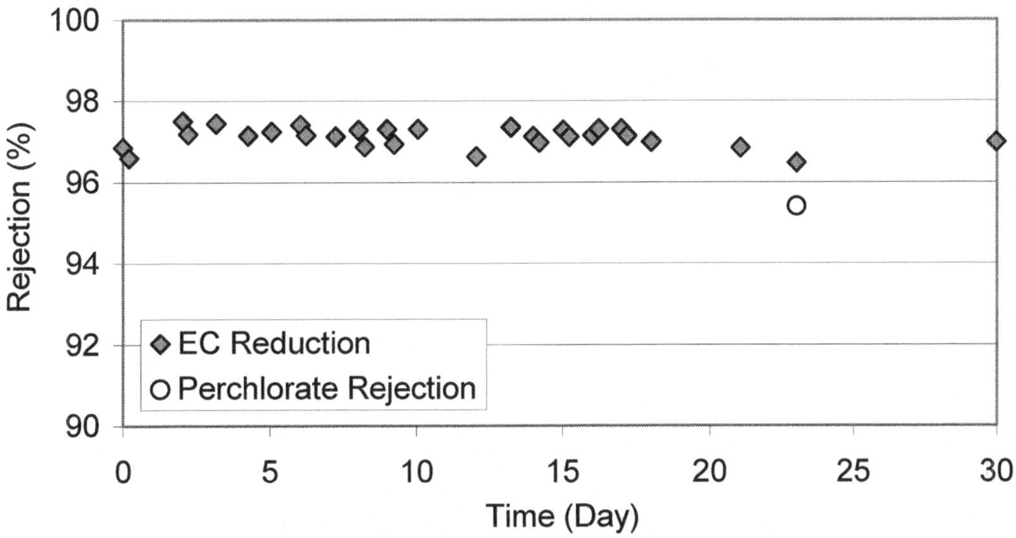

Figure 11.10 Salt rejection based on EC for nanofiltration membrane (TFC-S) at 50% recycle

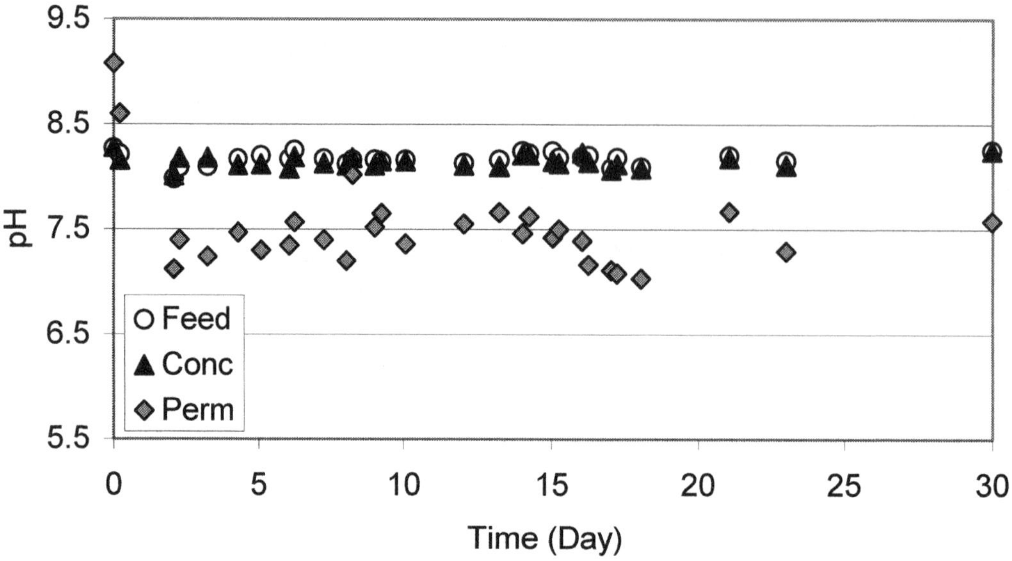

Figure 11.11 pH change for long term test with nanofiltration membrane at 50% recycle

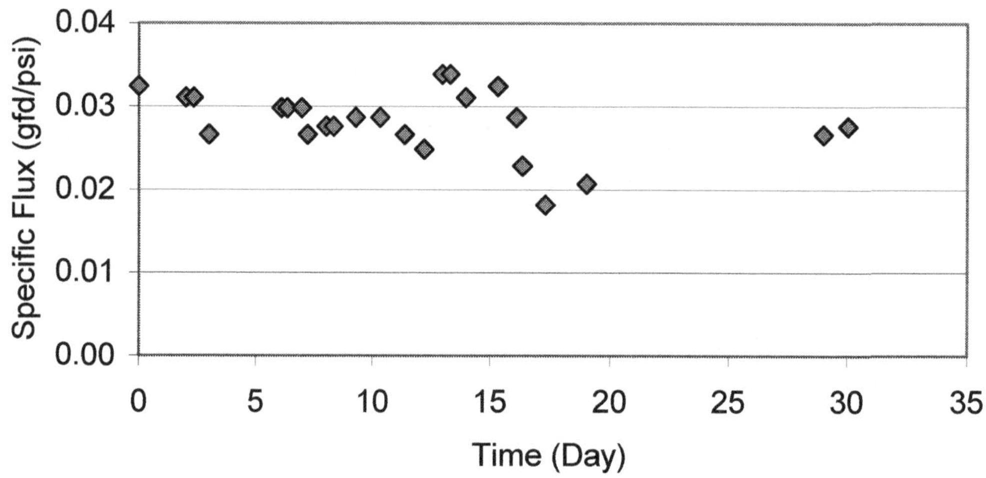

Figure 11.12 Specific flux change for long term test with RO membrane at 1 gpm feed rate

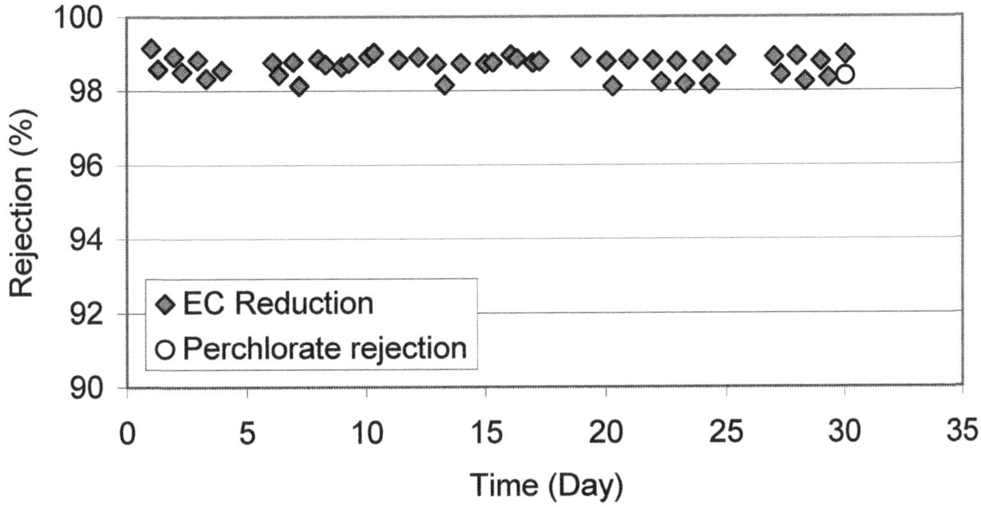

Figure 11.13 Rejection for long term test with RO membrane at 1 gpm feed rate

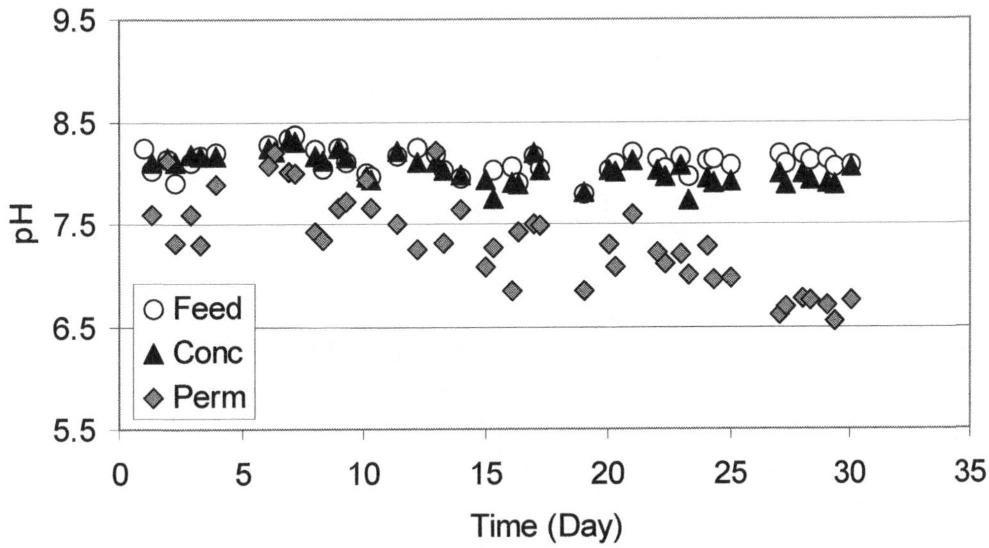

Figure 11.14 pH for long term test with RO membrane at 1 gpm feed rate

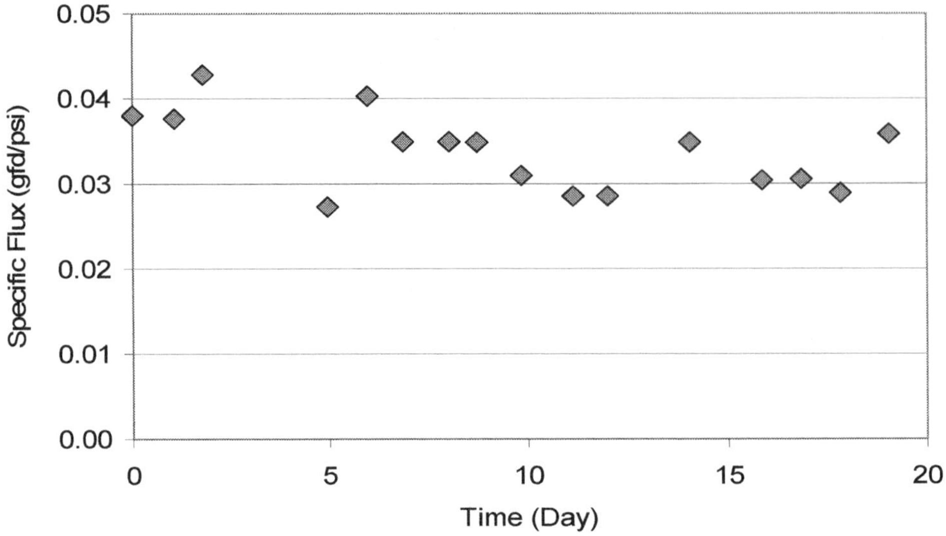

Figure 11.15 Specific flux change for long term test with RO membrane at higher feed rate

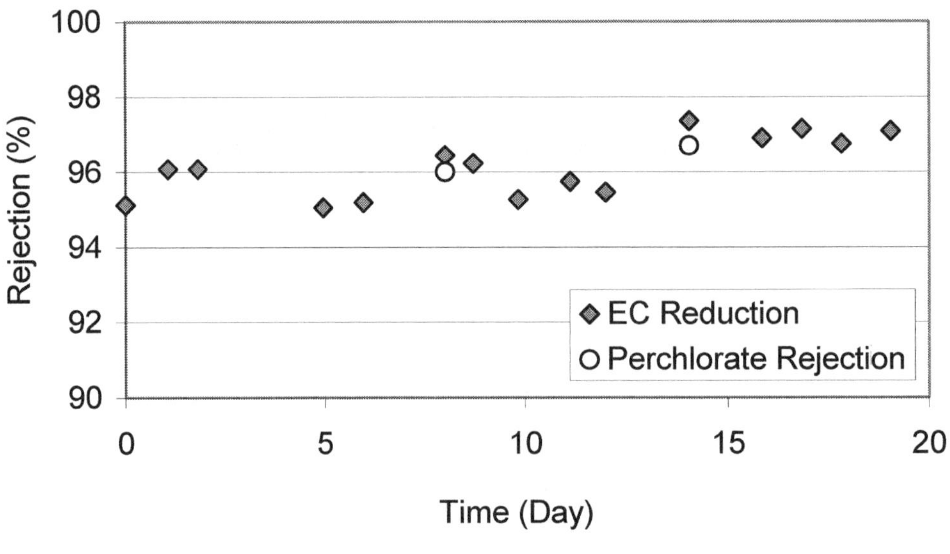

Figure 11.16 Salt rejection for long term test with RO membrane at high feed rate

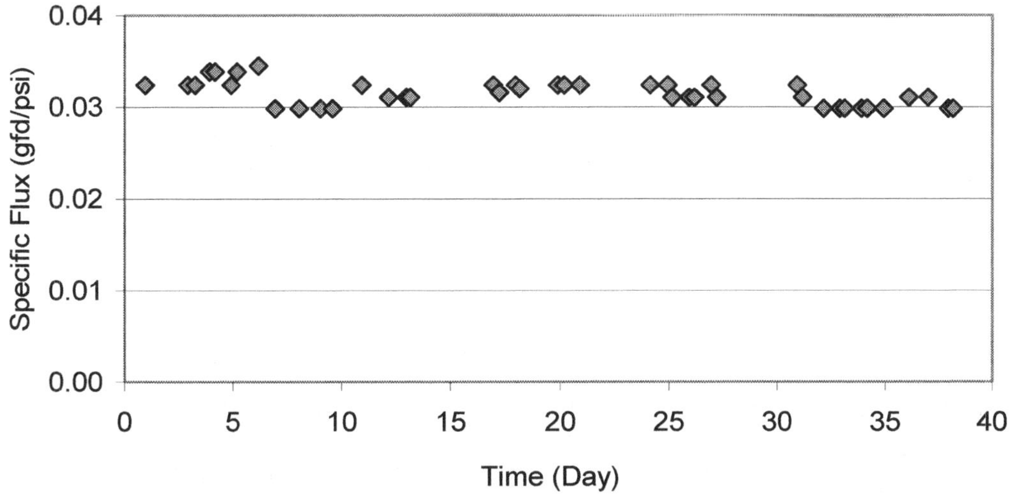

Figure 11.17 Specific flux change for long term test with RO at 50% recycle (1 gpm feed rate)

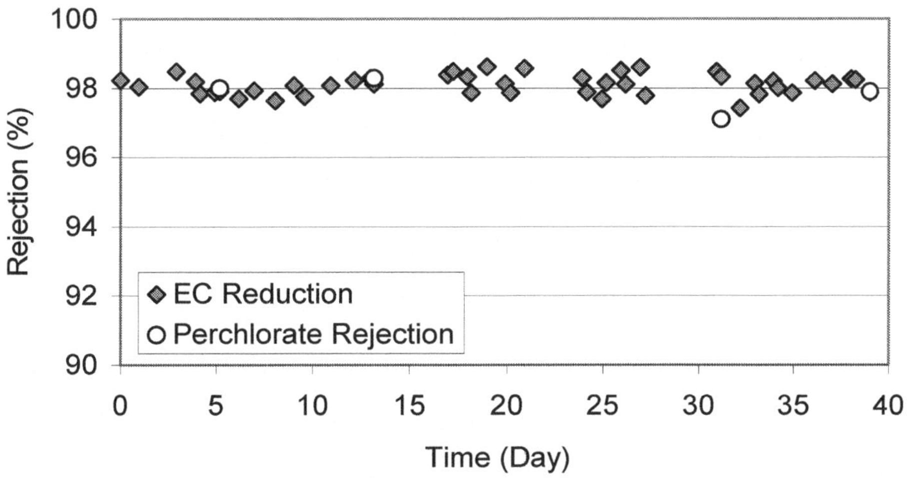

Figure 11.18 Salt rejection based on EC for RO at 50% recycle at 1 gpm feed rate

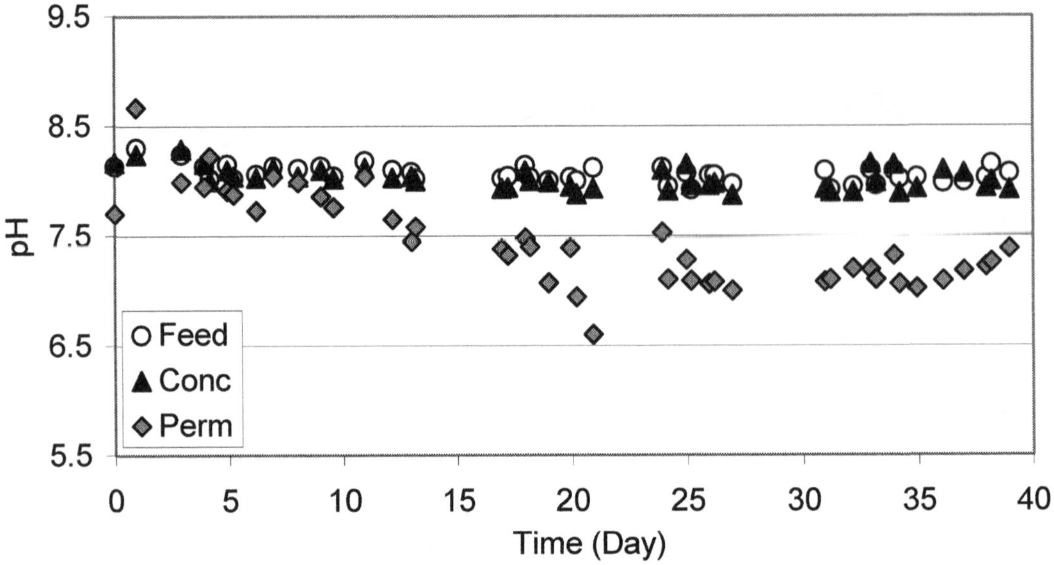

Figure 11.19 pH for long term test with RO at 50% recycle at 1 gpm feed rate

CHAPTER 12
CONCLUSIONS AND RECOMMEDATIONS FOR FUTURE RESEARCH

CONCLUSIONS

One of the significant findings of this research was that known mechanisms (e.g., solution diffusion, electrostatic interaction (repulsion), and steric (size) exclusion) for solute transport (rejection) of a target ion (in this case ClO_4^-) were demonstrated by several measurement techniques (e.g., hindered diffusion measurement, streaming potential measurement, and PEG rejection determination). Another significant finding of this research was that perchlorate transport was predicted well by a non-equilibrium thermodynamic model. The transport parameters used in the model were determined by independent measurements (and calculations with minimum assumptions). For example, the molecular transport coefficient was obtained by diffusion cell measurement under varying pH and conductivity. The conclusions of the associated experimental results below support the objectives of this research and research hypotheses.

- The changes of membrane surface charge, becoming more negative, were found to enhance perchlorate rejection as pH increased and conductivity decreased. This is because electrostatic repulsion significantly occurred between the perchlorate anion and the negatively charged membranes (ESNA, MX07, and GM) as pH increased and conductivity decreased. The data obtained from ZP measurement supported the above results that the ESNA, MX07, and GM membranes acquired more negative charge when pH increased regardless of the salt type at the same conductivity of 30 mS/m. The ZP values of the membranes became less negative with increasing conductivity regardless of salt type.
- Changes of membrane pore size, becoming smaller in the presence of co- and counter ions, were found to not significantly enhance perchlorate rejection. This is because the ionic radius of the perchlorate ion is still much smaller than the pore size of the membranes (ESNA, MX07, and GM). Thus, steric exclusion did not significantly affect perchlorate transport for the membranes. However, the relatively small pore size ESNA membrane showed much higher perchlorate rejection than the relatively large pore size MX07 and GM membranes. This indicates that steric exclusion still slightly affected perchlorate transport. Experimental results showed that PEG rejection by the membranes follows the order, $KCl < K_2SO_4 < CaCl_2$, due to ion binding (such as with Ca^{2+}) and increasing ions ($Cl^- < K^+ < SO_4^{2-}$) at the interface of the membranes.
- It was proven that electrostatic repulsion did not significantly occur between an uncharged inorganic species (in this case As (III)) and negatively charged membranes. As (III) rejection only slightly varied at pH 4, 6, and 8 and 30 mS/m, 60 mS/m, and 115 mS/m for the membranes (ESNA and MX07) since the As (III) exists primarily as an uncharged species below pH 9.13 based on its pK_a of 9.13. However, As (III) rejection significantly increased at pH 10 where As (III) is anionic.
- The changes of membrane surface charge, becoming more negative, were found to reduce hindered diffusion and partitioning for the MX07 and GM membranes as pH

increased and conductivity decreased. The measured hindered diffusion and partitioning were used to predict perchlorate transport dominated by diffusion driving force. The predicted perchlorate flux by both diffusion and convection decreased with increasing pH and increased with increasing conductivity. The predicted perchlorate flux by convection was much greater than that by diffusion regardless of pH and conductivity conditions. These results are consistent with the theory that solute flux is dominated by convection for a relatively large pore size membrane.

- An increase of recovery and J_o/k ratio resulted in a decrease of perchlorate rejection for the two RO membranes (LFC-1 and TFC-ULP) and two NF membranes (TFC-S and ESNA). For the RO membranes, an assumption was made that an increase of recovery and J_o/k ratio significantly affected solution diffusion due to concentration polarization. For the NF membranes, an increase of concentration polarization of solutes with increasing recovery and J_o/k ratio also significantly affected membrane surface charge and membrane pore size, which significantly increased perchlorate transport.
- Based on results of four different membranes at varying recoveries with natural waters, calcite ($CaCO_3$) scale was dominant at the membrane surface and detected by measurements (XRD, FTIR, SEM). Also, antiscalants, composed of phosphonate groups, were effective in the inhibition of calcite formation.
- The concentration of solute at the membrane surface can be predicted by a semiempirical constant (K). Considering results for each ion and four membranes, most K values of ions with RO membranes approached 1, while values for NF membranes were smaller than RO membranes.
- In an evaluation of costs and water quality of full vs. split stream treatment with respect to corrosion potential and effects on water cost associated with design and operational parameters, split RO treatment was shown to be more cost effective than full NF treatment. Also, split RO stream treatment reduced the corrosion potential of RO full stream treament, providing a cost-effective alternative to, and better water quality including rejection of perchlorate than, NF full stream treament.
- Considering design and operational parameters such as maintenance cost, interest rate, power cost, chemical cost, plant life, and membrane replacement, each parameter affected the product water cost.
- Perchlorate by a UF membrane modified with cationic surfactant was greater than expected, based mostly on steric/size exclusion as a result of a decrease in the membrane pore size. However, a limitation in use of a cationic surfactant to modify a negatively charged UF membrane for perchlorate rejection was that a significant amount of flux-decrease also occurred. As expected, perchlorate transport was not affected by anionic surfactant conditioning, which did not significantly influence membrane pore size or membrane surface charge.
- Since only short-term rejection tests were performed, long-term performance should be identified for future research.
- In the single element tests, the RO (LFC-1) and NF (TFC-S and ESNA) membranes showed effective perchlorate rejection regardless of the influent concentration, however, tight UF (GM) proved ineffective for removing perchlorate.

- While RO results can be explained by simple solution-diffusion as a rejection mechanism for ions including perchlorate, electrostatic rejection appeared to be the dominant mechanism of rejection for the NF membranes.
- For the long term tests, the specific flux declined by 25% for NF without recycle, which indicates fouling, and perchlorate rejection decreased from 97% to 92% after fouling. With 50% recycle, the NF membrane did not foul withing 30 days of operation under similar conditions compared without recycle. Also, for the long term tests with and without the recycle, the RO membrane did not show any signs of fouling for all conditions tested, and consistent perchlorate rejection was observed.
- EC reduction correlated well with perchlorate rejection, and can be used as a good indicator for perchlorate rejection.

RECOMMENDATIONS FOR FUTURE RESEARCH

While RO results can be explained by simple solution-diffusion as a rejection mechanism for perchlorate (and other ions), this study has demonstrated that there are two potential mechanisms for perchlorate rejection by NF membranes: steric (size) and electrostatic (charge exclusion). Given that perchlorate has a molecular mass of 100, effective rejection has been shown by (negatively) charged NF membranes having lower MMCOs (steric rejection and electrostatic rejection) and higher MMCOs (electrostatic rejection).

Although this short-term bench-testing study can predict perchlorate transport through the RO, NF, and UF membranes, to accurately model full-scale operation, a pilot-testing study is strongly recommended in order to address long-term perchlorate rejection and scale-up issues. The future research can be focused to: (i) confirm perchlorate removal efficiencies that were determined at the bench-scale, (ii) optimize membrane system operational conditions, to minimize generation of the concentrate stream and to evaluate concentrate treatment methods, (iii) investigate pretreatment, (iv) identify long-term performance as affected by membrane fouling, (v) identify effects of a long term oxidizing potential of perchlorate on flux-decline and rejection, and (vi) identify and address important concentrate stream issues. These recommended studies can be fulfilled through an intensive pilot-testing program interfaced with focused and supporting bench testing

REFERENCES

Amjad. 1985. Application of Antiscalants to Control Calcium Sulfate Scaling in Reverse Osmosis Systems. *Desalination*, 54:253.

Amjad. 1988. Calcium Sulfate Dihydrate (Gypsum) Scale Formation on Heat Exchanger Surfaces: The Influence of Scale Inhibitors. *Jour. Colloid and Interface Science*, 123:523.

APHA, AWWA, and WEF. 1998. *Standard Methods for the Examination of Water and Wastewater*. 20th ed. Washington, DC: APHA, AWWA, and WEF.

Awwa Research Foundation. 1996. Lyonnaise des Eaux, & Water Research Commission of South Africa. *Water Treatment Membrane Processes* (J. Mallevialle et al., editors). McGraw-Hill, New York.

Bauza, C.R., Sanchez, C.R., and Garcia, V.B. 1990. On Cation and Water Transport Numbers in Cellulose Acetate Membranes. *Jour. Non-Equilib. Thermodyn.*, 15:383.

Bellamy, L.J. 1975. *The Infra-red Spectra of Complex Molecules*, Chapman and Hall, London.

Bettinelli and Spezia. 1998. Determination of Trace Elements in Power Plant Emissions by Inductively Coupled Plasma Mass Spectrometry: Comparison With Other Spectrometric Techniques. *Microchemical Journal*, 59:203.

Boerlage and Kennedy. 1999. $BaSO_4$ Solubility Prediction in Reverse Osmosis Membrane Systems. *Jour. Membrane Sci.*, 159:47.

Booth, S.D.J., Cushing, R.S., Hassen, E., and Ford, L. 2000. Evaluating Electrodialysis Reversal for Perchlorate Treatment, *AWWA Annual Conference* in Denver, Colorado.

Bowen, W.R., and Cao, X. 1998. Electrokinetic Effects in Membrane Pores and the Determination of Zeta Potential. *Jour. Membrane Sci.*, 140:267.

Bowen, W.R., and Mukhtar, H. 1996. Characterisation and Prediction of Separation Performance of Nanofiltration Membranes. *Jour. Membrane Sci.*, 112:263.

Braghetta, A., DiGiano, F.A., and Ball, W.P. 1997. Nanofiltration of Natural Organic Matter: pH and Ionic Strength Effects. *Jour. Envir. Engrg.*, ASCE, 123(7):628.

Brandhuber, P. 1999. The Application of Pressure Driven Membrane Filtration as a Means of Arsenic Removal for Public Drinking Water Supply Systems. Ph.D. dissertation, Department of Civil, Environmental, and Architectural Engineering, University of Colorado at Boulder.

Brandhuber, P., and Amy, G. 1998. Alternative Methods for Membrane Filtration of Arsenic From Drinking Water. *Desalination*, 117:1.

Bromley. 1973. Thermodynamic Properties of Strong Electrolytes in Aqueous Solutions. *AIChE Journal*, 19:313.

Brown, J.C., Sun, L., Raskin, L.M., Chee-Sanford, J.C., and Lin, R. 2000. Removal of Perchlorate in Biologically Active Carbon Adsorption Systems, *AWWA Inorganic Contaminants Workshop* in Albuquerque, New Mexico.

Cadotte, J., Forester, R., Kim, M., Petersen, R., and Stocker, T. 1988. Nanofiltration Membranes Broaden the Use of Membrane Separation Technology, *Desalination*, 70:77.

Caldwell, D.J., King, J.H., Kinkead, E.R., Narayanan, L., and Mattie, D.R. 1995. Results of a Fourteen Day Oral-Dosing Toxicity Study of Ammonium Perchlorate. In: *Proceedings of the 1995 JANNAF Safety and Environmental Protection Subcommittee Meeting, Volume 1,* December, Tampa, FL, Calumbia, MD: Chemical Propulsion Information Agency, Joint Army, Navy, NASA, Air Force (JANNAF) interagency propulsion committed publication 634.

Causserand, C., Nystrom, M., and Aimar, P. 1994. Study of Streaming Potential of Clean and Fouled Ultrafiltration Membranes. *Jour. Membrane Sci.*, 88:211.

Chapman-Wilbert M., Delagah, S., and Pellegrino, J. 1999. Variance of Streaming Potential Measurements. *Jour. Membrane Sci.*, 161:247.

Chellam, and Jacangelo. 1997. Effect of Pretreatment on Surface Water Nanofiltration. *Jour. AWWA*, 89(10):77.

Childress, A., and Elimelech, M. 1996. Effect of Solution Chemistry on the Surface Charge of Polymeric Reverse Osmosis and Nanofiltration Membranes. *Jour. Membrane Sci.*, 119:253.

Cho, J., Amy, G., and Pellegrino, J. 1999. Membrane Filtration of Natural Organic Matter: Initial Comparison of Rejection and Flux Decline Characteristics With Ultrafiltration and Nanofiltration Membranes. *Wat. Res.*, 33(11):2517.

Cho, J., Amy, G., and Pellegrino, J. 2000. Membrane Filtration of Natural Organic Matter: Factors and Mechanisms Affecting Rejection and Flux-Decline With Charged Ultrafiltration (UF) Membrane. *Jour. Membrane Sci.*, 164:89.

Cho, J., Amy, G., Pellegrino, J., and Yoon, Y. 1998. Characterization of Clean and Natural Organic Matter (NOM) Fouled NF and UF Membranes, and Foulants Characterization. *Desalination*, 118:101.

Clark, M.M., and Lucas, P. 1998. Diffusion and Partitioning of Humic Acid in a Porous Ultrafiltration Membrane. *Jour. Membrane Sci.*, 143:13.

Darton. 2000. Membrane Chemical Research: Centuries Apart. *Desalination*, 132:121.

Darwish, M.A., Abdel, J.M., and Aly, G.S. 1989. Technical and Economical Comparison Between Large Capacity MSF and RO Desalting Plants. *Desalination*, 76:281.

Deen, W.M. 1987. Hindered Transport of Large Molecules in Liquid-Filled Pores. *Jour. AIChE*, 33(9):1409.

Deen, W.M., and Smith, F.G. 1982. Hindered Diffusion of Synthetic Polyelectrolytes in Charged Microporous Membranes. *Jour. Membrane Sci.*, 12:217.

Dodrill, Donna M., and Edwards, Marc. 1995. Corrosion Control on the Basis of Utility Experience. *Jour. AWWA*, 7:74.

Drak, and Glucina. 2000. Laboratory Technique for Predicting the Scaling Propensity of RO Feed Waters. *Proceedings of the Conference on Membranes in Drinking and Industrial Water Production*, L'Aquila, Italy, 2:483.

Drela, and Falewicz. 1998. New Rapid Test for Evaluation of Scale Inhibitors. *Wat. Res.*, 32(10):3188.

Eaton, A., Haghani, A., Cox, N., and Wong, E. 1998. A Comparison of the Ruggedness of Different Analytical Methods for Perchlorate, *Proceedings 1998 AWWA WQTC*, San Diego, CA.

Edwards, Marc, Schock, Michael R., and Meyer, Travis E. 1996. Alkalinity, pH and Copper Corrosion By-product Release. *Jour. AWWA*, 3:81.

Edzwald, J.K., Becker, W.C., and Wattier, K.L. 1985. Surrogate Parameters for Monitoring Organic Matter and THM Precursors. *Jour. AWWA*, 77:122.

Elimelech, M., Chen, W.H., and Waypa, J.J. 1994. Measuring the Zeta (Electrokinetic) Potential of Reverse Osmosis Membranes by a Streaming Potential Analyzer. *Desalination*, 95:269.

Elimelech, M., and O'Melia, C.R. 1990. Effect of Electrolyte Type on the Electrophoretic Mobility of Polystyrene Latex Colloids. *Colloids Surf.*, 44:165.

Ericsson, Bernt, and Hallmans, Bergt. 1991. Membrane Filtration as a Pre-Treatment Method. *Desalination*, 82:249.

Eriksson, P. 1988. Water and Salt Transport Through Two Types of Polyamide Composite Membranes. *Jour. Membrane Sci.*, 36:297.

Fairbrother, F. and Mastin, H. 1924. Studies in Electro-Endosmosis. *Jour. Chem. Soc.*, 75 (1924) 2318.

Garba, Y., Taha, S., Gondrexon, N., Cabon, J., and Dorange, G. 2000. Mechanisms Involved in Cadmium Salts Transport Through a Nanofiltration Membrane: Characterization and Distribution. *Jour. Membrane Sci.*, 168:135.

Gekas, V., Persson, K.M., Wahlgren, M., and Bjorn, Sivik. 1992. Contact Angles of Ultrafiltration Membranes and Their Possible Correlation to Membrane Performance. *Jour. Membrane Sci.*, 72:293.

Gonzalez, and Lafont. 1994. A Corrosion Inhibition Study of a Carbon Steel in Neutral Chloride Solutions by Zinc Salt/Phosphonic Acid Association. *Corrosion Science*, 37(11):1823-1837.

Harrar, and Locke. 1982. Field Tests of Organic Additives for Scale Control at the Salton Sea Geothermal Field. *Society of Petroleum Engineers Journal*, 22(1):17.

Harrington, P., and Shen, Y. 1998. Improved Analysis of Perchlorate in Water. *Proceedings 1998 AWWA WQTC*, San Diego, CA.

Hasson, and Semiat. 1998. Suppression of $CaCO_3$ Scale Suppression by Anti-Scalants. *Desalination*, 118:285.

Hautman, D.P., and Munch, D.J. 1999. Determination of Perchlorate in Drinking Water Using Ion Chromatography. *EPA Method*, 314.0.

Her, N., Amy, G., and Jarusutthirak, C. 2000. Seasonal Variations of Nanofiltration (NF) Foulants: Identification and Control. *Desalination*, 132:143.

Hong, S., and Elimelech, M. 1997. Chemical and Physical Aspects of Natural Organic Matter (NOM) Fouling of Nanofiltration Membranes. *Jour. Membrane Sci.*, 132:159.

Huheey, J.E. 1983. *Inorganic Chemistry*. Harper & Row, New York.

Kedem, O., and Katchalsky, A. 1958. Thermodynamic Analysis of the Permeability of Biological Membranes to Non-Electrolytes. *Jour. Biochimica and Biophysica*, 27:229.

Klepetsanis, and Koutsoukos. 1999. Calcium Carbonate Deposition on Cellulose. *J. of Colloid and Interface Science*, 224:56.

Lee, and Kim. 1999. Analysis of $CaSO_4$ Scale formation Mechanism in Various Nanofiltration Modules. *Jour. Membrane Science*, 164:89.

Lee, C.C., and Lin, Shun Dar. 1999. *Handbook of Environmental Engineering Calculation*, McGraw Hill.

Libutti, B.L., Knudsen, J.G., and Mueller, R.W. 1984. The Effects of Antiscalants on Fouling by Cooling Water. *Materials Performance*, 23:11.

Lide, D.R., ed. 1995. *CRC Handbook of Chemistry and Physics*. 76th ed. CRC Press.

Lisdonk, and Rietman. 2001. The Influence of Concentration Polarization on the Risk of Scaling in Spiral Wound Membrane Systems. *Proceedings of AWWA Membrane Technology Conference*, San Antonio, TX.

Logan, B.E., Kim, K., Miller, J., and LaPoint, D. 2000. Designing Systems to Biologically Treat Perchlorate-Contaminated Water. *AWWA Inorganic Contaminants Workshop* in Albuquerque, New Mexico.

Logan, and Rey. 1986. Scale Control in Multiple Stage Flash Evaporators. *Material Performance*, 25(6):38.

Mallevialle, J., Odendaal, P.E., and Wiesner, M.R. 1996. *Water Treatment Membrane Processes.* McGraw-Hill.

Malone, D.M., and Anderson, J.L. 1978. Hindered Diffusion of Particles Through Small Pores. *Chem, Eng. Sci.*, 33:1429.

Malone, D.M., and Anderson, J.L. 1977. Diffusional Boundary-Layer Resistance for Membranes With Low Porosity. *Jour. AIChE*, 23:177.

Mannisto, P.T., Ranta, T., and Leppaluoto, J. 1979. Effects of Methylmercaptoimidazole (MMI), Propylthiouracil (PTU), Potassium Perchlorate ($KClO_4$) and Potassium Iodide (KCl) on the Serum Concentrations of Thyrotrophin (TSH) and Thyroid Hormones in the Rat. *Acta Endocrinol*, 91:271.

Masselin, I., Chasseray, X., Chevalier, M., Laine, J., and Lemordant, D. 2000. Determination of the Porosity to Thickness Ratio $A_k/\Delta x$ for UF and MF Membranes by Diffusion Experiments. *Jour. Membrane Sci.*, 172:125.

Morel, G., Graciaa, A., and Lachaise, J. 1991. Enhanced Nitrate Ultrafiltration by Cationic Surfactant. *Jour. Membrane Sci.*, 56:1.

Morel, G., Ouazzani, N., Graciaa, A., and Lachaise, J. 1997. Surfactant Modified Ultrafiltration for Nitrate Ion Removal. *Jour. Membrane Sci.*, 134:47.

Mulder, M. 1991. *Basic Principles of Membrane Technology*, Kluwer Academic Publishers.

NACE Standard Test Method. 1995. Laboratory screening tests to determine the ability of scale inhibitors to prevent the precipitation of calcium sulphate and calcium carbonate from solution. TM0374-95, Item No. 21208.

Nakayama. 2000. Inhibitory Effects of Nitrilotris (Methylenephosphonic Acid) on Cathodic Reactions of Steels in Saturated $Ca(OH)_2$ Solutions. *Corrosion Science*, 42:1897-1920.

Nancollas, and Liu. 1975. Crystal Growth and Dissolution of Barium Sulfate. *Society of Petroleum Engineers Journal*, 15(6):509.

Nightingale, E.R. Jr. 1959. Phenomenological Theory of Ion Solvation: Effective Radii of Hydrated Ions. *Jour. Phys. Chem.*, 63:1381.

Nystrom, M., Kaipia, L., and Luque, S. 1995. Fouling and Retention of Nanofiltration Membranes. *Jour. Membrane Sci.*, 98:249.

Okazaki, I., Ohya, H., Semenova, S.I., Aihara, M., and Negishi, Y. 1998. Study on Molecular Weight Cut-Off Performance of Asymmetric Aromatic Polyimide Membrane "Effect of the Additive Agents." *Jour. Membrane Sci.*, 141:277.

Orvinim, and Speziali. 1998. Applicability and Limits of Instrumental Neutron Activation Analysis: State of the Art in the Year 2000. *Microchemical Journal*, 59:160.

Pabalan, and Pitzer. 1987. Thermodynamics of Concentrated Electrolyte Mixtures and the Prediction of Mineral Solubilities to High Temperatures for Mixtures in the System $Na\text{-}K\text{-}Mg\text{-}Cl\text{-}SO_4\text{-}OH\text{-}H_2O$. *Geochimica et Cosmochimica Acta*, 51:2429.

Porter, M.C. 1972. Concentration Polarization With Membrane Ultrafiltration. *Ind. Eng. Chem. Prod. Res. Develop.* 11(3):234.

Pusch, W. 1990. Performance of RO Membrane in Correlation With Membrane Structure, Transport Mechanisms of Matter and Module Design (Fouling). *Desalination*, 77:35.

Rautenbach, R. and Groschl, A. 1990. Separation Potential of Nanofiltration Membranes. *Desalination*, 77:73.

Rosa, M., and De Phino, M. 1997. Membrane Surface Characterization by Contact Angle Measurements Using the Immersed Method. *Jour. Membrane Sci.*, 131:167.

Shao, J. and Baltus, R.E. 2000. Effect of Solute Concentration on Hindered Diffusion in Porous Membranes. *Jour. AIChE*, 46(7):1307.

Shigan, S.A. 1963. Substantiatioin of the Maximum Permissible Concentration of Ammonium Perchlorate in Water of Reservoirs. *Gig. Saint*, 28:8.

Spiegler, K.S., and Kedem, O. 1966. Thermodynamics of Hyperfiltration (Reverse Osmosis): Criteria for Efficient Membranes. *Desalination*, 1:311.

Strathmann, H. 1992. *Membrane Handbook*, W. S. Ho, K. K. Sirkar, editors, Van Nostrand Reinhold.

Stumm, W. 1992. *Chemistry of the Solid-Water Interface*, Wiley, New York.

Sudmalis, and Sheikholeslami. 2000. Coprecipitation of $CaCO_3$ and $CaSO_4$. *The Canadian Journal of Chemical Engineering*, 78: 21.

Taylor, James S., Mulford, L.A., Duranceau, S.J,. and Barrett, W.M. 1989. Cost and Performance of a Membrane Pilot Plant. *Jour. AWWA*, 11:52.

Timmer, J.M.K., Van der Horst, H.C., and Robbertsen, T. 1993. Transport of Lactic Acid Through Reverse Osmosis and Nanofiltration Membranes. *Jour. Membrane Sci.*, 85:205.

Urase, T., Oh, J., and Yamamoto, K. 1998. Effect of pH on Rejection of Different Species of Arsenic by Nanofiltration. *Desalination,* 117:11.

U.S. EPA. 1998. Perchlorate Environmental Contamination: Toxicological Review and Risk Characterization Based on Emerging Information. *EPA report*, NCEA-1-0503.

Van der Horst, H.C., Timmer, J.M.K., Robbertsen, T., and Leeders, J. 1995. Use of Nanofiltration and Demineralization in the Dairy Industry: Model for Mass Transport. *Jour. Membrane Sci.*, 104:205.

Waypa, J.J., Elimelech, M., and Hering, J.G. 1997. Arsenic Removal by RO and NF Membranes. *Jour. AWWA*, 89:102.

Wiesner, and Aptel. 1996. *Water Treatment Membrane Processes*, McGraw-Hill, Chapter 4, 1.

Wilbert, M.C., Delagah, S., and Pellegrino, J. 1999. Variance of Streaming Potential Measurements. *Jour. Membrane Sci.*, 161:247.

Wilf, M., and Klinko, K. 1997a. Effect of New Pretreatment Methods and Improved Membrane Performance on Design of RO Seawater Systems. *Hydranautics technical paper.*

Wilf, M., and Klinko, K. 1997b. Optimization of Seawater RO Systems Design. *Hydranautics technical paper.*

Wilf, M., and Klinko, K. 1994. Performance of Commercial Seawater Membrane. *Desalination*, 96:456.

Yoon, Y. 2001. Rejection of Perchlorate by Reverse Osmosis (RO), Nanofiltration (NF), and Ultrafiltration (UF) Membranes: Mechanisms and Modeling. Ph. D. dissertation, Dept. of Civil, Environmental, and Architectural Engineering, University of Colorado at Boulder.

Yoon, Y., Amy, G., and Liang, S. 2000. Effect of Zeta Potential on Perchlorate Rejection by Negatively Charged Nanofiltration Membranes. *AWWA Inorganic Contaminants Workshop* in Albuquerque, New Mexico.

Zhang, W., and Hallstrom, B. 1990. Membrane Characterization Using the Contact Angle Technique: I Methodology of the Captive Bubble Technique. *Desalination*, 79:1.

Zhu, H., and Nystrom, M. 1998. Cleaning Results Characterized by Flux, Streaming Potential and FTIR Measurements. *Colloids and Surface*, 138:309.

Zhu, Xiaohua, and Elimelech, Menachem. 1997. Colloidal Fouling of Reverse Osmosis Membranes: Measurements and Fouling Mechanisms. *Environ. Sci. Technol.*, 31:3654.

Zuddas, and Mucci. 1998. Kinetics of Calcite Precipitation From Seawater: II. The Influence of the Ionic Strength. *Geochimica et Cosmochimica Acta*, 62(5):757.

ABBREVIATIONS

A	cross-sectional area (m^2)
C_a	Ca^{2+} concentration in the treated sample after precipitation (mg/L)
C_A	concentration in the fluid adjacent on the feed surface of the membrane in diffusion cell (mol m^{-3})
C_{avg}	average bulk fluid interfacial concentration between feed and permeate side (mol m^{-3})
C_b	feed (bulk) concentration (mol m^{-3})
C_b	Ca^{2+} concentration in the blank after precipitation (mg/L)
C_B	concentration in the fluid adjacent to the permeate surface of the membrane in diffusion cell (mol m^{-3})
Cc	Ca^{2+} concentration in the blank before precipitation (mg/L)
c_i	concentration of ion (mol m^{-3})
C_{if}	PEG concentration in initial feed fluid (mol m^{-3})
C_m	feed concentration of the membrane surface (mol m^{-3})
C_p	permeate concentration (mol m^{-3})
C_{pp}	PEG concentration in permeate (mol m^{-3})
C_s^{feed}	total solute concentration in feed side (mol m^{-3})
C_s^{perm}	total solute concentration in permeate side (mol m^{-3})
CMC	critical micelle concentration (mM)
D	diffusion coefficient of solute in water ($m^2\ s^{-1}$)
D_{eff}	effective diffusion coefficient ($cm^2\ s^{-1}$)
D_p	intrapore diffusion coefficient in membrane pores ($cm^2\ s^{-1}$)
g	gravitational acceleration (9.8 m/s^2, or 32.2 ft/s^2)
h	channel height (m)
H	partition coefficient (dimensionless)
H	total head (m or ft)
j	constant at 20 °C (0.163 for SI units, 2.525×10^{-4} for British uints)
J	pure water flux (m s^{-1})
j_i	ion flux (mol $m^{-2}\ s^{-1}$)
J_s	solute flux (mol $m^{-2}\ s^{-1}$)
J_v	solvent flux (m^3/m^2-s)
k	boundary layer back-diffusion mass transport coefficient (m s^{-1})
K	semiempirical constant (dimensionless)
K_{sp}	solubility product (dimensionless)
l	thickness of the separating layer of the membrane (m)
L	channel length (m)
L_p	filtration coefficient (solvent permeability) ($m^3\ m^{-1}\ s^{-1}\ Pa^{-1}$)

P_d	gage pressure at discharge (N/m^2 or lb/in^2)
P_s	gage pressure at suction (N/m^2 or lb/in^2)
P_w	power requirement (KW: m·m^3/min or hp:ft·gal/min)
PF	concentration polarization factor (dimensionless)
ΔP	differential pressure (Pa)
Q	ion activity product (dimensionless)
Q	discharge (m^3/mim or gpm)
Q_f	flow rate of feed stream (m^3/mim or gpm)
Q_p	flow rate of permeate stream (m^3/mim or gpm)
r	recovery (%)
R	gas constant (J mole^{-1} K^{-1})
R	ohmic resistance of electrolyte and channel (Ω)
R_{real}	real rejection fraction (dimensionless)
SI	Saturation Index (Dimensionless)
t	time (s)
T	absolute temperature (K)
U_b	average cross-flow velocity of bulk fluid (m s^{-1})
ΔU	streaming potential (mV)
V_A	volume of feed side of the membrane in diffusion cell (m^3)
V_B	volume of permeate side of the membrane in diffusion cell (m^3)
V_d	velocity in discharge (m/s or ft/s)
V_s	velocity in suction nozzles (m/s or ft/s)
x	distance normal to membrane face (m)
z_d	elevation of discharge gage (m)
z_i	valance of ion (dimensionless)
z_s	elevation of suction gage (m)

Greek symbols:

δ	thickness of the boundary layer (m)
$\Delta\Pi$	osmotic pressure gradient (Pa)
σ	molecular reflection coefficient (dimensionless)
ε	dielectric constant (dimensionless)
ε_o	permittivity of free space (m^{-1} Ω^{-1})
γ	specific weight of water (N/m^3 or lb/ft^2)
ω	molecular transport coefficient (mol m^{-2} s^{-1} Pa^{-1})
ξ	zeta potential (mV)
η	viscosity of solution (Pa s)
ψ	electrical potential gradient (V)

IWA Publishing's authorised EU representative for General Product Safety Regulations is Diane D'Arras, 15 rue Duret, 75116 Paris, France, e-mail: safety@iwap.co.uk.

Printed and bound by CPI Group (UK) Ltd, Croydon, CR0 4YY
18/09/2026
02218938-0005